Model-Based Tracking
Control of Nonlinear Systems

CRC Series: Modern Mechanics and Mathematics

Series Editors: David Gao and Martin Ostoja-Starzewski

PUBLISHED TITLES

CRC Series: Modern Mechanics and Mathematics

Model-Based Tracking Control of Nonlinear Systems

Elżbieta Jarzębowska

Warsaw University of Technology

Warsaw, Poland

CRC Press
Taylor & Francis Group
Boca Raton London New York

CRC Press is an imprint of the
Taylor & Francis Group, an **informa** business

A CHAPMAN & HALL BOOK

CRC Press
Taylor & Francis Group
6000 Broken Sound Parkway NW, Suite 300
Boca Raton, FL 33487-2742

First issued in paperback 2019

© 2012 by Taylor & Francis Group, LLC
CRC Press is an imprint of Taylor & Francis Group, an Informa business

No claim to original U.S. Government works

ISBN: 978-1-4398-1981-4 (hbk)
ISBN: 978-0-367-38110-3 (pbk)

Library of Congress Cataloging-in-Publication Data

Jarzebowska, Elzbieta.
    Model-based tracking control of nonlinear systems / Elzbieta Jarzebowska.
        p. cm. -- (CRC series, Modern mechanics and mathematics)
    Includes bibliographical references and index.
    ISBN 978-1-4398-1981-4 (hardback)
    1. Nonlinear systems--Automatic control--Mathematics. 2. Automatic tracking--Mathematics. 3. Nonlinear systems--Mathematical models. 4. Control theory. I. Title.

    TJ213.J36 2012
    629.8'36--dc23   2012017617

Visit the Taylor & Francis Web site at
http://www.taylorandfrancis.com

and the CRC Press Web site at
http://www.crcpress.com

To my Parents, who equipped me with passion and courage for life, and to

the memory of Prof. Roman Gutowski, who introduced me to mechanics.

Contents

Preface

The book presents model-based control methods and techniques for non-linear, specifically constrained, systems. It focuses on constructive control design methods with an emphasis on modeling constrained systems, generating dynamic control models, and designing tracking control algorithms for them.

Actually, an active research geared by applications continues on dynamics and control of constrained systems. It is reflected by numerous research papers, monographs, and research reports. Many of them are listed at the end of each book chapter, but it is impossible to make the list complete. The book is not aimed at the survey of existing modeling, tracking, and stabilization design methods and algorithms. It offers some generalization of a tracking control design for constrained mechanical systems for which constraints can be of the programmed type and of arbitrary order. This generalization is developed throughout the book in accordance with the three main steps of a control design project, i.e., model building, controller design, and a controller implementation. The book content focuses on model building and, based upon this model that consists of the generalized programmed motion equations, on a presentation of new tracking control strategy architecture.

The author would like to thank the editors at Taylor & Francis for their support in the book edition; Karol Pietrak, a Ph.D. candidate at Warsaw University of Technology, Warsaw, Poland, for excellent figure drawings in the book, and Maria Sanjuan-Janiec for the original book cover design.

The Author

Elżbieta M. Jarzębowska is currently with the Institute of Aeronautics and Applied Mechanics at the Power and Aeronautical Engineering Department, Warsaw University of Technology, Warsaw, Poland. She received the B.S., M.S., and Ph.D., D.Sc. degrees in mechanical engineering, control and mechanics of constrained systems from the Warsaw University of Technology.

Her fields of research expertise and teaching include dynamics modeling and analysis of multibody systems, nonlinear control of multibody systems including nonholonomic, underactuated, unmaned aerial vehicles (UAV), wheeled robotic systems, and geometric control theory.

Professor Jarzębowska was involved in research projects for Automotive Research Center and Engineering Research Center for Reconfigurable Machining Systems at the University of Michigan, Ann Arbor, Michigan. She also gained valuable experience when working for Ford Motor Company Research Laboratories, Dearborn, Michigan.

She is a member of American Society of Mechanical Engineers (ASME), Institute of Electrical and Electronics Engineers (IEEE), Gesellschaft für Angewandte Mathematik und Mechanik (GAMM), International Federation for the Promotion of Mechanism and Machine (IFToMM), Science Technical Committee of Mechatronics, and International Society for Advanced Research (SAR).

Her hobbies are psychology, swimming, yachting, and traveling.

1

Introduction

This book addresses dynamics modeling and control design methods for nonlinear mechanical systems. The premise of the book is that the two topics are closely related and should be studied together. Most systems of a practical interest are designed to perform various tasks, to work and serve a man. In order to execute work and services, these systems have to be controlled. Tasks and work that are to be completed as well as design and operation conditions are specified by task-based constraints or by other means. We may say, though, that most systems are constrained systems, and referring to nonlinear systems we mean that they are either holonomic or nonholonomic. These are mobile robots and vehicles, aircraft, underactuated manipulators, underwater vessels, space robots, a free-floating astronaut, multifinger grippers, and others. The examples also indicate that many systems are of interest to both mechanics and control. Robotics is an excellent example of a domain where systems are subjected to a variety of constraints, either material or task based. A robotic system can be viewed as a prototype of a system subjected to programmed constraints. Programmed constraints, roughly speaking, are motion specifications put on a system by a design or control engineer to obtain its desired behavior.

To develop dynamic models for constrained mechanical systems, methods of analytical mechanics are applied. From the control perspective, a question about a control algorithm design arises, i.e., how to make a system move according to a specified programmed constraint. Then, two aspects of research on constrained mechanical systems are examined in this book:

1. Theoretical aspect — dynamics modeling of constrained nonlinear systems
2. Practical aspect — design of control algorithms capable of tracking predefined motions

Both aspects inspire this book. The theoretical motivation involves a desire to generalize analytical mechanics results that concern constrained mechanical systems modeling and to accommodate them for control applications. The practical motivation is to bring to bear the mechanics models of constrained systems to design control algorithms.

This book follows an approach that system modeling cannot be separated from a control algorithm generation. Otherwise, control theory would be a sterile discipline without connections to physical systems. It is the author's belief that the process of modeling is as central in control problems as is control theory. A control algorithm design does not begin when a model is submitted to a control engineer; it begins from the model formulation. The full power of analytical dynamics and control theory may develop when both disciplines are put together and viewed as an integrated tool. Excellent control design will not remedy a poor model, and vice versa. Such an interdisciplinary approach coincides with current trends in science and engineering practice and is reflected in the following three main stages of a control design project:

1. Kinematics or dynamics model formulation
2. Control algorithm design according to a specified goal
3. Implementation of a control algorithm for a real system

The content of this book covers the first two stages of the control design project. It is intended to support the modeling stage and demonstrate how it may facilitate the overall control design. However, the book refers to implementation problems through the selection of control algorithms and respect of control constraints.

As the title of this book clearly states, we are interested in model-based control, meaning control based on dynamic models of systems. There are important reasons to formulate the problem of motion control, especially tracking, for nonlinear systems at the dynamic level. First, this is the level at which control actually takes place in practice. Designing controllers at the dynamic level usually leads to significant improvements in performance and implementability and can help in the early identification and resolution of difficulties. Second, unmodeled dynamics, friction, and disturbances can be taken into account at that level (Kane and Levinson 1985; Kwatny and Blankenship 2000; Lewis et al. 2004; Spong and Vidyasagar 1989). Also, for massive wheeled robots that operate at high speeds, dynamics-based control strategies are necessary to obtain realistic control results (see, e.g., De Luca et al. 2002). The third reason, the most significant from the perspective of the book, is that we consider constrained systems, and the constraints may be put on dynamic properties of systems. This explains why we do not address control at the kinematic level. It is interesting to consider tracking for a holonomically constrained system in this regard; the kinematic control problem in this case is trivial, but the dynamic control problem is still quite challenging (Lewis et al. 2004). Finally, many contributions to the kinematic control problems for first order nonholonomic systems can be found in the literature (e.g., Bloch 2003; Lewis et al. 1994; De Luca et al. 2002; Murray et al. 1994). A few textbooks address control of constrained systems at the dynamic level systematically, and there is no unified control theoretic approach to systems with nonholonomic constraints that may be of order higher than one. We believe that this book fills this gap to a certain extent.

The book is not intended to be one more monograph including the collection of almost all concepts, ideas, definitions, and theorems on nonlinear control. It is focused on constructive modeling and control methods developed to support the control design process. For this reason, many descriptive concepts of nonlinear control are not presented.

1.1 Scope and Outline

The scope of the book is presented in Figure 1.1.

The interdisciplinary nature of this book influenced the composition and exposition of its content. It is organized according to the control design project. It addresses kinematic and dynamic modeling methods first. Chapter 2 starts from the basics in nonlinear systems modeling. Modeling systems is not based on receipts. This is the art of modeling, i.e., how to generate an adequate system model. There are, however, some guidelines that help the model building. In this book, the prime goal of modeling is to make it control oriented—to make it transformable into a dynamic control model easily, which is basic for a controller design. In Section 2.2, we address constraints that are imposed on mechanical systems. Specifically, we introduce a concept of a programmed constraint, which may specify tasks. In Section 2.3, motion

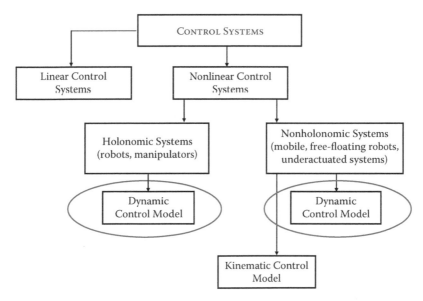

FIGURE 1.1 (SEE COLOR INSERT.)
The scope of the book—dynamic control models for nonlinear systems.

equations, i.e., dynamic models for nonlinear systems are derived. Some of these motion equations, e.g., Lagrange's equations are a standard part of the material presented in classical textbooks in analytical mechanics. Other equations presented, e.g., Maggi's or Nielsen's equations or the Boltzmann–Hamel equations in quasi-coordinates are not so widely used for control applications. We present them because we find them useful and underexploited for control. Selected kinds of equations are derived then based on their potential applicability to control. Chapter 2 closes with some latest modeling method in Section 2.4. It consists in an introduction to control theory new dynamic equations for constrained systems, i.e., generalized programmed motion equations (GPME). They enable us to generate equations of motion for systems with high order constraints. Our approach to modeling nonlinear systems is based on variational principles, because coordinate-based language is much more related to the physical meaning of control problems.

The novelty of Chapter 2 consists in the formulation of a unified representation of constraints and the formulation of kinematic and dynamic models for mechanical systems subjected to first and high order nonholonomic constraints.

Chapter 3 is devoted to an introduction to basic control concepts, ideas, and definitions nonlinear control theory uses. In nine sections of this chapter, we go through a classification of control problems and control properties of nonlinear systems. We emphasize a distinction between control properties of linear and nonlinear systems. In Sections 3.4 and 3.5, kinematic and dynamic control models for nonlinear systems are formulated. Kinematic control models are not the focus of the book; however, some results are cited to demonstrate control challenges that constrained systems may generate. Sections 3.6 to 3.9 address existing model-based control methods for nonlinear systems. Advanced readers may skip the material presented in Chapter 3.

Chapter 4 addresses stabilization strategies for nonlinear systems.

Chapter 5 presents existing model-based tracking control algorithms for nonlinear systems, which are subjected to first order constraints or are underactuated. The goal of the exposition of this material is twofold. First, the existing control algorithms for nonlinear systems are discussed from the point of view of the scope of their applicability; second, they prepare a reader for Chapter 7, where a formulation of a unified model-based tracking control strategy for a general class of constrained systems is presented. This formulation is an extension of the formulations already used in nonlinear control theory.

Chapter 6 delivers basics in path-following strategies for nonlinear systems.

Chapter 7 presents the most advanced material of the book. It uses the results from Chapters 2 and 5 to develop the model reference tracking strategy for programmed motion. It is detailed in Section 7.1. Architecture of this tracking strategy is new; however, it can utilize control laws that are known in nonlinear control. It means that we do not present any new control law for motion or force tracking, but the strategy that enables employing controllers depending on control purposes is new. It enables tracking any motion provided that it may be specified by a constraint equation. Controllers that are the most suitable for programmed motion tracking are also discussed in Chapter 7. Sections 7.2 to 7.5 present modifications and extensions of the model reference tracking control strategy. Examples of control motions of robotic systems illustrate the application of the tracking strategy in a case when system dynamics are completely known (Section 7.2), when the system dynamics are uncertain (Section 7.3), and when learning tracking is applied (Section 7.4). In Section 7.5, we present control-oriented modeling and control design with the aid of quasi-coordinates. In Sections 7.6 and 7.7, we present extensions, modifications, and other applications of the model reference tracking control strategy for programmed motion.

All chapters are illustrated with examples. Two leading examples are selected. They are detailed throughout the book sections to illustrate the theoretical development of the presented material. Examples are conceptually and computationally simple, and they represent two classes of mechanical systems relevant to practical applications. The first is a unicycle model that is a nonholonomic system model, and the second is a two-link planar manipulator model, which is a holonomic system. They are selected according to two criteria. First, we formulate for them the same control goals as reported in the literature and execute them, getting the same simulation results, using the strategy proposed in the book. Second, we demonstrate specific control goals that may be achieved only using the strategy presented in the book. Additionally, we demonstrate that using the model reference tracking control strategy, we may switch between controllers to obtain a desired tracking precision or select them for a specific nonlinear system.

Each chapter of the book ends with a list of references. Selection of references was a difficult task, as it was impossible to refer to all important and outstanding works, papers, and monographs in the fields of analytical mechanics and nonlinear control. The reason is that these fields are of great interest for both scientists and engineers, and a very active research continues. Thus, the author decided to include only references that were most relevant to the book content and apologizes for those she omits to cite. The references were selected according to the following:

1. References that treat both modeling and control of nonlinear systems.
2. References that present some significant milestones in one of the fields.
3. References published during the last 30 years if they do not satisfy criterion 2.

The book is addressed to graduate and postgraduate students as well as to control engineers and researchers. A basic course in analytical mechanics and linear control theory is appreciated.

1.2 Mechanics and Nonlinear Control

Nonlinear mechanical systems may be subjected to constraints, which may be holonomic or nonholonomic. Systems with either constraint are referred to as holonomic or nonholonomic. In this book, we address dynamics modeling and control of both kinds of systems, because both are relevant in applications. Holonomic system dynamics and control can be considered solved problems, at least theoretically (see, e.g., Isidori 1989; Lewis et al. 2004; Slotine and Li 1991; Spong and Vidyasagar 1989). However, nonholonomic system dynamics and control is still a challenge, and there are many open problems for them in both areas. The history of mechanics of nonholonomic systems is very rich. An incomplete list of literature in this area includes Appell (1911, 1947, 1953), Beghin (1947), Bloch (2003), Boltzmann (1902), Dobronravov (1970), Gaponov (1952), Gutowski (1972), Hamel (1904, 1967), Jarzębowska (2002, 2005), Jarzębowska and Lewandowski (2006), Kamman and Huston (1984), Kane and Levinson (1985, 1996), Lancos (1986), Layton (1998), Moon (1998), Nejmark and Fufaev (1972), Nielsen (1935), Papastavridis (2002), Pars (1965) Tzenoff (1924), Udwadia and Kalaba (1996), Vershik (1984), Voronets (1901), Vranceanu (1929), and Yun and Sarkar (1998).

At the beginning of the twentieth century, it was observed that for some systems, like electromechanical systems, equations of motion do not have the form of Lagrange's equations (Carvallo 1902; Gaponov 1952). A new trend to leave Lagrange's equations approach started. In this book, special attention is paid to results obtained within this trend by Nielsen, Tzenoff, Mangeron, and Deleanu (Mangeron and Deleanu 1962; Nielsen 1935; Tzenoff 1924). Their work motivated the author to develop new results, which can be applied to control (Jarzębowska 2002, 2005, 2006, 2007). These new results are presented in Chapters 2 and 7. Methods of classical analytical mechanics are applied to systems with nonholonomic constraints of first order and Appell's equations—to systems with second order constraints. Such systems are commonly referred to as Lagrangian and Hamiltonian systems. We consider systems with constraints that can be of high order and of non-material type, like task-based, which we refer to as programmed constraints. Robots, for example, are designed to do work, and requirements on their performance may be specified by a control engineer using the task-based constraints (see, e.g., Jarzębowska, 2007, and references therein). We go beyond the scope of classical analytical mechanics.

Recently, another approach to mechanics has been developed. This is the geometric approach (see, e.g., Bloch 2003; Brockett 1983; Isidori 1989; Lafferiere and Sussmann 1991; Marsden and Ratiu 1992; Sontag 1990) that emphasizes the role of symmetry and reduction. Both Hamiltonian and Lagrangian viewpoints benefit greatly from this renewed attention to geometric aspects of mechanical systems. Their geometric structure leads to stronger control algorithms than those obtained for generic nonlinear systems. As opposed to the more common language of variational principles or symplectic geometry, the geometric treatment relies on notions from Riemannian geometry (Marsden and Ratiu 1992). These are effective tools, as numerous contributions in modeling (Bloch 2003), stabilization and controllability (Brockett 1983), and dynamic feedback linearization (Isidori 1989) attest.

Similarly, nonlinear control theory today is also a well-developed field. Beginning in the 1970s, the results of numerous authors such as Brockett (1983), Hermann, Isidori, Krener, Sussmann (Isidori 1989; Lafferiere and Sussmann 1991; Sontag 1990 and references therein) have brought methods of differential geometry to bear on nonlinear control problems. Now, many books describe nonlinear control in a geometric light (Bloch 2003; Isidori 1989; Nijmeijer and van der Schaft 1990; Sontag 1990) and many others. We use some ideas, concepts, and definitions from geometric mechanics to describe control properties of nonholonomic control systems with the task-based constraints. Lagrangian or Hamiltonian models as provided by analytical mechanics are utilized in control theory. Using the language of geometric mechanics, these models may have symmetry properties, which can significantly facilitate control design. Also, many of these models are of the Chaplygin type, can be transformed to the power or chained forms, or to their extensions, and are driftless. We demonstrate that systems with the task-based constraints may not admit these properties. Modeling and control design for them may be a challenge.

Control theory has had a fruitful association with analytical mechanics from its birth. This historical relationship was confirmed during the past four decades with the emergence of a geometric theory for nonlinear control systems closely linked to the modern geometric formulation of analytical mechanics. The shared evolution of these fields reflected the needs for solving practical problems.

In the light of a variety of intensive studies on constrained mechanical systems from both analytical mechanics and control perspectives, research on systems with non-material nonholonomic constraints, referred to as programmed constraints is limited (Appell 1911; Galiulin 1971; Gutowski 1971; Korieniev 1964; Zotov and Tomofeyev 1992; Zotov 2003). In Soltakhanov et al. (2009), the concept of the programmed constraint and the formulation of high order constraint equations are presented following the literature; however, neither constructive modeling methods nor control models with applications are presented for them. The reason is not the lack of such constraints, because they are encountered very often, for example, in robotic systems design, operation, safety, and performance specifications. The reason is that there has been no general method to model systems with arbitrary order equality nonholonomic constraints. Consequently, there has been no attempt to describe tasks by equations of constraints. An exception is a position constraint, which is always presented in an equation form but is not merged into a constrained dynamic model (see, e.g., Kwatny and

Blankenship 2000; Murray et al. 1994). Of course, non-material constraints are satisfied by an action of appropriately selected control algorithms. This is usually done by taking the constraints into account when a controller is designed, for example, a constraint on the trajectory curvature put to make this trajectory available for a car-like vehicle (in De Luca et al. 2002, and references therein).

Elegant methods for modeling mechanical systems with first order non-holonomic constraints are generated using either variational or geometric mechanics. Nonlinear control theory takes advantage of both formulations. This book presents a method for modeling systems subjected to constraints of arbitrary order.

To have a closer look at constrained systems from the point of view of mechanics and nonlinear control theory, let us consider a couple of examples. They illustrate formulations of the constraints on system motions. They are simple yet instructive, and they demonstrate how one can formulate material and task-based constraint equations based on motion specifications. Many of these motion specifications are formulated in forms of constraint equations for the first time.

Example 1.1: Two-Wheeled Mobile Platform

The two-wheeled mobile platform presented in Figure 1.2 is a nonholonomic system with powered wheels. Other systems with powered wheels are car-like vehicles, scooters, or bikes.

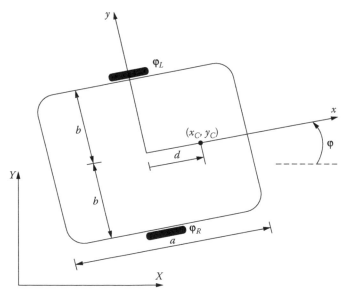

FIGURE 1.2
A two-wheeled mobile platform.

Motion of the platform may be specified by a coordinate vector $q \in R^5$, with $q = (q_1, q_2)$, $q_1 \in R^3$, $q_2 \in R^2$, and $q_1 = (x, y, \ )$, $q_2 = (\ _r, \ _l)$. Motors control the two wheels independently. Angles $\ , \ _r, \ _l$ denote the heading angle and wheel angles due to rolling for the right and left wheel, respectively. The distance between the wheels is equal to $2b$, and we assume that the robot mass center C is located at a distance d from the geometric center O. Material constraint equations for the mobile platform specify the condition that it does not slip sideways, and its driving wheels do not slip. They have the form

$$\dot{y}_C \cos \ - \dot{x}_C \sin \ - \ \dot{}\, d = 0,$$

$$\dot{x}_C \cos \ + \dot{y}_C \sin \ + \ \dot{}\, b = \ \dot{}\,_r, \tag{1.1}$$

$$\ = \frac{r}{2b}(\ _r - \ _l) + c_1,$$

where $\dot{x}_C, \dot{y}_C$- are the velocity components of the mass center C, and c_1 is the integration constant. Angles $\ , \ _r, \ _l$ can be defined in such a way that c_1 may be taken to be equal to zero.

The constraints in Equation (1.1) are referred to as material constraints. They consist of two nonholonomic and one holonomic equations of constraints.

The number of coordinates is $n = 5$, the number of the constraint equations is $k = 3$, the number of degrees of freedom is $n - k = 2$, and the number of control inputs is 2. There may be control torques τ_r, τ_l. Then, the mobile platform may be controlled at a kinematic or dynamic level. However, for heavy car-like vehicles, operating at high speeds, satisfactory control results can be achieved by model-based control strategies (De Luca et al. 2002).

Example 1.2: Snake-Board

The snake-board presented in Figure 1.3 is a nonholonomic system with passive wheels. A rider has to use body cyclic motions and leg and hand muscles to ride the vehicle. By coupling a twisting of the human torso with an appropriate turning of the wheel assemblies (the passive wheel assemblies can pivot freely about a vertical axis), where the turning is controlled by the rider's foot movements, the rider can generate a snake-like locomotion pattern without having to kick off the ground. Motion of the snake-board may be specified by a coordinate vector $q \in R^5$, $q = (x, y, \theta, \psi, \ )$, where (x, y, θ) are position variables and $(\psi, \ )$ are shape variables.

Equations of nonholonomic constraints for the snake-board that come from a condition of rolling its wheels without slipping are

$$-\dot{x}\sin(\theta + \ ) + \dot{y}\cos(\theta + \ ) - \dot{\theta}l\cos \ = 0,$$

$$-\dot{x}\sin(\theta - \ ) + \dot{y}\cos(\theta - \ ) + \dot{\theta}l\cos \ = 0. \tag{1.2}$$

In this example, the number of coordinates is $n = 5$, the number of the constraint equations is $k = 2$, the number of degrees of freedom is $n - k = 3$, and the number of control inputs is 2. There may be control torques τ_ψ, τ that represent the twisting of the human torso and the rider's foot movements. Control at the dynamic

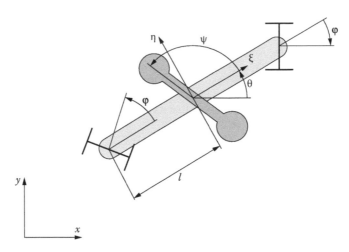

FIGURE 1.3
Model of a snake-board.

level is needed for the snake-board. More about the snake-board dynamics and
control can be found in Lewis et al. (1994).

Example 1.3: Roller-Racer

The roller-racer motion is very different from the snake-board, where the motion
of the rider is essential for the system propulsion. The roller-racer rider does not
have to move his body. The propulsion and steering of the roller-racer come from
a rotary motion at the joint that connects its segments. The joint torque is the only
control of the system; see Figure 1.4.

The motion of the roller-racer is usually specified by $q \in R^4$, where $q = (x, y, \theta, \psi)$
and (x, y, θ) are position variables, and ψ is a shape variable. Equations of nonholonomic constraints are

$$\dot{x}\sin\theta - \dot{y}\cos\theta = 0,$$

$$-\dot{x}\sin\psi + \dot{y}\cos\psi + \dot{\theta}l_1\cos(\theta - \psi) + l_2\dot{\psi} = 0. \tag{1.3}$$

The number of coordinates is $n = 4$, the number of the constraint equations
is $k = 2$, the number of degrees of freedom is $n - k = 2$, and there is only one
control input τ_ψ. Then, the purely kinematic analysis and a controller design
for the roller-racer are not possible. The roller-racer may be controlled at the
dynamic level only (see, e.g., Tsakiris and Krishnaprasad 1998; Jarzębowska and
Lewandowski 2006).

A roller-racer rider learns what value of a forward velocity is enough to ride
quite a smooth undulatory motion that permits maneuvering. Also, the rider learns
how fast his or her rotary motion at the rotary joint should be to execute this

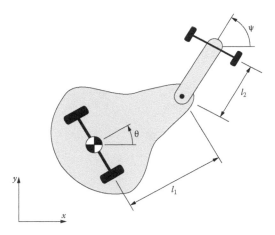

FIGURE 1.4
Model of a roller-racer.

maneuvering velocity. It is straightforward to think about the forward velocity and the angular velocity of the roller-racer to be the "natural coordinates" from the rider's perspective. It is also natural from the control perspective. Motion specification in quasi-velocities seems to be more "natural", that is,

$$\omega_1 = v = \dot{x}\cos\theta + \dot{y}\sin\theta,$$

$$\omega_2 = \dot{\theta}$$

(1.4)

for the maneuvering and

$$\omega_3 = \dot{x}\sin\theta - \dot{y}\cos\theta = 0,$$

$$\omega_4 = -\dot{x}\sin\psi + \dot{y}\cos\psi + l_1\dot{\theta}\cos(\psi - \theta) + l_2\dot{\psi} = 0$$

(1.5)

for the satisfaction of the nonholonomic constraints.

This example indicates that the selection of coordinates different than generalized coordinates may be useful for control-oriented modeling.

Example 1.4: Snake-Like Robot

The snake-like robot is a longer snake than the roller-racer. Its model is presented in Figure 1.5.

The robot may locomote by bending its body using the side force against sideslip.

The nonholonomic constraint equations that come from a condition of rolling its wheels without slipping have the same form as for wheeled vehicles with passive wheels, that is, $A(q)\dot{q} = 0$, where A is a $(n + 2 \times n + 2)$ matrix. It is possible to control the snake-like robot at a dynamic level only (see, e.g., Prautsch and Mita 1999).

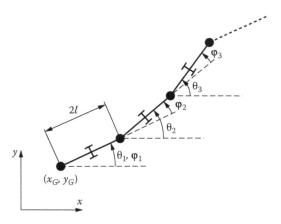

FIGURE 1.5
Model of a snake-like robot.

Example 1.5: Two-Link Planar Manipulator

Consider a model of a planar two-link manipulator presented in Figure 1.6. It moves in the horizontal plane (x,y). Two degrees of freedom are described by Θ_1, Θ_2.

It is a holonomic system; however, we may formulate a requirement that the manipulator end-effector is to move along a trajectory for which its curvature changes according to a specified function $\Phi^* = \frac{d\Phi(t)}{dt}$. In the task space (x,y), this constraint has the form

$$\ddot{x} = \frac{-\dot{\Phi}(\dot{x}^2 + \dot{y}^2)^{3/2}}{\dot{y}} - \frac{3(\dot{x}\ddot{x} + \dot{y}\ddot{y})(\dot{x}\ddot{y} - \dot{y}\ddot{x})}{\dot{y}(\dot{x}^2 + \dot{y}^2)} + \ddot{y}\frac{\dot{x}}{\dot{y}}. \tag{1.6}$$

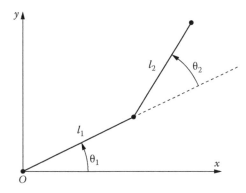

FIGURE 1.6
Model of a two-link planar manipulator.

In the joint coordinates (Θ_1, Θ_2), inserting $x = l_1\cos\Theta_1 + l_2\cos(\Theta_1+\Theta_2)$, $y = l_1\sin\Theta_1 + l_2\sin(\Theta_1+\Theta_2)$, and their time derivatives into Equation (1.6), it can be written as

$$F_2\ddot{\Theta}_1 + \ddot{\Theta}_2 - F_1 = 0, \qquad\qquad (1.7)$$

where

$$F_1 = \frac{A_\phi - A_1 - A_2 a_o}{a_2 + a_4 a_o}, \qquad F_2 = \frac{a_1 + a_2 + a_o(a_3 + a_4)}{a_2 + a_4 a_o},$$

and

$$A_\phi = \frac{-\Phi(a_5^2 + a_6^2)^2\left[\dot{\Phi}(a_5^2 + a_6^2) + 3\Phi(a_5 a_7 + a_6 a_8)\right]}{a_6(a_5 a_8 - a_7 a_6)},$$

$$A_1 = 3a_3\dot{\Theta}_1\ddot{\Theta}_1 + 3a_4(\ddot{\Theta}_1 + \ddot{\Theta}_2)(\dot{\Theta}_1 + \dot{\Theta}_2) - a_1\dot{\Theta}_1^3 - a_2(\dot{\Theta}_1 + \dot{\Theta}_2)^3,$$

$$A_2 = 3a_3\dot{\Theta}_1\ddot{\Theta}_1 + 3a_2(\ddot{\Theta}_1 + \ddot{\Theta}_2)(\dot{\Theta}_1 + \dot{\Theta}_2) + a_3\dot{\Theta}_1^3 + a_4(\dot{\Theta}_1 + \dot{\Theta}_2)^3, \quad a_o = a_5/a_6.$$

$$a_1 = -l_1\sin\Theta_1, \quad a_2 = -l_2\sin(\Theta_1+\Theta_2), \quad a_3 = -l_1\cos\Theta_1, \quad a_4 = -l_2\cos(\Theta_1+\Theta_2),$$

$$a_5 = a_1\dot{\Theta}_1 + a_2(\dot{\Theta}_1 + \dot{\Theta}_2), \quad a_6 = -a_3\dot{\Theta}_1 - a_4(\dot{\Theta}_1 + \dot{\Theta}_2),$$

$$a_7 = a_1\dot{\Theta}_1 + a_3\dot{\Theta}_1^2 + a_2(\ddot{\Theta}_1 + \ddot{\Theta}_2) + a_4(\dot{\Theta}_1 + \dot{\Theta}_2)^2,$$

$$a_8 = -a_3\ddot{\Theta}_1 + a_1\dot{\Theta}_1^2 - a_4(\ddot{\Theta}_1 + \ddot{\Theta}_2) + a_2(\dot{\Theta}_1 + \dot{\Theta}_2)^2.$$

Equation (1.7) specifies a task-based nonholonomic constraint equation of the third order. It can be thought of as a task that specifies writing on a surface, painting, or scribing.

Example 1.6: Free-Floating Space Manipulator

Consider a space two-link manipulator as shown in Figure 1.7 that consists of a base described by its moment of inertia J and ϕ, the orientation of the base relative to a fixed axis in the plane. Let θ_1 be the angle of the first link of mass m_1 and length l_1 relative to the base, and θ_2 the angle of the second link of mass m_2 and length l_2 relative to the first. For simplicity, we assume that link masses are concentrated at the ends of the links. In this model, the manipulator base is pinned to the ground at its center. Pinning the base permits the body to rotate freely but prevents translation. The law of conservation of the angular momentum for free-floating systems implies that moving the links causes the base body to rotate.

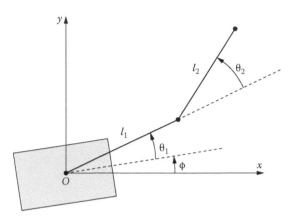

FIGURE 1.7
Model of a two-link space manipulator.

This conservation law is viewed as a nonholonomic constraint on the system and it has the form

$$\left[J+(m_1+m_2)l_1^2+m_2 l_2^2\right]\dot\phi+\left[(m_1+m_2)l_1^2+m_2 l_2^2\right]\dot\theta_1+$$

$$+m_2 l_2^2\dot\theta_2+m_2 l_1 l_2\cos\theta_2(2\dot\phi+2\dot\theta_1+\dot\theta_2)=0. \qquad (1.8)$$

Free-floating systems are considered nonholonomic with first order constraints. Constraint Equation (1.8) is of the same form as the constraint equation that originates from the condition of rolling wheels without slipping, i.e., $A(q)\dot q=0$.

Example 1.7: Ishlinsky's Example

Consider an example of a non-Chaplygin system with material constraints. One of such systems studied in the literature is a ball rolling on a rotating plate (Bloch 2003). Another nonholonomic non-Chaplygin system is the Ishlinsky example. For the first time, it is studied from the point of view of control theory in Jarzębowska and McClamroch (2000). From the point of mechanics, it is studied in (Nejmark and Fufaev 1972).

The Ishlinsky system consists of three uniform cylinders (see Figure 1.8). Two cylinders, the lower ones, each of radius a, can roll without slipping on a horizontal plane. The third cylinder, the upper one, of radius R can roll without slipping on the two lower cylinders. Because the lower cylinders have identical radii, the rotation axis of the upper cylinder always remains in a fixed horizontal plane. Let G denote the mass center of the upper cylinder and suppose that its coordinates are (x, y) and (ξ, η) with respect to the two selected fixed reference frames. The vertical coordinate of G is a constant $z=2a+R$. We denote by η the angle between the axis of rotation of the upper cylinder and the coordinate axis OX and by ϕ the roll angle of the upper cylinder. All the cylinders roll without slipping, so $\alpha=$ const., $\alpha\neq 0$, and $\alpha\neq\eta$. Consequently, the following generalized coordinates describe the motion of the three cylinders: $q_1=x$, $q_2=y$, $q_3=\eta$, $q_4=$ ₁, $q_5=$ ₂,

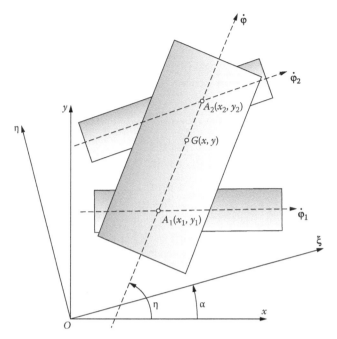

FIGURE 1.8
The Ishlinsky example.

$q_6 = $, where $_1$, $_2$ are roll angles of the two lower cylinders. Because the lower cylinders roll without slipping on the horizontal plane, the geometric relations $y_1 = a_1$, $\eta_2 = a_2$ hold. The upper cylinder also rolls without slipping on the two lower ones, giving

$$v_1 = v + \omega \times \rho_1, \quad v_2 = v + \omega \times \rho_2, \tag{1.9a}$$

where v is the translational velocity vector of G, v_1 and v_2 are the translational velocity vectors of A_1 and A_2, and ρ_1 and ρ_2 are the position vectors from G to A_1, and from G to A_2.

Conditions of Equation (1.9a) generate nonholonomic constraint equations for the upper cylinder motion. Scalar components of ρ_1 and ρ_2 can be expressed as

$$\rho_{1x} = (a_1 - y)ctg\eta, \; \rho_{1y} = (a_1 - y), \; \rho_{1z} = -R,$$

$$\rho_{2x} = \frac{\cos\eta}{\sin(\eta - \alpha)}[a_2 + x\sin\alpha - y\cos\alpha], \rho_{2y} = \frac{\sin\eta}{\sin(\eta - \alpha)}[a_2 + x\sin\alpha - y\cos\alpha], \; \rho_{2z} = -R,$$

and the components of ω, v_1, v_2, and v are

$$\omega_x = \dot{}\cos\eta, \; \omega_y = \dot{}\sin\eta, \; \omega_z = \dot{\eta}, \; v_x = \dot{x}, \; v_y = \dot{y}, \; v_z = \dot{z},$$

$$v_{1x} = 0, \; v_{1y} = 2a\dot{}_1, \; v_{1z} = 0, \; v_{2x} = -2a\dot{}_2\sin\alpha, \; v_{2y} = 2a\dot{}_2\cos\alpha, \; v_{2z} = 0.$$

Introducing the relations above into Equation (1.9a), we obtain the kinematic constraints

$$\dot{x} = \frac{2az\sin\eta}{w\sin\eta - z\sin(\eta-\alpha)} \,{}_1\sin\alpha + R\,{}^\cdot\sin\eta,$$

$$\dot{y} = \frac{2a\sin\eta}{w\sin\eta - z\sin(\eta-\alpha)}[w - z\cos\alpha] \,{}^\cdot_1 - R\,{}^\cdot\cos\eta,$$

$$\dot{\eta} = \frac{2a\sin\eta}{w\sin\eta - z\sin(\eta-\alpha)} \,{}_1\sin\alpha,$$

$$\dot{}_2 = \frac{\sin\eta}{\sin(\eta-\alpha)} \,\dot{}_1.$$

(1.9b)

Later, we explain that the set of first order nonholonomic equations (1.9b) is non-Chaplygin in contrast to the Chaplygin first order nonholonomic constraint equations (1.1), (1.3), or (1.8).

Example 1.8: Grioli's Example

According to Grioli's theorem (Grioli 1972), the necessary and sufficient condition for a body to perform the pseudoregular precession is

$$(p\dot{q} - \dot{p}q) + r(p^2 + q^2) - \lambda(p^2 + q^2)^{3/2} = 0,$$ (1.10)

where $\lambda = const.$, $p = \omega_\xi$, $q = \omega_\eta$, $r = \omega_\zeta$ are angular velocities of the body in the (ξ, η, ζ) frame fixed in the body, as presented in Figure 1.9. The velocities ω_ξ, ω_η, ω_ζ are quasi-velocities, which may be presented using Euler's angles , ψ, as follows:

$$\omega_\xi = \dot{\psi}\sin\ \sin\ +\ {}^\cdot\cos\ ,\quad \omega_\eta = \dot{\psi}\sin\ \cos\ -\ {}^\cdot\sin\ ,\quad \omega_\zeta = \dot{\psi}\cos\ +\ {}^\cdot.$$ (1.11)

Inserting Equation (1.11) into the Grioli condition (1.10), we obtain

$$\dot{\psi}\,{}^\cdot\sin\ -\ \ddot{\psi}\sin\ + 2\dot{\psi}\,{}^2\cos\ + \dot{\psi}^3\sin^2\ \cos\ -\lambda(\dot{\psi}^2\sin\ +\ {}^{\cdot\,2})^{3/2} = 0.$$ (1.12)

The constraint equation (1.12) is task based and it is second order nonholonomic.

Example 1.9: Galiulin and Korenev Example

Galiulin and Korienev worked on missiles control and guidance (Galiulin 1971; Korieniev 1964). A concept of a programmed motion appears in their works in the context of a trajectory tracking, i.e., tracking a moving target. The problem is illustrated in Figure 1.10.

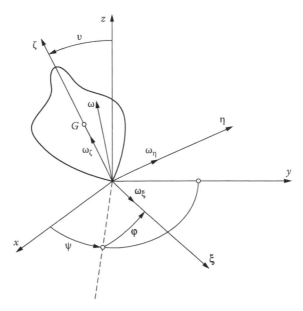

FIGURE 1.9
Motion of a body about a fixed point.

It can be formulated as follows: A target Q moves a prescribed motion $\xi(t)$ along OX. A follower, modeled as a particle, moves in the (x,y) plane in such a way that its velocity vector is directed toward Q. The constraint equation for the follower is

$$y\dot{x} + (\xi - x)\dot{y} = 0. \tag{1.13}$$

It may be nonholonomic, and it is task based. The follower moves along a curve of pursuit.

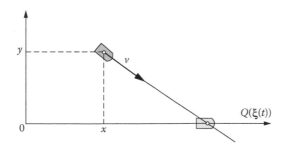

FIGURE 1.10
Tracking of a moving target.

Example 1.10: Motion along a Predefined Trajectory

Consider again the two-wheeled mobile platform and specify the task-based constraint for it. It may be a control goal, motion, or design demand, e.g., a desired trajectory equation:

$$B_1 = x^2 + y^2 - R(t) = 0, \quad R(t) = 0.2 + 0.01t \tag{1.14}$$

or a trajectory curvature rate of change

$$\ddot{x} = \frac{-\dot{\Phi}(\dot{x}^2 + \dot{y}^2)^{3/2}}{\dot{y}} - \frac{3(\dot{x}\ddot{x} + \dot{y}\ddot{y})(\dot{x}\ddot{y} - \dot{y}\ddot{x})}{\dot{y}(\dot{x}^2 + \dot{y}^2)} + \ddot{y}\frac{\dot{x}}{\dot{y}}. \tag{1.15}$$

Another constraint of the type as (1.15) may specify other motion properties. The set of constraints for the mobile platform consists then of Equations (1.1) and (1.15), i.e., the material constraints are to be supplemented by the task-based constraints.

Example 1.11: Modeling Constraints

For modeling of some classes of systems, e.g., robotic systems, other coordinates may be useful (see, e.g., de Jalon and Bayo 1994). Consider a four-bar mechanism and model its kinematics in natural coordinates, see Figure 1.11. They are the Cartesian coordinates for points 1 and 2, that is, (x_1, y_1, x_2, y_2).

Conditions that specify that the mechanism consists of rigid links result in three position constraint equations:

$$(x_1 - x_A)^2 + (y_1 - y_A)^2 - L_2^2 = 0,$$
$$(x_2 - x_1)^2 + (y_2 - y_1)^2 - L_3^2 = 0, \tag{1.16}$$
$$(x_2 - x_B)^2 + (y_2 - y_B)^2 - L_4^2 = 0.$$

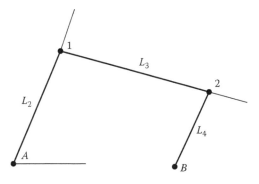

FIGURE 1.11
Modeling a four-bar mechanism in natural coordinates.

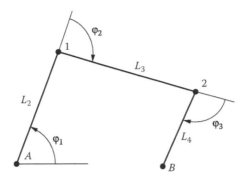

FIGURE 1.12
Modeling a four-bar mechanism in relative coordinates.

This mechanism may be modeled using relative coordinates, see Figure 1.12. They produce no constraints for open chains, but for the closed chain the constraint equations are

$$L_2 \cos \varphi_1 + L_3 \cos(\varphi_1 + \varphi_2) + L_4 \cos(\varphi_1 + \varphi_2 + \varphi_3) - AB = 0,$$
$$L_2 \sin \varphi_1 + L_3 \sin(\varphi_1 + \varphi_2) + L_4 \sin(\varphi_1 + \varphi_2 + \varphi_3) = 0. \tag{1.17}$$

From the examples above, we may conclude that constrained systems, specifically nonholonomic, are a large class of systems. From the perspective of mechanics and derivation of equations of motion, most of them belong to the class of first order nonholonomic systems. They may be approached by Lagrange's equations with multipliers, and usually these equations are used to generate dynamic control models for them. However, not all systems with task-based constraints may be approached by Lagrange's equations because they fail when the constraints are of an order higher than one. Also, we could see that a holonomic system, e.g., the planar manipulator, may be made nonholonomic by a constraint put upon its motion.

From the perspective of nonlinear control, the constrained systems differ and may not be approached by the same control strategies and algorithms. Some of them may be controlled at the kinematic level and the other at the dynamic level only. Their control properties depend upon the way they are designed and propelled, i.e., how many control inputs are available and whether their wheels, if there are any, are powered or not. They are divided into two control groups, which are treated separately, i.e., the group of fully actuated and the group of underactuated systems. Finally, based on the roller-racer example, we could see that the selection of the coordinates for control-oriented modeling may be significant.

Then, a system design, way of its propulsion, control goals, and other task-based or work-space constraints may determine the way of its modeling.

1.3 Role of Modeling in a Control Design Process

In Section 1.2, we presented examples that showed clearly that nonlinear systems that belonged to the same class as mechanical systems might differ from the point of view of nonlinear control. This is the basic reason to address system modeling in the context of a controller design. The steps of a model-based control design project are presented in the diagram in Figure 1.13. It specifies a more general concept of tracking control comparing to a traditional nonlinear control design in which a control goal is either a trajectory to follow or stabilization to equilibrium; compare, for example, to Kwatny and Blankenship (2000), Murray, Li, and Sastry (1994), and Yun and Sarkar (1998). It follows the more general approach to control developed in this book. Also, in Figure 1.13, motion equations of nonholonomic systems that furnish a dynamic control model are not specified. This is due to the fact that we discuss several possibilities of

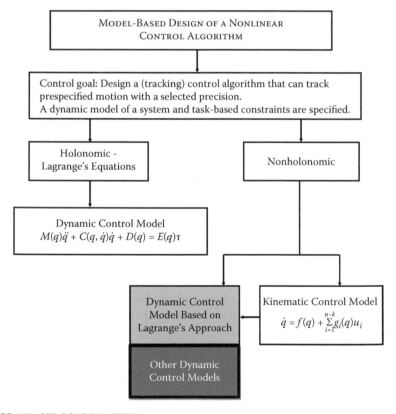

FIGURE 1.13 (SEE COLOR INSERT.)
Model-based control design for nonlinear systems.

the generation of this model based on a specific system and control goals. Figure 1.13 thus conforms to the control-oriented modeling approach to constrained systems.

References

Appell, P. 1911. Exemple de mouvement d'un point assujeti a une liason exprimee par une relation non lineaire entre les composantes de la vitesse. *Comptes Rendus*: 48–50.

Appell, P. 1947. *Traite de mecanique rationnelle.* Vol. 1, vol. 2. Paris: Gauthier-Villars.

Appell, P. 1953. *Traite de mecanique rationnelle.* Vol. 1, vol. 2. Paris: Gauthier-Villars.

Beghin, H. 1947. *Course de mecanique,* Paris.

Bloch, A.M. 2003. *Nonholonomic mechanics and control,* New York: Springer-Verlag.

Boltzmann, L. 1902. Sitzungsber., Vol. 3. Wien: Akad. Wiss.

Brockett, R.W. 1983. Asymptotic stability and feedback stabilization. In *Differential geometric control theory,* ed. R.W. Brockett, R.S. Millman, and H.J. Sussmann, Boston, MA: Birkhauser.

Carvallo, C. 1902. L'electricite de l'experience et ramenee au principe dex travaux virtuels. *Coll. Scienta* 19.

Chaplygin, S.A. 1897. About a motion of a heavy body on a horizontal plane. In *Izbrannye trudy klassiki nauki,* 363–375. Moscow: Nauka (in Russian).

de Jalon, J.G., and E. Bayo. 1994. *Kinematic and dynamic simulation of multibody systems.* Mech. Eng. Series. Berlin: Springer-Verlag.

De Luca, A., G. Oriolo, L. Paone, and P.R. Giordano. 2002. Experiments in visual feedback control of a wheeled mobile robot. In *Proc. IEEE Conf. Robot. Automat.* 2073–2078. Washington, DC.

Dobronravov, V.V. 1970. *Foundations of mechanics of non-holonomic systems.* Moscow: Vyschaja Shkola (in Russian).

Galiulin, A.C. 1971. *Design of systems for programmed motion.* Moscow: Nauka (in Russian).

Gaponov, V. 1952. Nonholonomic systems of S.A. Chaplygin and the theory of commutative electrical machinery. *Doklady Akademii Nauk SSSR* 87:401 (in Russian).

Grioli, G. 1972. Particular solutions in stereodynamics. In *Centro Internazionale Matematico Estivo,* 1–65, Roma (in Italian).

Gutowski, R. 1971. *Analytical mechanics,* Warsaw: PWN (in Polish).

Hamel, G. 1904. Die Lagrange-Eulersche gleichugen der mechanic. *Z. Math. Phys.* 50:1–57.

Hamel, G. 1967. *Theoretische mechanik.* Berlin: Springer-Verlag.

Isidori, A. 1989. *Nonlinear control systems.* 2nd ed. Berlin: Springer-Verlag.

Jarzębowska, E., and N.H. McClamroch. 2000. On nonlinear control of the Ishlinsky problem as an example of a nonholonomic non-Chaplygin system. In *Proc. Am. Contr. Conf.* 3249–3253. Chicago, IL.

Jarzębowska, E. 2002. On derivation of motion equations for systems with nonholonomic high-order program constraints. *Multibody Syst. Dyn.* 7(3):307–329.

Jarzębowska, E. 2005. Dynamics modeling of nonholonomic mechanical systems: Theory and applications. *Nonlin. Anal.* 63(5–7):185–197.

Jarzębowska, E. 2006. Control oriented dynamic formulation of robotic systems with program constraints. *Robotica* 24(1):61–73.

Jarzębowska, E., and R. Lewandowski. 2006. Modeling and control design using the Boltzmann-Hamel equations: A roller-racer example. In *Proc. 8th IFAC Symposium on Robot Contr. SYROCO*, Bologna, Italy.

Jarzębowska, E. 2007. Stabilizability and motion tracking conditions for nonholonomic control systems. *Mathematical Problems in Engineering.* Cairo: Hindawi.

Kamman, J.W., and R.L. Huston 1984. Dynamics of constrained multibody systems. *J. Appl. Mech.* 51:899–903.

Kane, T.R., and D.L. Levinson. 1985. *Dynamics—Theory and applications.* New York: McGraw-Hill.

Kane, T.R., and D.L. Levinson. 1996. The use of Kane's dynamical equations in robotics. *Int. J. Robot. Res.* 2(3):3–21.

Korieniev, G.V. 1964. *Introduction to mechanics of a controllable body.* Moscow: Nauka (in Russian).

Kwatny, H.G., and G.L. Blankenship. 2000. *Nonlinear control and analytical mechanics, a computational approach.* Boston: Birkhauser.

Lafferiere, G., and H. Sussmann. 1991. Motion planning for controllable systems without drift. In *Proc. IEEE Int. Conf. Robot. Automat.*, 1148–1153.

Lancos, C. 1986. *The variational principles of mechanics.* 4th ed. New York: Dover.

Layton, R.A. 1998. *Principles of analytical system dynamics.* New York: Springer-Verlag.

Lewis, A.D., J.P. Ostrowski, R.M. Murray, and J. Burdick. 1994. Nonholonomic mechanics and locomotion: The snakeboard example. In *IEEE Int. Conf. Robot. Automat.*, 2391–2400.

Lewis, F.L., C.T. Abdallah, and D.M. Dawson. 2004. *Control of robot manipulators.* New York: Marcel Dekker.

Mangeron, D., and S. Deleanu 1962. Sur une classe d'equations de la mecanique analytique ou sens de Tzenoff. *Comptes Rendus de l'Academie Bulgare des Sciences* 15:1–9.

Marsden, J.E., and T.S. Ratiu. 1992. *An introduction to mechanics and symmetry.* Texts in Appl. Math. 17, Springer-Verlag.

Moon, F.C. 1998. *Applied dynamics.* New York: John Wiley & Sons.

Murray, R.M., Z.X. Li, and S.S. Sastry. 1994. *A mathematical introduction to robotic manipulation.* Boca Raton, FL: CRC Press.

Nejmark, J.I., and N.A. Fufaev. 1972. *Dynamics of nonholonomic systems.* Providence, RI: American Mathematical Society.

Nielsen, J. 1935. *Vorlesungen uber elementare mechanik.* Berlin: Verlag von J. Springer.

Nijmeijer, H., and A. van der Schaft. 1990. *Nonlinear dynamical control systems.* New York: Springer-Verlag.

Papastavridis, J.G. 2002. *Analytical mechanics, a comprehensive treatise on the dynamics of constrained systems; for engineers, physicians, and mathematicians.* New York: Oxford University Press.

Pars, L.A. 1965. *Treatise of analytical dynamics.* London: W. Heinemann.

Prautsch, P., and T. Mita. 1999. Control and analysis of the gait of snake robots. In *Proc. IEEE Int. Conf. on Contr. Appl.* 502–507.

Slotine, J.J., and W. Li. 1991. *Applied nonlinear control.* Englewood Cliffs, NJ: Prentice Hall.

Soltakhanov, Sh., M.P. Yushkov and S.A. Zegzhda. 2009. *Mechanics of nonholonomic systems*. Berlin: Springer.

Sontag, E.D. 1990. *Mathematical control theory*. New York: Springer.

Spong, M.W., and M. Vidyasagar. 1989. *Robot control and dynamics*. New York: Wiley.

Tsakiris, D.P., and P.S. Krishnaprasad. 1998. Oscillations, SE(2)-snakes and motion control: A study of the roller-racer. *Technical Report, CDSS*, University of Maryland.

Tzenoff, J. 1924. Sur les equations du mouvement des systemes mecanique nonholonomes. *Mat. Ann.* 91.

Udwadia, F., and R. Kalaba. 1996. *Analytical dynamics—A new approach*. New York: Cambridge University Press.

Vershik, A.M. 1984. Classical and nonclassical dynamics with constraints. In *New in Global Analysis*, Moscow: Gos. Univ.

Voronets, P.V. 1901. On the equations of motion for nonholonomic systems. *Matematicheskii Sbornik* 22(4) (in Russian).

Vranceanu, G. 1929. Studio geometrice dei sistemi anolonomi. *Bull. Mat.*, 30.

Yun, X., and N. Sarkar. 1998. Unified formulation of robotic systems with holonomic and nonholonomic constraints. *IEEE Trans. Robot. Automat.* 14(4):640–650.

Zotov, Yu.K., and A.V. Tomofeyev. 1992. Controllability and stabilization of programmed motions of reversible mechanical and electromechanical systems. *J. Appl. Math. Mech.* (trans. from Russian), 56(6):873–880.

Zotov, Yu.K. 2003. Controllability and stabilization of programmed motions of an automobile-type transport robot. *J. Appl. Math. Mech.* (trans. from Russian), 67(3):303–327.

2

Dynamics Modeling of Constrained Systems

In model-based control system analysis and design, explicit dynamic equations of motion are required. Hence, a unified theoretical framework, including a systematic generation of equations of motion for nonlinear, especially nonholonomic, systems is needed. Modeling methods for nonlinear holonomic systems are well known and can be found in many classical textbooks in mechanics and control (e.g., Isidori 1989; Lewis et al. 2004; Murray et al. 1994; Nijmeijer and van der Schaft 1990; Pars 1965; Slotine and Li 1991; Udwadia and Kalaba 1996).

Modeling methods available for nonholonomic systems are presented in this chapter. Any systematic survey is provided; this can be found in Bloch (2003), de Jalon and Bayo (1994), Dobronravov (1970), Gutowski (1971), Gutowski and Radziszewski (1969), Hamel (1904, 1967), Kamman and Huston (1984), Kane and Levinson (1985), Lancos (1986), Layton (1998), Moon (1998), Nejmark and Fufaev (1972), Nielsen (1935), Papastavridis (1988, 1994, 1998, 2002), Pars (1965), Rong (1996a, 1996b, 1997), Soltakhanov et al. (2009), Tianling (1992), Tzenoff (1924), Udwadia and Kalaba (1996), Voronets (1901), Vranceanu (1929) and many others. Instead, we focus on constraint specifications and on ways they are managed in mechanics and control settings. Modeling methods presented in this chapter are the basis for design of nonlinear controllers, which is the topic of Chapters 5 and 7.

2.1 Introduction—Art of Modeling

Modeling systems may be completed using either vector or analytical mechanics methods. The latter ones proved to be more efficient for complex nonlinear systems, and they are the basis for a development of computer codes for automatic modeling, simulation, design, and control.

A modeling process starts from setting goals of modeling, careful analysis of a system to be modeled, and a further model destination. There are no receipts of how to develop a good model. However, there are some guidelines and experience gathered. From the perspective of this book, our main guideline is to make the model suitable for control applications, i.e., make it suitable to transfer it to a control model, either kinematic or dynamic.

Thus, we focus on demands required by the control design process and on control-oriented goals.

Modeling nonlinear systems is quite a challenging task. It is much less obvious because the response of a nonlinear system to one input signal does not reflect its response to another one. A frequency-domain description is not possible for nonlinear systems. With a controller design as a final goal of modeling, two points can be made. First, one should understand well the modeled system physical properties and the specification of a control task to obtain an accurate model for a controller design. Second, the system model is not just "mapping" of its properties; it should provide some characterization of uncertainties and judge unmodeled dynamics. A separate question is what is meant by an "accurate" model. There is no unique answer to this question. An "accurate" model means a set of different properties that depend upon modeling and control goals, available computer and motor powers, and many other factors. How do we develop this "accurate" model? The only answer known to the author is to use your experience, try to understand the way a system works and is controlled, and think about alternative approaches even when modeling looks standard.

In this section, we recall basic concepts and definitions used in analytical mechanics to develop system models. Let us start from a concept of coordinates and their selection.

2.1.1 Selection of Coordinates

A position of a system of n particles in space, in some selected reference frame, say (x,y,z) frame, may be determined by n position vectors p_i. Each vector may be presented by three coordinates so $3n$ scalar functions specify the system position. If three unit vectors e_x, e_y, e_z specify the frame, then coordinates of the particles can be written as

$$x_i = p_i e_x, \quad y_i = p_i e_y, \quad z_i = p_i e_z. \qquad i = 1,...,n \qquad (2.1)$$

However, in most cases, systems are not free to move. They are constrained.

Example 2.1

Consider a spherical pendulum presented in Figure 2.1. The link length L is assumed to be constant. Select the Cartesian coordinates x,y,z to specify the pendulum motion. These coordinates are not mutually independent. They are related by

$$x^2 + y^2 + z^2 - L^2 = 0. \qquad (2.2)$$

Relation (2.2) reflects the assumption that L is constant and it has to be satisfied during the pendulum motion.

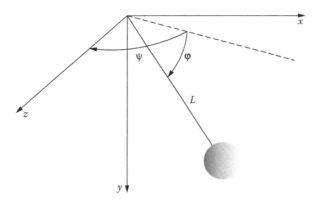

FIGURE 2.1
Model of a spherical pendulum.

When motion equations for the pendulum are generated in x,y,z coordinates, the dynamic model consists of differential-algebraic equations (DAEs) that are of index 3; for details about the DAE index, see Layton (1998). Then, they may not be the best coordinates from the control application perspective. We may select the generalized coordinates ψ and φ that uniquely specify the pendulum location at any time instant.

Example 2.2

Consider a four-bar mechanism as presented in Figure 2.2. It consists of links of constant length l_1, l_2, l_3.
To build such a mechanism, the following design conditions have to be satisfied

$$l_1 \cos \varphi_1 + l_2 \cos \varphi_2 - l_3 \cos \varphi_3 - d_1 = 0,$$
$$l_1 \sin \varphi_1 + l_2 \sin \varphi_2 - l_3 \sin \varphi_3 - d_2 = 0. \tag{2.3}$$

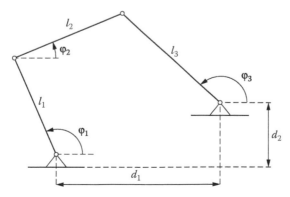

FIGURE 2.2
Model of a four-bar mechanism.

We may generalize Equation (2.2) saying that when the Cartesian coordinates specify a system motion, the following constraints may hold:

$$f_\alpha(t, x_1, y_1, z_1, ..., x_n, y_n, z_n) = 0, \alpha \leq 3n, \alpha = 1, ..., k \qquad (2.4)$$

They reflect geometric properties of a system. In Equation (2.4) α is the number of relations that must be satisfied. We assume that $\alpha \leq 3n$. When $\alpha = 3n$, the system motion is uniquely determined by $3n$ functions $x_i = x_i(t)$, $y_i = y_i(t)$, $z_i = z_i(t)$.

Also, Equation (2.3) may be generalized by

$$f_1(_1, _2, _3) = 0, \qquad f_2(_1, _2, _3) = 0. \qquad \alpha = 2 \qquad (2.5)$$

Similar examples of the constraint specifications are given in Chapter 1. Also, compare Figures 1.12 and 2.2 for various possibilities of angle selections.

The first observation then is that the Cartesian coordinates may not always be the most suitable to specify a system motion. For many multi-link systems, they are good for motion analysis and simulation; for details about modeling, see de Jalon and Bayo (1994).

We would like to have coordinates that minimize the number of auxiliary conditions that have to be handled in dynamic models and are related to control inputs.

Consider then a set of n particles constrained by (2.4). Then, $m = 3n - k$ out of the coordinates x_i, y_i, z_i, $i = 1, ..., n$, are mutually independent, and each of n functions x_i, y_i, z_i may be presented as a function of time and m new parameters $q_1(t), ..., q_m(t)$. The parameters are selected in such a way that Equations (2.4) are satisfied by all q_i, $i = 1, ..., m$, at any time instant t in a selected space. New parameters $q_1(t), ..., q_m(t)$ are referred to as generalized coordinates and may be formally introduced as

$$x_i = f_{1i}(t, q_1, ..., q_m), y_i = f_{2i}(t, q_1, ..., q_m), z_i = f_{3i}(t, q_1, ..., q_m). \qquad i=1, ..., n \qquad (2.6)$$

First, there is no rule of how to select the generalized coordinates. Quite often, they have a physical meaning of a translation or rotation. However, they do not need to have the physical interpretation. Their selection is due to a designer, either a mechanical or control engineer. It is necessary for the coordinates $q_1, ..., q_m$ that the functions in Equation (2.6) are uniquely determined, are continuous bounded and differentiable time functions, and the Jacobian of one of the m function combination is nonzero.

A concept of a configuration space is related to the set of coordinates $q_1, q_2, ..., q_m$. Similarly as for the triplet (x, y, z) that is represented by a point in the three-dimensional Euclidean space, we may consider the set of $q_1, q_2, ..., q_m$ as coordinates of a "particle" in a m-dimensional space. Analogously, we may generalize a concept of a curve and motion along the curve. Instead of a particle motion specified by $x = f_1(t)$, $y = f_2(t)$, $z = f_3(t)$, we may consider a set of equations $q_1 = q_1(t), ..., q_m = q_m(t)$. This set is equivalent to a map, in which the "particle" of m-dimensional space moves along some curve in this space.

The "particle" represents any system in the configuration space $D_m(q_1,...q_m)$. A rigid body location may be presented as a "particle" in D_6. The space D_6 is not a physical space for the body. It represents an injection between the body location and the point in D_6. The $D_{m+1}(q_i, t)$, $i = 1,...,m$ space, for which a "particle" coordinates are $(q_1,...,q_m, t)$, is referred to as $(m + 1)$ space-time and each curve $q_i = q_i(t)$, $i = 1,...,m$, is a trajectory in this space-time.

Another concept is a number of degrees of freedom. It is equal to m and to the smallest number of parameters that uniquely specify a system configuration in time t, in a selected reference frame. The pendulum from Example 2.1 has $m = n - k = 3 - 1 = 2$ degrees of freedom. The independent parameters q_1, q_2 for the pendulum may be selected as $q_1 = $, $q_2 = \psi$. For the four-bar mechanism from Example 2.2, the number of degrees of freedom is equal to $m = 3 - 2 = 1$, and one independent parameter may be selected as $q_1 = $ $_1$.

In many robotic systems, for their motion analysis, the selection of the independent coordinates, i.e., the generalized coordinates, is not always the best choice. As presented in Example 1.11, other coordinates may be more suitable (for more examples see de Jalon and Bayo 1994). In subsequent chapters, we demonstrate that the generalized coordinates may not be the best choice for control-oriented modeling. These are the reasons for referring to the concept of the number of degrees of freedom, because the selection of coordinates for a system modeling starts from the question about their number and whether we need an independent or dependent set of coordinates. The selection of dependent coordinates is followed, in most cases, by the increased number of equations a model consists of, and they may require further numerical support, i.e., numerical stabilization of the constraint equations.

2.1.2 Generalized Velocities and Quasi-Velocities

In most cases, generalized velocities are defined as time derivatives of generalized coordinates, i.e., they are a set of $\dot{q}_1,...,\dot{q}_m$, $m = n - k$. If a position vector of a particle or a selected body point is specified by r, which is presented by generalized coordinates as

$$r = r(t, q_1,...,q_m),\tag{2.7}$$

then the velocity vector is of the form

$$v = \sum_{r=1}^{m} \frac{\partial r}{\partial q_r}\dot{q}_r + \frac{\partial r}{\partial t}.\tag{2.8}$$

Sometimes, the coordinates $q_1,...,q_m$ are referred to as Lagrange's variables. Similarly, the time derivative of the generalized velocity is referred to as the generalized acceleration.

In Kane and Levinson (1985), the generalized velocities are defined as

$$u_r = \sum_{l=1}^{n} a_{rl}\dot{q}_l + b_r, \qquad r = 1,...,m\tag{2.9}$$

where a_{rl} and b_r are functions of generalized coordinates and time. They have to be selected in such a way that the velocities $\dot{q}_l$, $l = 1, ..., n$ can be computed from Equations (2.9).

In this book, the generalized velocities are defined as $\dot{q}_1, ..., \dot{q}_m$, $m = n - k$.

As we showed in Chapter 1, in Example 1.3, sometimes for the motion description it is more convenient to select a system velocity as a variable, e.g., control variable. This velocity is a combination of generalized velocities, and it is referred to as a quasi-velocity. Quasi-coordinates and quasi-velocities were introduced first to derive equations of motion referred to as the Boltzmann–Hamel equations, for details see Nejmark and Fufaev (1972). Relations between the generalized velocities and quasi-velocities are assumed linear and nonintegrable, i.e.,

$$\omega_r = \omega_r(t, q_\sigma, \dot{q}_\sigma). \qquad \sigma, r = 1, ..., n. \tag{2.10}$$

Also, it is assumed that inverse transformations can be computed for Equation (2.10) in the considered state space, i.e.,

$$\dot{q}_\lambda = \dot{q}_\lambda(t, q_\sigma, \omega_r). \qquad \lambda = 1, ..., n \tag{2.11}$$

Quasi-coordinates can be introduced by the following relations:

$$d\pi_r = \sum_{\sigma=1}^{n} \frac{\partial \omega_r}{\partial \dot{q}_\sigma} dq_\sigma \qquad r = 1, ..., n \tag{2.12}$$

and the assumption that Equations (2.12) are nonintegrable holds. Based on Equations (2.10) through (2.12), the generalized coordinates are related to the quasi-coordinates as

$$dq_\lambda = \sum_{\mu=1}^{n} \frac{\partial \dot{q}_\lambda}{\partial \omega_\mu} d\pi_\mu. \qquad \lambda = 1, ..., n \tag{2.13}$$

Based on Equations (2.12) and (2.13), variations of quasi-coordinates and generalized coordinates are related as

$$\delta\pi_r = \sum_{\sigma=1}^{n} \frac{\partial \omega_r}{\partial \dot{q}_\sigma} \delta q_\sigma, \quad \delta q_\lambda = \sum_{\mu=1}^{n} \frac{\partial \dot{q}_\lambda}{\partial \omega_\mu} \delta\pi_\mu \qquad r, \lambda = 1, ..., n \tag{2.14}$$

For our further theoretical development, consider an arbitrary function $= (q_\sigma), \sigma = 1,...,n$, from C^2 and compute its total differential using Equation (2.13), i.e.,

$$d = \sum_{\lambda=1}^{n} \frac{\partial}{\partial q_\lambda} dq_\lambda = \sum_{\lambda=1}^{n} \frac{\partial}{\partial q_\lambda} \sum_{\mu=1}^{n} \frac{\partial \dot{q}_\lambda}{\partial \omega_\mu} d\pi_\mu = \sum_{\mu=1}^{n} \left(\sum_{\lambda=1}^{n} \frac{\partial}{\partial q_\lambda} \frac{\partial \dot{q}_\lambda}{\partial \omega_\mu} \right) d\pi_\mu.$$

Denoting by

$$\frac{\partial}{\partial \pi_\mu} = \sum_{\lambda=1}^{n} \frac{\partial}{\partial q_\lambda} \frac{\partial \dot{q}_\lambda}{\partial \omega_\mu}, \qquad \mu = 1,...,n \qquad (2.15)$$

the total differential of $= (q_\sigma)$ becomes

$$d = \sum_{\mu=1}^{n} \frac{\partial}{\partial \pi_\mu} d\pi_\mu. \qquad (2.16)$$

Relation (2.15) is the operator formula of the form

$$\frac{\partial \cdot}{\partial \pi_\mu} = \sum_{\lambda=1}^{n} \frac{\partial \cdot}{\partial q_\lambda} \frac{\partial \dot{q}_\lambda}{\partial \omega_\mu}, \qquad \mu = 1,...,n \qquad (2.17)$$

Based on Equations (2.17), the inverse formula is

$$\frac{\partial \cdot}{\partial q_\lambda} = \sum_{r=1}^{n} \frac{\partial \cdot}{\partial \pi_r} \frac{\partial \omega_r}{\partial \dot{q}_\mu}, \qquad \lambda = 1,...,n \qquad (2.18)$$

In Section 2.3.6, we use quasi-coordinates to derive the Boltzmann–Hamel equations.

2.2 Constrained Systems

The concept of constraints in classical mechanics is based on the assumption that constraints are given a priori, and they are put upon a mechanical system through other bodies or physical systems, i.e., when bodies are in contact with each other or roll without slipping. They are position and kinematic constraints, and they are "known" and "given" by nature (Dobronravov 1970; Gutowski 1971; Kamman and Huston 1984; Kane and Levinson 1985; Lancos 1986; Layton 1998; Moon 1998; Nejmark and Fufaev 1972; Papastavridis 1988, 1994, 1998, 2002; Pars 1965; Udwadia and Kalaba 1996). Also, as we demonstrated in Chapter 1, the constraint sources may originate from modeling, control, design, or task specifications.

One general classification divides constraints into unilateral constraints specified by inequalities, and equality constraints specified by equations. Constraints can be modeled as ideal or nonideal. We address ideal equality constraints.

2.2.1 Holonomic Constraints

Holonomic (also referred to as position or geometric) constraints are specified by algebraic equations of the form

$$_\alpha(t, q_1, \ldots, q_n) = 0. \qquad \alpha = 1, \ldots, a, \quad a < n. \tag{2.19}$$

We assume that functions $_\alpha$, $\alpha = 1, \ldots, a$ are defined on a $(n+1)$-dimensional manifold and have continuous derivatives up to the second order at least. In a vector notation Equations (2.19) have the form $A(t, q) = 0$, where q is an n-dimensional vector of generalized coordinates, and A is an a-dimensional vector. When, for example, $\alpha = 1$, $n = 3$, and Equations (2.19) does not depend on time, it takes the form $_1(q_1, q_2, q_3) = 0$. It may be interpreted as a stationary plane at which a particle has to stay during its motion. When $\alpha = 2$, $n = 3$, Equations (2.19) take the form $_1(q_1, q_2, q_3) = 0$, $_2(q_1, q_2, q_3) = 0$. In this case, the constraints specify a line in space. It is generated as the intersection of two surfaces $_1 = 0$ and $_2 = 0$.

Geometric constraints (2.19) are referred to as rheonomic when functions $_\alpha$, $\alpha = 1, \ldots, a$ depend on time explicitly and scleronomic when they do not. They belong to the group of material constraints because they mostly specify relations in system bodies, e.g., constant link length in a multilink system. Usually, material position constraint equations are of the form $A(q) = 0$.

Position constraints restrict velocities and accelerations of a system. By differentiating Equations (2.19) once and twice with respect to time, we obtain constraints on velocities and accelerations, that is,

$$\frac{d_\alpha}{dt} = \sum_{r=1}^{n} \frac{\partial_\alpha}{\partial q_r} \dot{q}_r + \frac{\partial_\alpha}{\partial t} = \sum_{r=1}^{n} \dot{q}_r grad_r {}_\alpha + \frac{\partial_\alpha}{\partial t} = 0, \tag{2.20}$$

$$\frac{d^2_\alpha}{dt} = \sum_{r=1}^{n} \frac{\partial_\alpha}{\partial q_r} \ddot{q}_r + \sum_{r=1}^{n} \dot{q}_r grad_r {}_\alpha \frac{d_\alpha}{dt} + \frac{d}{dt}\left(\frac{d_\alpha}{dt}\right) = 0. \tag{2.21}$$

The term $grad_r \; \alpha = \frac{\partial \; \alpha}{\partial q_r}$ is a constraint gradient at a point specified by coordinates q_r. It follows from Equations (2.20) and (2.21) that the position constraints restrict velocity and acceleration components in the direction of the constraint gradient. The inverse statement is not true.

Other examples of the position constraints are given in Example 1.11. The position constraints are not a problem from the modeling point of view. We may select a set of coordinates in such a way that they are identically satisfied, i.e., we may select a set of generalized coordinates.

2.2.2 Nonholonomic Constraints

Kinematic (velocity) constraints are specified by first order differential equations

$$_\beta(t, q_1, ..., q_n, \dot{q}_1, ..., \dot{q}_n) = 0. \qquad \beta = 1, ..., b, \quad b < n. \qquad (2.22)$$

We assume that functions $_\beta, \beta = 1, ..., b$ are defined on a $(2n+1)$ -dimensional manifold and have continuous derivatives. In a vector notation Equations (2.22) have the form $B_1(t, q, \dot{q}) = 0$, where B_1 is a b-dimensional vector. Most material kinematic constraints do not depend on time explicitly, and the form $B_1(q, \dot{q}) = 0$ is often used.

In many cases, kinematic constraints are linear in velocities

$$\sum_{\sigma=1}^{n} b_{\beta\sigma}(t, q_1, ..., q_n)\dot{q}_\sigma + b_{\beta o}(t, q_1, ..., q_n) = 0. \qquad \beta = 1, ..., b \qquad (2.23)$$

In a matrix notation (2.23) have the form $B_1(t, q)\dot{q} + b_1(t, q) = 0$, where $B_1(t, q)$ is a $(b \times n)$ matrix, and $b_1(t, q)$ is a b-dimensional vector.

Kinematic constraints (Equation 2.22 or Equation 2.23) also restrict accelerations. If the constraint equations (2.22) can be integrated with respect to time, we call them holonomic. Otherwise, they are called nonholonomic. The nonholonomic constraints do not restrict positions. They are first order differential equations, and we refer to them as first order constraints. Constraints (2.19) and (2.22) are referred to as material constraints in the classical mechanics setting (Gutowski 1971; Nejmark and Fufaev 1972).

Example 2.3

Consider a rolling falling disk on a plane (x, y) that may roll without slipping (Figure 2.3).

The disk location is specified by x, y that are the coordinates of a contact point between the disk and the plane, ψ is the rolling angle, φ - heading angle, α - yaw angle.

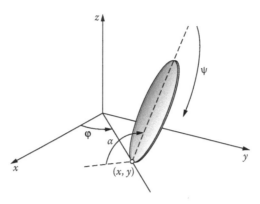

FIGURE 2.3
Model of a rolling falling disk.

The assumption of rolling without slipping implies that the velocity of the contact point of the disk and the ground is equal to zero. It yields the following relations:

$$\dot{x} = r\dot{\psi}\cos \quad , \qquad \dot{y} = r\dot{\psi}\sin \quad . \tag{2.24}$$

The coordinates $x, y,$ $,\psi,\alpha$ may take arbitrary values during the disk motion, i.e., it may reach any location on the plane but the velocities have to satisfy Equations (2.24).

The classification of the constraints provided in this section is not the only one that exists in the literature. In both mechanics and control, there are constrained systems referred to as Chaplygin (Bloch 2003; Kolmanovsky and McClamroch 1995) and non-Chaplygin, e.g., one is analyzed in Jarzębowska and McClamroch (2000).

Nonholonomic constraint equations (2.22) can be written in the form

$$\dot{q}_i = \sum_{j=1}^{m} {}_{ij}(q_1, ..., q_n)\dot{q}_j, \qquad i = m+1, ..., n, \, m = n-k \tag{2.25}$$

where the first m out of n generalized velocities in the number equal to the number of system's degrees of freedom, are regarded as independent, $(n-m)$ velocities are dependent, and $_{ij}(q_1, ..., q_n)$ are smooth functions of their arguments. Generalized velocities can be partitioned in this way at least locally to obtain Equations (2.25).

Following the early work of Chaplygin (1897), if constraint functions satisfy certain symmetry properties, namely that they are cyclic in the last $(n-m)$ generalized coordinates, we obtain Chaplygin nonholonomic constraint equations in the form

$$\dot{q}_i = \sum_{j=1}^{m} {}_{ij}(q_1, ..., q_m)\dot{q}_j. \qquad i = m+1, ..., n \tag{2.26}$$

From the control point of view, Equations (2.25) and (2.26) can be presented in the control form by viewing the independent generalized velocities as inputs. Then Equations (2.25) take the form

$$\dot{q}_j = u_j,$$

$$\dot{q}_i = \sum_{j=1}^{m} {}_{ij}(q_1,...,q_n)u_j. \qquad i = m+1,...,n \qquad (2.27)$$

Often, the nonlinear control form (2.27) is presented as

$$\dot{y}_i = u_i,$$

$$\dot{z} = \sum_{i=1}^{m} \tilde{g}_i(z,y)u_i, \qquad m \geq 2 \qquad (2.28)$$

and the vector q is partitioned as $q=(z,y)$, where $z=(q_{m+1},...,q_n)$ is a $(n-m)$-dimensional vector referred to as the fiber vector, and $y=(q_1,...,q_m)$ is a m-dimensional base vector. If the Chaplygin assumption holds, Equations (2.27) can be presented in the nonlinear control form as

$$\dot{q}_j = u_j,$$

$$\dot{q}_i = \sum_{j=1}^{m} {}_{ij}(q_1,...,q_m)u_j. \qquad i = m+1,...,n \qquad (2.29)$$

Following Equations (2.28) and (2.29), a control system is Chaplygin if ${}_{ij}$ depend on the base vector but not on the fiber vector. Most systems with material nonholonomic constraints are Chaplygin, and for this reason the kinematic control model (2.29) is a focus of many theoretic control studies (e.g., Bloch 2003; Bushnell et al. 1995; Campion, d'Andrea-Novel, and Bastin 1991; Samson 1990, 1995; Scheuer and Laugier 1998). The nonholonomic constraint equations (2.25) or, equivalently, (2.27) that cannot be written in the form of (2.26) or, equivalently (2.29), are referred to as non-Chaplygin nonholonomic systems. In Chapter 3, we get back to the discussion about the control properties of Chaplygin and non-Chaplygin systems.

2.2.3 Programmed Constraints

Material constraints are a significant class of motion limitations in engineering practice, but there are many problems for which constraints are formulated in a different way. For example, in design or operation problems, constraints are formulated before a system is designed. Tasks are excellent examples of such constraints. They are specified first, and then we develop dynamic or control models of a system with the task-based constraints. Generally, sources of these constraints are not in other bodies. Such constraints are referred to as non-material and may arise as performance, design, operation, control, or safety requirements. They can be formulated in the form of algebraic or differential equations, or inequalities. The earliest formulation of the non-material constraints known to the author was given by Appell (1911). He described them as constraints "that can be realized not through a direct contact." Similar ideas were introduced by Mieszczerski at the beginning of the twentieth century. Beghin developed a concept of servo-constraints (Beghin 1947). These new "constraint sources" were a motivation to specify constraints by the formulations like

$$_\beta(t, q_1, ..., q_n, \dot{q}_1, ..., \dot{q}_n) = 0. \qquad \beta = 1, ..., k, \quad k < n \qquad (2.30)$$

In the vector form, Equation (2.30) can be written as $B_1(t, q, \dot{q}) = 0$, where B_1 is a k-dimensional vector.

Later works that concern non-material position constraints called "programmed" are by Korieniev and Galliulin (see Example 1.9). The concept of programmed constraints is presented in Gutowski (1971) and Gutowski and Radziszewski (1969), and a control problem for a system with position programmed constraints is addressed in Zotov and Tomofeyev (1992) and Zotov (2003). Moreover, there are a lot of examples of specifications of restrictions on system motions which may be viewed as non-material constraints. The Grioli example, presented in Chapter 1, Example 1.8, specifies a non-material nonholonomic constraint equation of the second order. In biomechanical modeling, it was found that the third order time derivative of a position coordinate influences the smoothness of a limb motion (Hogan 1987). For underactuated systems, a second order nonholonomic non-material constraint originates from control or design requirements (e.g., Oriolo and Nakamura 1991; Reyhanoglu et al. 1996). In navigation of wheeled mobile robots, to avoid wheel slippage and mechanical shock during motion, a dynamic constraint put on the acceleration has to be taken into account (Koh and Cho 1999). In path planning problems, for car-like robots, to secure motion smoothness two additional constraints are put upon trajectory curvature and its time derivative (Koh and Cho 1999; Oriolo, de Luca, and Vendittelli 2002; Scheuer and Laugier 1998; Shekl and Lumelsky 1998). They are constraint equations of the second and third orders, respectively. Driving and task constraints are examples of non-material constraints (de Jalon and Bayo 1994; Murray et al. 1994; Nikravesh 1988). A position constraint equation that specifies a trajectory may also be viewed as non-material; however, it is not presented in this form in a control setting (see, e.g., Fukao et al. 2000; Koh and Cho 1999). Other non-material constraints on the space vehicle velocity are reported in Vafa (1991). None of the non-material constraint reported above is formulated in the programmed equation form.

The overview of the constraint classifications in mechanics and a variety of requirements on system's motions reported in the literature can be summarized as follows:

1. Many problems are formulated as synthesis problems, and motion requirements may be viewed as non-material constraints imposed on a system before it is designed and put into operation.
2. Constraints that specify motion requirements may be of orders higher than one or two.
3. Non-material constraints may arise in modeling and analysis of electromechanical and biomechanical systems.
4. No unified approach to the specification of non-material constraints or any other unified constraint has been formulated in classical mechanics.

These conclusions are a motivation to present an extended concept of a constraint (Jarzębowska 2002, 2007b).

Definition 2.1: A programmed constraint is any requirement put on a physical system motion specified by an equation.

Definition 2.2: A programmed motion is a system motion that satisfies a programmed constraint.

A system can be subjected to both material and programmed constraints. Programmed constraints do not have to be satisfied during all motion of a system. Examples 1.8, 1.9, and 1.10 provide formulations of the programmed constraints.

2.3 Equations of Motion for Systems with First Order Constraints

Classical analytical mechanics is a discipline with its roots in the eighteenth and nineteenth century works of Lagrange, Hamilton, Jacobi, and many others. Variational principles underlie the basic theory of analytical mechanics, and in contrast to geometric mechanics, it is based on the variational approach. Within the classical mechanics framework, work, energy, and constraints are the unifying concepts in generalizing the principles of analytical mechanics to encompass multidisciplinary systems (see, e.g., a systematic exposition of the principles of analytical dynamics applied to the modeling of multidisciplinary systems in Layton 1998).

Classical analytical mechanics provides modeling methods for systems with first order nonholonomic constraints, which are usually material (see, e.g., Appell 1911, 1941, 1953; Bloch 2003; Boltzmann 1902; de Jalon and

Bayo 1994; Dobronravov 1970; Gutowski 1971; Hamel 1904, 1967; Kamman and Huston 1984; Kane and Levinson 1985; Kwatny and Blankenship 2000; Lancos 1986; Layton 1998; Moon 1998; Nejmark and Fufaev 1972; Nielsen 1935; Papastavridis 1988, 1994, 1998, 2002; Pars 1965; Rong 1996a, 1996b, 1997; Soltakhanov et al. 2009; Tianling 1992; Udwadia and Kalaba 1996; Voronets 1901; Vranceanu 1929). Appell-Gibbs equations can also be applied to systems with second order constraints (Appell 1953; Gutowski 1972; Nejmark and Fufaev 1972). At the beginning of the twentieth century, a trend of leaving the Lagrange's equations approach began. It was accompanied by efforts to eliminate unknown constraint reaction forces from equations of motion.

Historically, the problem of elimination of unknown constraint reaction forces from equations of motion was solved by Chaplygin (1897). In 1901, Voronetz obtained equations for nonholonomic systems for a more general case in which the kinetic energy, potential energy, and coefficients of constraint equations might depend on all generalized coordinates as well as on time (Voronets 1901). In 1898, Volterra proposed equations of motion in the kinematic forms, later called quasi-velocities (Nejmark and Fufaev 1972). In 1901, Maggi derived equations of motion with no constraint reaction forces in a form that yields Volterra's equations (Nejmark and Fufaev 1972). Appell's equations were derived in 1899 on the basis of the principle of the least constraint. Boltzmann (1902) and Hamel (1904) derived a form of equations in quasi-coordinates, which coincide for holonomic systems. Later versions of the equations were developed by Tzenoff (1924), Vranceanu (1929), and Schouten. For more details, see Nejmark and Fufaev (1972), Papastavridis (2002), and references therein. Some later works deal with nonlinear nonholonomic constraints but of the first order (Rong 1996a, 1996b, 1997). Between 1910 and 1930, dynamics of nonlinear nonholonomic systems made a significant progress. Appell, Chetaev, and Hamel derived equations for mechanical systems subjected to constraint equations nonlinear in velocities. Some other efforts to leave the Lagrange's equations approach resulted in equations obtained by Nielsen (1935) and Tzenoff (1924).

In this section, we derive equations of motion of the Lagrange type and their modifications that result from the elimination of the constraint reaction forces. All these equations may be applied to systems subjected to first order constraints. Also, we derive equations of motion in quasi-coordinates. In Section 2.4, we generalize the equations of motion for constrained systems.

2.3.1 D'Alembert Principle

Consider a system of N particles subjected to position constraints given by Equations (2.4).

A virtual displacement $\delta r_i(\delta x_i, \delta y_i, \delta z_i)$, $i = 1,...,N$, is a vector that satisfies the relations

$$\sum_{i=1}^{N} \delta r_i \ grad_i f_\alpha = 0, \qquad \alpha = 1,...,a \tag{2.31}$$

where a symbol δ indicates variation. Relation (2.31) may be presented as

$$\sum_{i=1}^{N} \left(\frac{\partial f_\alpha}{\partial x_i} \delta x_i + \frac{\partial f_\alpha}{\partial y_i} \delta y_i + \frac{\partial f_\alpha}{\partial z_i} \delta z_i \right) = 0. \qquad \alpha = 1,...,a. \tag{2.32}$$

The virtual displacement concept may be interpreted geometrically. For a constrained particle located on the plane $f(t,x,y,z) = 0$, the following holds:

$$\frac{df}{dt} = \frac{\partial f}{\partial x} \dot{x} + \frac{\partial f}{\partial y} \dot{y} + \frac{\partial f}{\partial z} \dot{z} + \frac{\partial f}{\partial t} = v_i \ grad \ f + \frac{\partial f}{\partial t} = 0. \tag{2.33a}$$

According to Equations (2.31), δr_i satisfies

$$\delta r_i \ grad \ f = 0. \tag{2.33b}$$

From the comparison of Equations (2.33a) and (2.33b), it follows that for the scleronomic constraints, virtual displacements δr_i are proportional to the velocities v_i, that is, to the velocities permitted by the constraints. The virtual displacement is then an arbitrary displacement tangent to the stationary plane $f(t,x,y,z) = 0$ at the selected point, for which $t = 0$.

For the rheonomic constraints, the virtual displacement is not proportional to the particle velocity. Then, according to Equation (2.33b), the virtual displacement vector is directed along the allowable velocity vector at the time t, at which the plane is actually "frozen."

Introducing the generalized coordinates by relations

$$r_i = r_i(t, q_1,...,q_n), \qquad i = 1,...,N, \tag{2.34}$$

the virtual displacements may be presented as

$$\delta r_i = \sum_{r=1}^{n} \frac{\partial r_i}{\partial q_r} \delta q_r, \qquad i = 1,...,N. \tag{2.35}$$

Also, it is important to note that δq_i is not the same as dq_i. The virtual displacement satisfies the constraints only, and the dq_i satisfies both the velocity constraints and the equations of motion.

Example 2.4

Let the constraint equation (2.4) has the form

$$f = x^2 + y^2 + z^2 - R^2 = 0. \tag{2.36}$$

It specifies a condition that a particle is to remain on a sphere of radius R. According to Equation (2.31),

$$\delta r \, grad f = \frac{\partial f}{\partial x} \delta x + \frac{\partial f}{\partial y} \delta y + \frac{\partial f}{\partial z} \delta z = 2(x \delta x + y \delta y + z \delta z) = 0. \tag{2.37}$$

The selection of arbitrary x and y, and determination of z from Equation (2.37), results in a vector r at the point (x,y,z).

Consider now the same set of N particles, on which forces F_i, $i = 1,...N$, may act. For constrained systems, the constraints are replaced by constraint reaction forces and the second Newton law may be written for such systems as

$$m_i a_i = F_i + R_i, \qquad i = 1,...N \tag{2.38}$$

where a_i denotes acceleration vectors for the system particles.

Usually, in mechanics we are interested in the system motion, not in the reaction forces R_i. Reactions are "auxiliary" quantities that have to be taken into account in order to derive equations of motion based on Newton's second law. They are not given a priori—they only have to assure the system motion according to the constraints.

A class of constraints, referred to as ideal, can be distinguished. Constraints are ideal if the work done by the constraint reaction forces is equal to zero, i.e.,

$$\sum_{i=1}^{N} R_i \delta r_i = 0. \tag{2.39}$$

In this book, we consider ideal constraints only.

In nonlinear control, the same assumption about the constraints is adopted. However, the constraints are viewed in a different way in a control setting, i.e., we are not interested in directions a system cannot move but in the directions it can. We address the constraint reformulation in Chapter 3.

Example 2.5

Consider a rigid body, for which distances between its particles remain constants, that is,

$$(r_i - r_j)^2 = const, \qquad i, j = 1, ..., N \qquad (2.40)$$

Constraint reaction forces in the rigid body should enforce the constraints. According to Newton's third law, reactions R_{ij} satisfy $R_{ij} = -R_{ji}$, and due to an interaction,

$$R_{ij} = \lambda(r_i - r_j), \qquad \lambda = const. \qquad (2.41)$$

The virtual work due to reaction forces is equal to

$$R_{ij}\delta r_i + R_{ji}\delta r_j = R_{ij}(\delta r_i - \delta r_j) = \lambda(r_i - r_j)(\delta r_i - \delta r_j) = 0. \qquad (2.42)$$

Example 2.6

Consider a rigid body constrained by

$$f(x, y) = 0. \qquad (2.43)$$

The constraint specifies a demand that the body remains on the surface f during its motion. The surface is assumed to be smooth and frictionless. A reaction R is perpendicular to the surface, then $R\delta r = 0$. The constraint is ideal.

Example 2.7

Consider the rigid body from Example 2.6, for which the surface is not stationary. Then, $\delta r \neq 0$, but the constraints are still ideal, because $\delta r_i = \delta r_j$ for no slip, and $R_{ij} = -R_{ji}$. Then

$$R_{ij}\delta r^i + R_{ji}\delta r^j = \delta r^i(R_{ij} - R_{ji}) = 0. \qquad (2.44)$$

Let us now transform Equation (2.38) as follows:

$$F_i + R_i - m_i a_i = S_i = 0, \qquad i = 1, ..., N \qquad (2.45)$$

and denote $-m_i a_i = B_i$, $i = 1, ..., N$.

Vectors B_i, $i = 1, ... N$, are referred to as inertia forces. Then, Equation (2.45) takes the form

$$F_i + R_i + B_i = S_i = 0. \qquad i = 1, ..., N. \qquad (2.46)$$

Due to d'Alembert, the forces S_i satisfy the relation

$$\sum_{i=1}^{N} S_i \delta r_i = \sum_{i=1}^{N} (F_i + R_i + B_i) \delta r_i = 0. \tag{2.47}$$

If the constraints imposed upon a system are ideal, then based on (2.39), (2.47) takes the form

$$\sum_{i=1}^{N} (F_i + B_i) \delta r_i = \sum_{i=1}^{N} (F_i - m_i a_i) \delta r_i = 0. \tag{2.48}$$

Relation (2.48) provides an analytical formulation of d'Alembert's principle. It is the basis for the generation of equations of motion of constrained systems.

D'Alembert's principle can be formulated as follows:

If a system of particles is subjected to holonomic and nonholonomic equality and ideal constraints, the virtual work due to applied and inertia forces acting along a virtual displacement is equal to zero.

It is important to note that d'Alembert's principle is formulated based on the concept of the ideal constraints and their replacement by the constraint reaction forces. It may be proceeded in the other way: postulate d'Alembert's principle (2.48) and then there is no need to postulate to replace the constraints by reaction forces, because it results from d'Alembert's principle.

Consider a particle that may move on a smooth surface $f(x,y,z) = 0$. Based on d'Alembert's principle, we may write

$$(F_x - m\ddot{x})\delta x + (F_y - m\ddot{y})\delta y + (F_z - m\ddot{z})\delta z = 0. \tag{2.49}$$

The particle virtual displacement vector satisfies

$$\delta r \operatorname{grad} f = \frac{\partial f}{\partial x}\delta x + \frac{\partial f}{\partial y}\delta y + \frac{\partial f}{\partial z}\delta z = 0. \tag{2.50}$$

Multiplying (2.50) by some $\lambda = \lambda(t, x, y, z)$ and adding it to (2.49) results in

$$\left(F_x - m\ddot{x} + \lambda\frac{\partial f}{\partial x}\right)\delta x + \left(F_y - m\ddot{y} + \lambda\frac{\partial f}{\partial y}\right)\delta y + \left(F_z - m\ddot{z} + \lambda\frac{\partial f}{\partial z}\right)\delta z = 0. \tag{2.51}$$

Based on Equation (2.49), we may not conclude that $m\ddot{x} = F_x$, $m\ddot{y} = F_y$, $m\ddot{z} = F_z$ or terms in brackets in Equation (2.51) are all equal to zero, because x, y, and z are related by (2.50). One of them, e.g., x, may be determined from (2.50) when y and z are independent. Then, selecting $\lambda = \lambda^*$, such that,

$$F_x - m\ddot{x} + \lambda^* \frac{\partial f}{\partial x} = 0, \tag{2.52}$$

we obtain

$$\left(F_y - m\ddot{y} + \lambda^* \frac{\partial f}{\partial y} \right)\delta y + \left(F_z - m\ddot{z} + \lambda^* \frac{\partial f}{\partial z} \right)\delta z = 0,$$

Since δy and δz are mutually independent,

$$F_y - m\ddot{y} + \lambda^* \frac{\partial f}{\partial y} = 0, \qquad F_z - m\ddot{z} + \lambda^* \frac{\partial f}{\partial z} = 0. \tag{2.53}$$

Adding both sides of Equations (2.52) and (2.53), we obtain that

$$F - m\,a + \lambda^* grad f = 0.$$

We may introduce the notation

$$R = \lambda^* grad f, \tag{2.54}$$

which we refer to as the reaction of a smooth surface at the point (x,y,z). It follows from Equation (2.54) that the vector R is perpendicular to the surface $f(x,y,z) = 0$ at (x,y,z). Then

$$ma = F + R. \tag{2.55}$$

In this way, we demonstrated that starting from d'Alembert's principle, we arrived to the concept of the constraint reaction force and the Newton's second law for a constrained particle. The presented reasoning is referred to as the method of undetermined Lagrange multipliers. We use it to derive equations of motion for systems subjected to noholonomic constraints in Section 2.3.3.

D'Alembert's principle (2.48) does not assume any special constraint kind. It may then be applied to systems subjected to holonomic as well as nonholonomic constraints as long as they are ideal. When we would like to apply Equation (2.48) to a system subjected to nonholonomic constraints, we would have to define a virtual displacement for it.

For a system subjected to nonholonomic constraints, a virtual displacement concept is introduced as follows. The constraint equation (2.4) may be rewritten in the form

$$f_\beta(t, q_1, ..., q_n) = 0, \qquad \beta = 1, ..., b \tag{2.56}$$

and virtual displacements $\delta r_i(q_1, ..., q_n)$ satisfies

$$\sum_{i=1}^{N} \delta r_i grad_i f_\beta = \sum_{i=1}^{n} \frac{\partial f_\beta}{\partial q_i} \delta q_i = 0. \tag{2.57}$$

When a system is subjected to nonholonomic constraints, linear in velocities

$$\overset{*}{\beta} = \sum_{r=1}^{n} H_{\beta r}\dot{q}_r + D_\beta = 0, \qquad \beta = 1,...,b \tag{2.58}$$

with $H_{\beta r}$ being functions of time and coordinates, we may rewrite (2.56) in the form of Equation (2.58):

$$\overset{*}{\beta} = \frac{df_\beta}{dt} = \sum_{i=1}^{n} \frac{\partial f_\beta}{\partial q_i}\dot{q}_i + \frac{\partial f_\beta}{\partial t}. \qquad \beta = 1,...,b \tag{2.59}$$

The following hold for Equations (2.56) and (2.59):

$$\frac{\partial \overset{*}{\beta}}{\partial \dot{q}_i} = \frac{\partial f_\beta}{\partial q_i}, \qquad i{=}1,...,n,\ \beta = 1,...,b \tag{2.60}$$

so using (2.59), Equation (2.57) may be written as

$$\sum_{i=1}^{n} \frac{\partial f_\beta}{\partial q_i}\delta q_i = \sum_{i=1}^{n} \frac{\partial \overset{*}{\beta}}{\partial \dot{q}_i}\delta q_i = 0. \qquad \beta = 1,...,b \tag{2.61}$$

Relation (2.61) may be considered an equivalent definition of a virtual displacement for a system subjected to position constraints. We adopt then (2.61) as a definition of a virtual displacent for a system subjected to nonholonomic constraints. It is as follows:

$$\sum_{i=1}^{n} \frac{\partial \overset{\alpha}{}}{\partial \dot{q}_i}\delta q_i = \sum_{i=1}^{n} H_{\beta_i}\delta q_i = 0. \qquad \beta = 1,...,b \tag{2.62}$$

We may say then, that for a system subjected to material nonholonomic constraints of the form

$$\overset{}{\beta}(t,q_i,\dot{q}_i) = 0, \qquad i{=}1,...,n,\ \beta = 1,...,b \tag{2.63}$$

vectors of virtual displacements are those whose coordinates δq_i satisfy the relations

$$\sum_{i=1}^{n} \frac{\partial \overset{\alpha}{}}{\partial \dot{q}_i}\delta q_i = 0. \qquad \beta = 1,...,b \tag{2.64}$$

The definition above is proposed by Appell and Chetaev. It applies to non-linear nonholonomic constraints of the first order in q. If $\dot{q}_i$, $i = 1, ..., n$, is linear in (2.63) or is not present, definition (2.64) coincides with (2.62) or (2.61). The virtual displacement that satisfies Equation (2.64) is referred to as the virtual displacement in the Appell–Chetaev sense (Nejmark and Fufaev 1972).

2.3.2 Lagrange's Equations for Holonomic Systems

To derive equations of motion for a system subjected to holonomic constraints, we start from Equation (2.48), which we transform as follows. First, transform the right-hand side term:

$$\sum_{i=1}^{n} m_i a_i \delta r_i = \sum_{i=1}^{n} F_i \delta r_i \tag{2.65}$$

using the identity

$$m_i a_i \delta r_i = m_i \dot{v}_i \delta r_i = \frac{d}{dt}(m_i v_i \delta r_i) - m_i v_i (\delta r_i)' = \frac{d}{dt}(m_i v_i \delta r_i) - m_i v_i \delta v_i.$$

It is true for holonomic systems, for which $d\delta \equiv \delta d$. Also, we have that

$$v_i \delta v_i = \frac{1}{2} \delta (v_i \, v_i) = \frac{1}{2} \delta (v_i)^2$$

and Equation (2.65) takes the form

$$\frac{d}{dt} \sum_{i=1}^{n} m_i v_i \delta r_i = \delta \frac{1}{2} \sum_{i=1}^{n} m_i (v_i)^2 + \sum_{i=1}^{n} F_i \delta r_i \tag{2.66}$$

The right-hand side terms in (2.66) are

$$\delta T = \delta \frac{1}{2} \sum_{i=1}^{n} m_i (v_i)^2 \tag{2.67}$$

and

$$\sum_{i=1}^{n} F_i \delta r_i = \delta' W, \tag{2.68}$$

in which T is the kinetic energy of a system, and W is referred to as a virtual work of applied forces due to virtual displacements. The notation $\delta'W$ indicates that it is not a variation of any function (an exception is the case of a virtual work of potential forces).

Based on (2.67) and (2.68), Equation (2.66) takes the form

$$\frac{d}{dt} \sum_{i=1}^{n} m_i v_i \delta r_i = \delta T + \delta' W. \tag{2.69}$$

Equation (2.69) is referred to as the central form of the Lagrange equation. When applied forces are potential, i.e., $F_i = -grad V$, then $\delta'W = -\delta V$ and the right-hand side of Equation (2.69) is

$$\delta T + \delta' W = \delta T - \delta V = \delta(T - V) = \delta L \tag{2.70}$$

A function $L = T - V$ is known as the Lagrange function or just Lagrangian. Using the notion of L, Equation (2.69) may be written as

$$\frac{d}{dt} \sum_{i=1}^{n} m_i v_i \delta r_i = \delta L. \tag{2.71}$$

Introducing the generalized coordinates $r_i = r_i(t, q_1, ..., q_n)$, $i = 1, ..., N$ and the virtual displacements (2.35), the virtual work $\delta'W$ due to applied forces may be written as

$$\delta' W = \sum_{i=1}^{n} F_i \sum_{r=1}^{n} \frac{\partial r_i}{\partial q_r} \delta q_r = \sum_{r=1}^{n} \delta q_r \sum_{i=1}^{n} F_i \frac{\partial r_i}{\partial q_r}.$$

Quantities

$$Q_r = \sum_{i=1}^{n} F_i \frac{\partial r_i}{\partial q_r}, \qquad r = 1, ..., n \tag{2.72}$$

are referred to as generalized forces.

The left-hand side of (2.69) may be transformed as follows. First, use (2.35) to obtain

$$\sum_{i=1}^{N} m_i v_i \delta r_i = \sum_{i=1}^{N} m_i v_i \sum_{i=1}^{n} \frac{\partial r_i}{\partial q_r} \delta q_r = \sum_{r=1}^{n} \delta q_r \sum_{i=1}^{N} m_i v_i \frac{\partial r_i}{\partial q_r} \tag{2.73}$$

and introducing the generalized momenta as

$$p_r = \sum_{i=1}^{N} m_i v_i \frac{\partial r_i}{\partial q_r}, \qquad r = 1,...,n \tag{2.74}$$

the right-hand side of Equation (2.73) takes the form

$$\sum_{i=1}^{N} m_i v_i \delta r_i = \sum_{r=1}^{n} p_r \delta q_r. \tag{2.75}$$

Also, based on (2.8), we may write that

$$\frac{\partial r_i}{\partial q_r} = \frac{\partial v_i}{\partial \dot{q}_r}, \qquad i = 1,...,N, \qquad r = 1,...,n$$

and present (2.74) as

$$p_r = \sum_{i=1}^{N} m_i v_i \frac{\partial v_i}{\partial \dot{q}_r} = \frac{\partial}{\partial \dot{q}_r} \frac{1}{2} \sum_{i=1}^{N} m_i v_i v_i = \frac{\partial T}{\partial \dot{q}_r}. \qquad r = 1,...,n \tag{2.76}$$

Inserting (2.75) and (2.72) into (2.69), we obtain

$$\frac{d}{dt} \sum_{r=1}^{n} p_r \delta q_r = \delta T + \sum_{r=1}^{n} Q_r \delta q_r. \tag{2.77}$$

Specifically, for applied forces F_i, for which $F_i = -grad V$, Equation (2.77) takes the form

$$\frac{d}{dt} \sum_{r=1}^{n} p_r \delta q_r = \delta L. \tag{2.78}$$

In Equations (2.77) and (2.78), the kinetic energy and Lagrangian are both specified in generalized coordinates, i.e., $T = T(t,q_r,\dot{q}_r)$, $L = L(t,q_r,\dot{q}_r)$, $r = 1,...,n$. The variation of the kinetic energy T may be determined as

$$\delta T = \sum_{r=1}^{n} \left(\frac{\partial T}{\partial q_r} \delta q_r + \frac{\partial T}{\partial \dot{q}_r} \delta \dot{q}_r \right). \tag{2.79}$$

Inserting Equations (2.76) and (2.79) into (2.77), we obtain

$$\frac{d}{dt}\sum_{r=1}^{n}\frac{\partial T}{\partial \dot{q}_r}\delta q_r = \sum_{r=1}^{n}\left(\frac{\partial T}{\partial q_r}\delta q_r + \frac{\partial T}{\partial \dot{q}_r}\delta \dot{q}_r + Q_r\delta q_r\right),$$

and using the commutative rule $d\delta \equiv \delta d$ yields

$$\sum_{r=1}^{n}\frac{d}{dt}\left(\frac{\partial T}{\partial \dot{q}_r}\right)\delta q_r + \sum_{r=1}^{n}\frac{\partial T}{\partial \dot{q}_r}\delta \dot{q}_r = \sum_{r=1}^{n}\frac{\partial T}{\partial \dot{q}_r}\delta \dot{q}_r + \sum_{r=1}^{n}\left(\frac{\partial T}{\partial q_r}+Q_r\right)\delta q_r. \qquad (2.80)$$

After rearrangement of the terms in Equation (2.80), we may rewrite it in the form

$$\sum_{r=1}^{n}\left[\frac{d}{dt}\left(\frac{\partial T}{\partial \dot{q}_r}\right) - \frac{\partial T}{\partial q_r} - Q_r\right]\delta q_r = 0. \qquad (2.81)$$

The system we consider is holonomic, and the variations $\delta q_r, r = 1,...,n$, are independent. Hence,

$$\frac{d}{dt}\left(\frac{\partial T}{\partial \dot{q}_r}\right) - \frac{\partial T}{\partial q_r} = Q_r. \qquad r = 1,...,n \qquad (2.82)$$

Equations (2.82) are Lagrange's equations specified in generalized coordinates. In the case of potential forces, $F_i = -grad V$, $L = T\text{-}V$, and Lagrange's equations may be written as

$$\frac{d}{dt}\left(\frac{\partial L}{\partial \dot{q}_r}\right) - \frac{\partial L}{\partial q_r} = 0. \qquad r = 1,...,n \qquad (2.83)$$

Example 2.8

Derive equations of motion for a system presented in Figure 2.4. The disks are assumed to be thin and uniform, their radiuses are R and r, and masses M and m, respectively. The bigger disk may rotate about the mass center S of a smaller one.

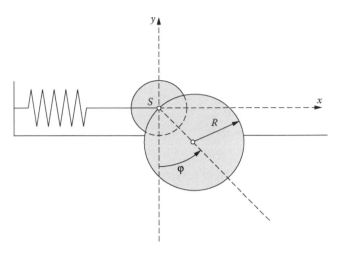

FIGURE 2.4
A system of two disks.

Generalized coordinates are selected as $q_1 = x$, $q_2 = \varphi$. The spring constant is equal to c. The smaller disk may roll on a smooth surface.

To apply Equations (2.83) to the two-disk system, we proceed as follows. First, determine the kinetic energy of a smaller disk T_m and a bigger one T_M, then the potential energy V of the system, and its Lagrangian L:

$$T_m = \frac{1}{2}m\dot{x}^2 + \frac{1}{2}\cdot\frac{1}{2}mr^2\left(\frac{\dot{x}}{r}\right)^2 = \frac{3}{4}m\dot{x}^2, \quad T_M = \frac{1}{2}M(\dot{x}^2 + \dot{\varphi}^2 R^2 + 2\dot{x}\dot{\varphi}R\cos\varphi) + \frac{1}{4}MR^2\dot{\varphi}^2,$$
(2.84)

$$V = MgR(1 - \cos\varphi) + \frac{1}{2}cx^2,$$
(2.85)

$$L = \frac{1}{2}(M + \frac{3}{2}m)\dot{x}^2 + \frac{3}{4}MR^2\dot{\varphi}^2 + MR\cos\varphi\,\dot{x}\dot{\varphi} - MgR(1 - \cos\varphi) - \frac{1}{2}cx^2.$$
(2.86)

Next, calculate all derivatives as Equations (2.83) indicate and insert them into it to obtain

$$\left(M + \frac{3}{2}m\right)\ddot{x} + MR\cos\varphi\,\ddot{\varphi} - MR\sin\varphi\,\dot{\varphi}^2 + cx = 0,$$
(2.87)

$$\frac{3}{2}MR\ddot{\varphi} + MR\cos\varphi\,\ddot{x} + MgR\sin\varphi = 0.$$

2.3.3 Lagrange's Equations for First Order Nonholonomic Systems

Consider a system subjected to first order nonholonomic constraints (2.22). Assume that it is the Appell–Chetaev system. Then, variations δq_r are related by

$$\sum_{r=1}^{n} \frac{\partial \overset{\beta}{}}{\partial \dot{q}_r} \delta q_r = 0. \qquad \beta = 1,...,b \qquad (2.88)$$

To derive equations of motion of a system subjected to the constraints (2.22), we take advantage of the method of undetermined Lagrange multipliers (see Section 2.3.1). Relations (2.88) multiplied by λ_β, $\beta = 1,...,b$, summed from 1 to b,

$$\sum_{\beta=1}^{b} \lambda_\beta \sum_{r=1}^{n} \frac{\partial \overset{\beta}{}}{\partial \dot{q}_r} \delta q_r = 0$$

and subtracted from the Lagrange's equations (2.81) yield

$$\sum_{r=1}^{n} \left[\frac{d}{dt}\left(\frac{\partial T}{\partial \dot{q}_r}\right) - \frac{\partial T}{\partial q_r} - Q_r - \sum_{\beta=1}^{b} \lambda_\beta \frac{\partial \overset{\beta}{}}{\partial \dot{q}_r} \right] \delta q_r = 0. \qquad (2.89)$$

Next we select β multipliers λ, $\beta = 1,...,b$, in such a way that the sums in brackets multiplied by δq_β are all equal to zeros. Based on (2.88), variations $\delta q_{\beta+1},...,\delta q_n$ may be regarded as independent. Then, the sums in the brackets multiplied by these variations are also equal to zeros. Thus, we obtain equations of motion of the system with nonholonomic constraints in the form

$$\frac{d}{dt}\left(\frac{\partial T}{\partial \dot{q}_r}\right) - \frac{\partial T}{\partial q_r} = Q_r + \sum_{\beta=1}^{b} \lambda_\beta \frac{\partial \overset{\beta}{}}{\partial \dot{q}_r}. \qquad r = 1,...,n. \qquad (2.90)$$

Equations (2.90) and equations of constraints $\,_\beta(t, q_\sigma, \dot{q}_\sigma) = 0$, $\sigma = 1,...,n$, are a set of $(n + b)$ equations for n unknown coordinates q_r and b multipliers λ. Specifically, for the constraints in the form (2.58), Equations (2.90) have the form

$$\frac{d}{dt}\left(\frac{\partial T}{\partial \dot{q}_r}\right) - \frac{\partial T}{\partial q_r} = Q_r + \sum_{\beta=1}^{b} \lambda_\beta H_{\beta r}. \qquad r = 1,...,n. \qquad (2.91)$$

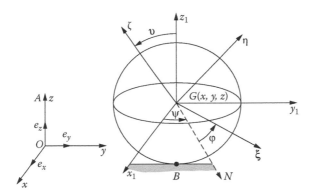

FIGURE 2.5
Model of a rolling ball.

Example 2.9

Consider a ball rolling and spinning without slipping on a plane surface. The ball mass is m, its radius is r (see Figure 2.5).

Select generalized coordinated $q_1 = \quad , q_2 = \psi, q_3 = \quad , q_4 = x, q_5 = y, q_6 = a =$ const. to specify the ball motion. A condition of rolling without slipping generates the vector equation in the form

$$v_B = v_C + \omega \times a = 0, \tag{2.92}$$

with $a = CA = -a\, e_z, v_C = \dot{x} e_x + \dot{y} e_y$, and ω being the angular velocity vector of the ball.

Inserting the kinematic relations:

$$\omega_x = \dot{}\, \sin\ \sin\psi + \dot{}\, \cos\psi, \omega_y = -\dot{}\, \sin\ \cos\psi + \dot{}\, \sin\psi \tag{2.93}$$

into Equation (2.92), we transform it to

$$(\dot{x} - a\,\omega_y) e_x + (\dot{y} + a\,\omega_x) e_y = 0. \tag{2.94}$$

Using Equations (2.93) again, the constraints (2.94) can be presented as

$$-a\sin\ \cos\psi\,\dot{} - a\sin\psi\,\dot{} - \dot{x} = 0,$$
$$a\sin\ \sin\psi\,\dot{} + a\cos\psi\,\dot{} + \dot{y} = 0. \tag{2.95}$$

To obtain equations of motion for the ball, we follow Equations (2.91):

$$\frac{d}{dt}\left(\frac{\partial L}{\partial \dot{}}\right) - \frac{\partial L}{\partial} = \lambda_1 H_{11} + \lambda_2 H_{21}, \quad \frac{d}{dt}\left(\frac{\partial L}{\partial \dot{\psi}}\right) - \frac{\partial L}{\partial \psi} = \lambda_1 H_{12} + \lambda_2 H_{22},$$

$$\frac{d}{dt}\left(\frac{\partial L}{\partial \dot{}}\right) - \frac{\partial L}{\partial} = \lambda_1 H_{13} + \lambda_2 H_{23}, \quad \frac{d}{dt}\left(\frac{\partial L}{\partial \dot{x}}\right) - \frac{\partial L}{\partial x} = \lambda_1 H_{14} + \lambda_2 H_{24}, \qquad (2.96)$$

$$\frac{d}{dt}\left(\frac{\partial L}{\partial \dot{y}}\right) - \frac{\partial L}{\partial y} = \lambda_1 H_{15} + \lambda_2 H_{25}.$$

From the constraint equations (2.95), we obtain

$$H_{11} = -a\sin \ \cos\psi, \ H_{12} = 0, \ H_{13} = a\sin\psi, \ H_{14} = -1, \ H_{15} = 0$$
$$H_{21} = -a\sin \ \sin\psi, \ H_{22} = 0, \ H_{23} = -a\cos\psi, \ H_{24} = 0, \ H_{25} = -1. \qquad (2.97)$$

For the ball $L = T$ and $I = 2ma^2/5$. Then

$$L = \frac{1}{2}m(\dot{x}^2 + \dot{y}^2) + \frac{1}{2}I(\dot{}^2 + \dot{\psi}^2 + \ ^2 + 2\dot{}\ \dot{\psi}\cos \ ). \qquad (2.98)$$

Performing all differentiations as indicated by Equations (2.96), we obtain the equations of motion for the ball:

$$I \ddot{} + I \ddot{\psi}\cos \ - I \dot{\psi} \ \dot{}\sin \ = -\lambda_1 a\sin \ \cos\psi - \lambda_2 a\sin \ \sin\psi,$$
$$I \ddot{\psi} + I \ddot{}\cos \ - I \dot{} \ \dot{}\sin \ = \lambda_1 \cdot 0 + \lambda_2 \cdot 0,$$
$$I \ddot{} + I \dot{} \ \dot{\psi}\sin \ = \lambda_1 a\sin\psi - \lambda_2 a\cos\psi, \qquad (2.99)$$
$$m\ddot{x} = -\lambda_1 + \lambda_2 \cdot 0,$$
$$m\ddot{y} = \lambda_1 \cdot 0 - \lambda_2.$$

The constraint equations (2.95) and the equation of motion (2.99) are the set of seven equations, from which $,\psi, ,x,y,\lambda_1,\lambda_2$ may be determined.

2.3.4 Maggi's Equations

Lagrange's equations with multipliers (2.91) are usually applied to derive equations of motion for systems subjected to first order nonholonomic constraints in both mechanics and control settings. However, the elimination of the multipliers requires some preprocessing work, when equations are for control applications. Maggi's equations are free of the multipliers, i.e., of the constraint reaction forces. They are used rather rarely, yet they may compete with Lagrange's equations specifically for control applications.

In subsequent sections, we develop a unified dynamic modeling frame-work for constrained systems. We will demonstrate that Maggi's as well as Lagrange's equations are peculiar cases of the generalized programmed motion equations.

To eliminate multipliers from Equations (2.90), we assume that b out of n first order time derivatives of generalized coordinates can be locally solved in terms of the rest $(n-b)$ of the first order derivatives such that $\dot{q} = (\dot{q}_\beta, \dot{q}_\mu)$, $\dot{q}_\beta \in R^b$, $\dot{q}_\mu \in R^{n-b}$. Then the constraint equations (2.22) take the form

$$\dot{q}_\beta = g_\beta^{(1)}(t, q_1, ..., q_n, \dot{q}_{b+1}, ..., \dot{q}_n). \qquad \beta = 1, ..., b. \tag{2.100}$$

Now, take (2.81) and rewrite it in the form

$$\sum_{\beta=1}^{b}\left[\frac{d}{dt}\left(\frac{\partial T}{\partial \dot{q}_\beta}\right) - \frac{\partial T}{\partial q_\beta} - Q_\beta\right]\delta q_\beta + \sum_{\mu=b+1}^{n}\left[\frac{d}{dt}\left(\frac{\partial T}{\partial \dot{q}_\mu}\right) - \frac{\partial T}{\partial q_\mu} - Q_\mu\right]\delta q_\mu = 0. \tag{2.101}$$

Based on (2.64), variations δq_β satisfy

$$\delta q_\beta = \sum_{\mu=b+1}^{n}\frac{\partial g_\beta^{(1)}}{\partial \dot{q}_\mu}\delta q_\mu. \tag{2.102}$$

Inserting (2.102) into (2.101) results in

$$\sum_{\mu=b+1}^{n}\left\{\frac{d}{dt}\left(\frac{\partial T}{\partial \dot{q}_\mu}\right) - \frac{\partial T}{\partial q_\mu} - Q_\mu + \sum_{\beta=1}^{b}\left[\frac{d}{dt}\left(\frac{\partial T}{\partial \dot{q}_\beta}\right) - \frac{\partial T}{\partial q_\beta} - Q_\beta\right]\frac{\partial g_\beta^{(1)}}{\partial \dot{q}_\mu}\right\}\delta q_\mu = 0. \tag{2.103}$$

Variations δq_μ, $\mu = b + 1, ..., n$, are independent, so we obtain equations in the form

$$\frac{d}{dt}\left(\frac{\partial T}{\partial \dot{q}_\mu}\right) - \frac{\partial T}{\partial q_\mu} - Q_\mu + \sum_{\beta=1}^{b}\left[\frac{d}{dt}\left(\frac{\partial T}{\partial \dot{q}_\beta}\right) - \frac{\partial T}{\partial q_\beta} - Q_\beta\right]\frac{\partial g_\beta^{(1)}}{\partial \dot{q}_\mu} = 0. \qquad \mu = b + 1, ..., n \tag{2.104}$$

Equations (2.104) are referred to as Maggi's equations for a system subjected to the constraints (2.100) (Jarzębowska 2007b). Maggi's equations may be presented in a different form. Denote by $T^{(c)}$ the kinetic energy of a system, in which generalized velocities $\dot{q}_\beta \in R^k$ are replaced by right-hand sides of (2.100). Then, $T^{(c)}$ has the form

$$T^{(c)} = T\left[t, q_1, ..., q_n, g_1^{(1)}(t, q_1, ..., q_n, \dot{q}_{b+1}, ..., \dot{q}_n), ..., g_\beta^{(1)}(t, q_1, ..., q_n, \dot{q}_{k+1}, ..., \dot{q}_n), \dot{q}_{b+1}, ..., \dot{q}_n\right]$$

from which the following hold:

$$\frac{\partial T}{\partial \dot{q}_\mu} = \frac{\partial T^{(c)}}{\partial \dot{q}_\mu} - \sum_{\beta=1}^{b} \frac{\partial T}{\partial \dot{q}_\beta} \frac{\partial g_\beta^{(1)}}{\partial \dot{q}_\mu}, \quad \frac{\partial T}{\partial q_\mu} = \frac{\partial T^{(c)}}{\partial q_\mu} - \sum_{\beta=1}^{b} \frac{\partial T}{\partial \dot{q}_\beta} \frac{\partial g_\beta^{(1)}}{\partial q_\mu}. \quad \mu = b+1,\dots,n \quad (2.105)$$

Inserting (2.105) into (2.104), we obtain

$$\frac{d}{dt}\left(\frac{\partial T^{(c)}}{\partial \dot{q}_\mu} - \sum_{\beta=1}^{b} \frac{\partial T}{\partial \dot{q}_\beta} \frac{\partial g_\beta^{(1)}}{\partial \dot{q}_\mu} \right) - \frac{\partial T^{(c)}}{\partial q_\mu} + \sum_{\beta=1}^{b} \frac{\partial T}{\partial \dot{q}_\beta} \frac{\partial g_\beta^{(1)}}{\partial q_\mu} - Q_\mu + \sum_{\beta=1}^{b} \left[\frac{d}{dt}\left(\frac{\partial T}{\partial \dot{q}_\beta} \right) - \frac{\partial T}{\partial q_\beta} - Q_\beta \right] \frac{\partial g_\beta^{(1)}}{\partial \dot{q}_\mu} = 0$$

or introducing $Q_\mu^{(c)} = Q_\mu + \sum_{\beta=1}^{b} Q_\beta \frac{\partial g_\beta^{(1)}}{\partial \dot{q}_\mu}$ and rearranging terms, Maggi's equations take the form

$$\frac{d}{dt}\left(\frac{\partial T^{(c)}}{\partial \dot{q}_\mu} \right) - \frac{\partial T^{(c)}}{\partial q_\mu} - \sum_{\beta=1}^{b} \frac{\partial T}{\partial \dot{q}_\beta} \left[\frac{d}{dt}\left(\frac{\partial g_\beta^{(1)}}{\partial \dot{q}_\mu} \right) - \frac{\partial g_\beta^{(1)}}{\partial q_\mu} \right] - \sum_{\beta=1}^{b} \frac{\partial T}{\partial \dot{q}_\beta} \frac{\partial g_\beta^{(1)}}{\partial \dot{q}_\mu} = Q_\mu. \quad (2.106)$$

In classical textbooks on analytical mechanics, Maggi's equations are presented rather rarely, and they are derived in a different form; compare to Nejmark and Fufaev (1972). First, it is assumed that the nonholonomic constraint equations are linear in velocities. Then, in order to eliminate Lagrange multipliers, r parameters $\dot{e}_1,\dots,\dot{e}_r$ in a number equal to the number of degrees of freedom of a system are introduced. All generalized velocities, $\dot{q}_i$, $i = 1,\dots,n$, are presented by the parameters as

$$\dot{q}_\sigma = \sum_{i=1}^{r} C_{i\sigma}\dot{e}_i + D_\sigma. \qquad \sigma = 1,\dots,n \qquad (2.107)$$

$C_{i\sigma}$ and D_σ are functions of time and coordinates q_i, $i = 1,\dots,n$. Parameters $\dot{e}_1,\dots,\dot{e}_r$ are referred to as kinematic parameters. When $r = n$, Equations (2.107) define transformations from $\dot{q}_i$ to $\dot{e}_i$. Comparing Equations (2.100) and (2.107), it can be seen that $\dot{e}_i$ play the role of independent velocities $\dot{q}_\mu$. Based on Equations (2.107),

$$C_{i\sigma} = \frac{\partial \dot{q}_\sigma}{\partial \dot{e}_i}, \qquad \delta q_\sigma = \sum_{i=1}^{r} C_{i\sigma}\delta e_i, \qquad \sigma = 1,\dots,n, \, i = 1,\dots,r \qquad (2.108)$$

and $\delta q_\sigma = \sum_{i=1}^{r} \frac{\partial \dot{q}_\sigma}{\partial \dot{e}_i} \delta e_i$. Relations (2.108) inserted into Equation (2.81) yield

$$\sum_{\sigma=1}^{n} \left[\frac{d}{dt}\left(\frac{\partial T}{\partial \dot{q}_\sigma}\right) - \frac{\partial T}{\partial q_\sigma} \right] \sum_{i=1}^{r} C_{i\sigma}\delta e_i = \sum_{\sigma=1}^{n} Q_\sigma \sum_{i=1}^{r} C_{i\sigma}\delta e_i.$$

Variations δe_i, $i = 1,...,r$, are independent, and we obtain

$$\sum_{\sigma=1}^{n} \left[\frac{d}{dt}\left(\frac{\partial T}{\partial \dot{q}_\sigma}\right) - \frac{\partial T}{\partial q_\sigma} \right] C_{i\sigma} = \sum_{\sigma=1}^{n} Q_\sigma C_{i\sigma}. \qquad i = 1,...,r \qquad (2.109)$$

Denoting by

$$\Phi_i = \sum_{\sigma=1}^{n} Q_\sigma C_{i\sigma}, \qquad i = 1,...,r \qquad (2.110)$$

Equations (2.109) may be presented as

$$\sum_{\sigma=1}^{n} \left[\frac{d}{dt}\left(\frac{\partial T}{\partial \dot{q}_\sigma}\right) - \frac{\partial T}{\partial q_\sigma} \right] C_{i\sigma} = \Phi_i, \qquad i = 1,...,r. \qquad (2.111)$$

Equations (2.111) are referred to as Maggi's equations. Together with (2.107), they are a set of $(n + r)$ equations for unknown $q_1,...,q_n$, and $\dot{e}_1,...,\dot{e}_r$.

2.3.5 Nielsen's Equations

Nielsen's equations can be derived from Lagrange's equations or from the Jourdain principle (Nielsen 1935; Jarzębowska 2007b). They apply to systems with first order nonholonomic constraints $_\beta(t,q_\sigma,\dot{q}_\sigma) = 0$, $\beta = 1,...,k$, $\sigma = 1,...,n$. They may be obtained in different ways, and in this section we demonstrate one of them. In Section 2.4, we will obtain Nielsen's equations as a peculiar case of a general form of equations of motion.

Now, take Lagrange equations in which we transform the term $\frac{d}{dt}\left(\frac{\partial T}{\partial \dot{q}_\sigma}\right)$ as follows:

$$\frac{d}{dt}\left(\frac{\partial T}{\partial \dot{q}_\sigma}\right) = \frac{\partial^2 T}{\partial \dot{q}_\sigma \partial t} + \sum_{\lambda=1}^{n} \frac{\partial^2 T}{\partial \dot{q}_\sigma \partial q_\lambda} \dot{q}_\lambda + \sum_{\lambda=1}^{n} \frac{\partial^2 T}{\partial \dot{q}_\sigma \partial \dot{q}_\lambda} \ddot{q}_\lambda, \qquad \sigma = 1,...,n \quad (2.112)$$

Also, we have that

$$\dot{T} = \frac{\partial T}{\partial t} + \sum_{\lambda=1}^{n} \frac{\partial T}{\partial q_\lambda} \dot{q}_\lambda + \sum_{\lambda=1}^{n} \frac{\partial T}{\partial \dot{q}_\lambda} \ddot{q}_\lambda , \tag{2.113}$$

and

$$\frac{\partial \dot{T}}{\partial \dot{q}_\sigma} = \frac{\partial^2 T}{\partial t \partial \dot{q}_\sigma} + \sum_{\lambda=1}^{n} \frac{\partial^2 T}{\partial q_\lambda \partial \dot{q}_\sigma} \dot{q}_\lambda + \sum_{\lambda=1}^{n} \frac{\partial^2 T}{\partial \dot{q}_\lambda \partial \dot{q}_\sigma} \ddot{q}_\lambda + \frac{\partial T}{\partial q_\sigma} . \tag{2.114}$$

Comparing Equations (2.112) and (2.114), we may write that

$$\frac{d}{dt}\left(\frac{\partial T}{\partial \dot{q}_\sigma} \right) = \frac{\partial \dot{T}}{\partial \dot{q}_\sigma} - \frac{\partial T}{\partial q_\sigma} . \tag{2.115}$$

Relations (2.115) inserted into (2.81) yield

$$\sum_{\sigma=1}^{n} \left(\frac{\partial \dot{T}}{\partial \dot{q}_\sigma} - 2\frac{\partial T}{\partial q_\sigma} - Q_\sigma \right) \delta q_\sigma = 0 \tag{2.116}$$

or

$$\sum_{\beta=1}^{k} \left(\frac{\partial \dot{T}}{\partial \dot{q}_\beta} - 2\frac{\partial T}{\partial q_\beta} - Q_\beta \right) \delta q_\beta + \sum_{\mu=k+1}^{n} \left(\frac{\partial \dot{T}}{\partial \dot{q}_\mu} - 2\frac{\partial T}{\partial q_\mu} - Q_\mu \right) \delta q_\mu = 0. \tag{2.117}$$

Based on the constraints (2.100), variations δq_β satisfy

$$\delta q_\beta = \sum_{\mu=k+1}^{n} \frac{\partial g_\beta^{(1)}}{\partial \dot{q}_\mu} \delta q_\mu .$$

Then (2.117) becomes

$$\sum_{\beta=1}^{k} \left(\frac{\partial \dot{T}}{\partial \dot{q}_\beta} - 2\frac{\partial T}{\partial q_\beta} - Q_\beta \right) \sum_{\mu=k+1}^{n} \frac{\partial g_\beta^{(1)}}{\partial \dot{q}_\mu} \delta q_\mu + \sum_{\mu=k+1}^{n} \left(\frac{\partial \dot{T}}{\partial \dot{q}_\mu} - 2\frac{\partial T}{\partial q_\mu} - Q_\mu \right) \delta q_\mu = 0. \tag{2.118}$$

Variations $\delta q_\mu, \mu = k+1,...,n$, are independent, so we rewrite (2.118) in the form

$$\frac{\partial \dot{T}}{\partial \dot{q}_\mu} - 2\frac{\partial T}{\partial q_\mu} - Q_\mu + \sum_{\beta=1}^{k} \left(\frac{\partial \dot{T}}{\partial \dot{q}_\beta} - 2\frac{\partial T}{\partial q_\beta} - Q_\beta \right) \frac{\partial g_\beta^{(1)}}{\partial \dot{q}_\mu} = 0. \qquad \mu = k+1,...,n \tag{2.119}$$

We refer to Equations (2.119) as Nielsen's equations in Maggi's form (Jarzębowska 2007b). Together with k nonholonomic constraints (2.100), they are a set of $(n - k) + k$ equations for n unknown q'_σ s.

Equations (2.119) can be transformed into an equivalent form. Let us denote by $T^{(c)}$ the kinetic energy function in which the right-hand sides of Equations (2.100) replace generalized velocities $\dot{q}_\beta$. $T^{(c)}$ is then expressed in terms of generalized coordinates q_σ, $\sigma = 1, ..., n$, and independent generalized velocities $\dot{q}_\mu$, $\mu = k + 1, ..., n$. Also let $\dot{T}^{(c)}$ denote the time derivative of T, in which the right-hand side of Equations (2.100) replace $\dot{q}_\beta$ and the right-hand side of

$$\ddot{q}_\beta = \gamma_\beta^{(2)}(t, q_1, ..., q_n, \dot{q}_{k+1}, ..., \dot{q}_n, \ddot{q}_{k+1}, ..., \ddot{q}_n), \qquad \beta = 1, ..., k \qquad (2.120)$$

replace $\ddot{q}_\beta$. We have then

$$T^{(c)} = T(t, q_\sigma, g_\beta^{(1)}(t, q_\sigma, \dot{q}_\mu), \dot{q}_\mu), \qquad (2.121)$$

$$\dot{T}^{(c)} = \dot{T}(t, q_\sigma, g_\beta^{(1)}(t, q_\sigma, \dot{q}_\mu), \dot{q}_\mu, \gamma_\beta^{(2)}(t, q_\sigma, \dot{q}_\mu, \ddot{q}_\mu), \ddot{q}_\mu)). \qquad (2.122)$$

Based on (2.100) and (2.120) through (2.122), the following hold:

$$\frac{\partial T^{(c)}}{\partial q_\mu} = \frac{\partial T}{\partial q_\mu} + \sum_{\beta=1}^{k} \frac{\partial T}{\partial \dot{q}_\beta} \frac{\partial g_\beta^{(1)}}{\partial q_\mu}, \qquad (2.123a)$$

$$\frac{\partial \dot{T}^{(c)}}{\partial \dot{q}_\mu} = \frac{\partial \dot{T}}{\partial \dot{q}_\mu} + \sum_{\beta=1}^{k} \frac{\partial \dot{T}}{\partial \dot{q}_\beta} \frac{\partial g_\beta^{(1)}}{\partial \dot{q}_\mu} + \sum_{\beta=1}^{k} \frac{\partial \dot{T}}{\partial \ddot{q}_\beta} \frac{\partial \gamma_\beta^{(2)}}{\partial \dot{q}_\mu}. \qquad (2.123b)$$

From (2.123b), we obtain that

$$\frac{\partial \dot{T}}{\partial \dot{q}_\mu} = \frac{\partial \dot{T}^{(c)}}{\partial \dot{q}_\mu} - \sum_{\beta=1}^{k} \frac{\partial \dot{T}}{\partial \dot{q}_\beta} \frac{\partial g_\beta^{(1)}}{\partial \dot{q}_\mu} - \sum_{\beta=1}^{k} \frac{\partial \dot{T}}{\partial \ddot{q}_\beta} \frac{\partial \gamma_\beta^{(2)}}{\partial \dot{q}_\mu} \qquad (2.123c)$$

and based on (2.113),

$$\frac{\partial \dot{T}}{\partial \ddot{q}_\sigma} = \frac{\partial T}{\partial \dot{q}_\sigma}. \qquad \sigma = 1, ..., n \qquad (2.124)$$

Relation (2.124) substituted into the third term on the right-hand side of (2.123c) yields

$$\frac{\partial \dot{T}}{\partial \dot{q}_\mu} = \frac{\partial \dot{T}^{(c)}}{\partial \dot{q}_\mu} - \sum_{\beta=1}^{k} \frac{\partial \dot{T}}{\partial \dot{q}_\beta} \frac{\partial g_\beta^{(1)}}{\partial \dot{q}_\mu} - \sum_{\beta=1}^{k} \frac{\partial T}{\partial \dot{q}_\beta} \frac{\partial \gamma_\beta^{(2)}}{\partial \dot{q}_\mu}. \tag{2.125}$$

Based on Equation (2.123a), we obtain that

$$\frac{\partial T}{\partial q_\mu} = \frac{\partial T^{(c)}}{\partial q_\mu} - \sum_{\beta=1}^{k} \frac{\partial T}{\partial \dot{q}_\beta} \frac{\partial g_\beta^{(1)}}{\partial q_\mu}. \tag{2.126}$$

Substitution of Equations (2.125) and (2.126) into (2.119) yields

$$\frac{\partial \dot{T}^{(c)}}{\partial \dot{q}_\mu} - 2\frac{\partial T^{(c)}}{\partial q_\mu} - \sum_{\beta=1}^{k} \frac{\partial T}{\partial \dot{q}_\beta} \left(\frac{\partial \gamma_\beta^{(2)}}{\partial \dot{q}_\mu} - 2\frac{\partial g_\beta^{(1)}}{\partial q_\mu} \right) - 2\sum_{\beta=1}^{k} \frac{\partial T}{\partial q_\beta} \frac{\partial g_\beta^{(1)}}{\partial \dot{q}_\mu} = Q_\mu^{(c)}, \quad \mu = k+1,...,n \tag{2.127}$$

where $Q_\mu^{(c)} = Q_\mu + \sum_{\beta=1}^{k} Q_\beta \frac{\partial g_\beta^{(1)}}{\partial \dot{q}_\mu}$. These are equations of motion of Nielsen's type for a system with the constraints (2.100). They are equivalent to Equations (2.119). The difference is that in Equations (2.127), the kinetic energy $T^{(c)}$ is a function of $q_\sigma, \sigma = 1,...,n$, and $\dot{q}_\mu, \mu = k+1,...,n$.

2.3.6 Equations of Motion in Quasi-Coordinates

In Chapter 1 in Example 1.3, we could see that the generalized coordinates might not always be the most suitable coordinates for motion description. Also, from the perspective of specifications of control goals or control constraints, it may be easier to use quasi-velocities.

In this section, we derive the Boltzmann–Hamel equations, i.e., the equations of motion in quasi-coordinates. In some textbooks, these equations are derived under an assumption that the relations between quasi-velocities and generalized velocities (2.10) are linear (Nejmark and Fufaev 1972). In this book, we relax this assumption and let these relations be nonlinear (Jarzębowska 2008b).

The derivation of the Boltzmann–Hamel equations with the assumption that the relations

$$\omega_r = \omega_r(t, q_\sigma, \dot{q}_\sigma). \qquad \sigma, r = 1,...,n, \tag{2.128}$$

may be nonlinear and a system is subjected to first order nonholonomic constraints starts from the principal form of the dynamics motion equation in the form

$$\frac{d}{dt}\sum_{\sigma=1}^{n} p_\sigma \delta q_\sigma = \delta T + \sum_{\sigma=1}^{n} Q_\sigma \delta q_\sigma + \sum_{\sigma=1}^{n} p_\sigma [(\delta q_\sigma)^{\cdot} - \delta \dot{q}_\sigma] \delta q_\sigma. \qquad (2.129)$$

Notice, that Equation (2.129) differs from (2.77). This is due to the fact that in (2.129), it is assumed that operations $d(\cdot)$ and $\delta(\cdot)$ do not commute. The commutability rule is not true for quasi-coordinates and quasi-velocities, in general. Using Equations (2.12), the left-hand side of (2.129) may be transformed to

$$\sum_{\lambda=1}^{n} p_\lambda \delta q_\lambda = \sum_{\lambda=1}^{n} p_\lambda \sum_{\mu=1}^{n} \frac{\partial \dot{q}_\lambda}{\partial \omega_\mu} \delta \pi_\mu = \sum_{\mu=1}^{n} \left(\sum_{\lambda=1}^{n} p_\lambda \frac{\partial \dot{q}_\lambda}{\partial \omega_\mu} \right) \delta \pi_\mu.$$

Denoting by

$$\tilde{p}_\mu = \sum_{\lambda=1}^{n} p_\lambda \frac{\partial \dot{q}_\lambda}{\partial \omega_\mu}, \qquad \mu = 1,...,n \qquad (2.130)$$

the following holds

$$\sum_{\lambda=1}^{n} p_\lambda \delta q_\lambda = \sum_{\mu=1}^{n} \tilde{p}_\mu \delta \pi_\mu. \qquad (2.131)$$

Now, denote by $\tilde{T}$ the kinetic energy of a system in which the generalized velocities are replaced by the quasi-velocities according to Equations (2.11):

$$T = T(t, q_\sigma, \dot{q}_\lambda(t, q_\sigma, \omega_r)) = \tilde{T}(t, q_\sigma, \omega_\sigma). \qquad \sigma, r, \lambda = 1,...,n, \qquad (2.132)$$

Based on Equations (2.14), the virtual work of external forces $Q_\lambda, \lambda = 1,...,n$, that correspond to virtual displacements δq_λ can be written as

$$\delta A' = \sum_{\lambda=1}^{n} Q_\lambda \delta q_\lambda = \sum_{\lambda=1}^{n} Q_\lambda \sum_{\mu=1}^{n} \frac{\partial \dot{q}_\lambda}{\partial \omega_\mu} \delta \pi_\mu = \sum_{\mu=1}^{n} \left(\sum_{\lambda=1}^{n} Q_\lambda \frac{\partial \dot{q}_\lambda}{\partial \omega_\mu} \right) \delta \pi_\mu.$$

Denoting by

$$\tilde{Q}_\mu = \sum_{\lambda=1}^{n} Q_\lambda \frac{\partial \dot{q}_\lambda}{\partial \omega_\mu}, \qquad \mu = 1,\ldots,n \tag{2.133}$$

we obtain that

$$\sum_{\lambda=1}^{n} Q_\lambda \delta q_\lambda = \sum_{\mu=1}^{n} \tilde{Q}_\mu \delta \pi_\mu. \tag{2.134}$$

Based on Equations (2.10) and (2.14), we have that

$$\delta \omega_r = \sum_{\lambda=1}^{n} \frac{\partial \omega_r}{\partial q_\lambda} \delta q_\lambda + \sum_{\lambda=1}^{n} \frac{\partial \omega_r}{\partial \dot{q}_\lambda} \delta \dot{q}_\lambda,$$

$$(\delta \pi_r)^\cdot = \sum_{\lambda=1}^{n} \left(\frac{\partial \omega_r}{\partial \dot{q}_\lambda} \delta q_\lambda \right)^\cdot = \sum_{\lambda=1}^{n} \frac{\partial \omega_r}{\partial \dot{q}_\lambda} (\delta q_\lambda)^\cdot + \sum_{\lambda=1}^{n} \frac{d}{dt}\left(\frac{\partial \omega_r}{\partial \dot{q}_\lambda} \right) \delta q_\lambda, \qquad \lambda = 1,\ldots,n$$

and hence,

$$(\delta \pi_r)^\cdot - \delta \omega_r = \sum_{\lambda=1}^{n} \frac{\partial \omega_r}{\partial \dot{q}_\lambda} [(\delta q_\lambda)^\cdot - \delta \dot{q}_\lambda] + \sum_{\lambda=1}^{n} \left[\frac{d}{dt}\left(\frac{\partial \omega_r}{\partial \dot{q}_\lambda} \right) - \frac{\partial \omega_r}{\partial q_\lambda} \right] \delta q_\lambda,$$

what may be written as

$$\sum_{\lambda=1}^{n} \frac{\partial \omega_r}{\partial \dot{q}_\lambda} [(\delta q_\lambda)^\cdot - \delta \dot{q}_\lambda] = (\delta \pi_r)^\cdot - \delta \omega_r - \sum_{\lambda=1}^{n} \left[\frac{d}{dt}\left(\frac{\partial \omega_r}{\partial \dot{q}_\lambda} \right) - \frac{\partial \omega_r}{\partial q_\lambda} \right] \delta q_\lambda. \quad r = 1,\ldots,n \tag{2.135}$$

Multiplying both sides of Equations (2.135) by $\tilde{p}_r$, summing it over n, and using the second of relations (2.14) we obtain

$$\sum_{r=1}^{n} \tilde{p}_r \sum_{\lambda=1}^{n} \frac{\partial \omega_r}{\partial \dot{q}_\lambda} [(\delta q_\lambda)^\cdot - \delta \dot{q}_\lambda] = \sum_{r=1}^{n} \tilde{p}_r [(\delta \pi_r)^\cdot - \delta \omega_r]$$

$$- \sum_{r=1}^{n} \sum_{\mu=1}^{n} \sum_{\lambda=1}^{n} \tilde{p}_r \frac{\partial \dot{q}_\lambda}{\partial \omega_\mu} \left[\frac{d}{dt}\left(\frac{\partial \omega_r}{\partial \dot{q}_\lambda} \right) - \frac{\partial \omega_r}{\partial q_\lambda} \right] \delta \pi_\mu.$$

Based on relations (2.130), $p_\lambda = \sum\limits_{r=1}^{n} \tilde{p}_r \frac{\partial \omega_r}{\partial \dot{q}_\lambda}$, $\lambda = 1, ..., n$; hence, we get

$$\sum_{\lambda=1}^{n} p_\lambda [(\delta q_\lambda)^{\boldsymbol{\cdot}} - \delta \dot{q}_\lambda] = \sum_{r=1}^{n} \tilde{p}_r [(\delta \pi_r)^{\boldsymbol{\cdot}} - \delta \omega_r] - \sum_{r=1}^{n} \sum_{\mu=1}^{n} \sum_{\lambda=1}^{n} \tilde{p}_r \frac{\partial \dot{q}_\lambda}{\partial \omega_\mu} \left[\frac{d}{dt} \left(\frac{\partial \omega_r}{\partial \dot{q}_\lambda} \right) - \frac{\partial \omega_r}{\partial q_\lambda} \right] \delta \pi_\mu .$$

(2.136)

By introducing the notation

$$W_\mu^r = \sum_{\lambda=1}^{n} \frac{\partial \dot{q}_\lambda}{\partial \omega_\mu} \left[\frac{d}{dt} \left(\frac{\partial \omega_r}{\partial \dot{q}_\lambda} \right) - \frac{\partial \omega_r}{\partial q_\lambda} \right], \qquad r, \mu = 1, ..., n, \qquad (2.137)$$

(2.136) may be written in the form

$$\sum_{\lambda=1}^{n} p_\lambda [(\delta q_\lambda)^{\boldsymbol{\cdot}} - \delta \dot{q}_\lambda] = \sum_{r=1}^{n} \tilde{p}_r \left\{ [(\delta \pi_r)^{\boldsymbol{\cdot}} - \delta \omega_r] - \sum_{\mu=1}^{n} W_\mu^r \delta \pi_\mu \right\}. \qquad (2.138)$$

Inserting (2.131), (2.132), (2.134), and (2.138) into Equation (2.129), we finally obtain

$$\frac{d}{dt} \sum_{\mu=1}^{n} \tilde{p}_\mu \delta \pi_\mu = \delta \tilde{T} + \sum_{\mu=1}^{n} \tilde{Q}_\mu \delta \pi_\mu + \sum_{r=1}^{n} \tilde{p}_r [(\delta \pi_r)^{\boldsymbol{\cdot}} - \delta \omega_r] - \sum_{r=1}^{n} \tilde{p}_r \sum_{\mu=1}^{n} W_\mu^r \delta \pi_\mu . \quad (2.139)$$

Equation (2.139) is the principal form of the equation of motion developed in quasi-coordinates with nonlinear relations between quasi-velocities and generalized velocities. When these relations are linear, Equation (2.139) coincides with the equation obtained in Nejmark and Fufaev (1972). Symbols W_μ^r may be regarded as the generalized Boltzmann symbols for the nonlinear relations between the quasi-velocities and generalized velocities.

To develop motion equations in the noninertial coordinates, we can start from Equation (2.139). First, we derive some relations useful in our development. Based on relations (2.132) and (2.130), we obtain

$$\frac{\partial \tilde{T}}{\partial \omega_\mu} = \sum_{\lambda=1}^{n} \frac{\partial T}{\partial \dot{q}_\lambda} \frac{\partial \dot{q}_\lambda}{\partial \omega_\mu} = \tilde{p}_\mu . \qquad (2.140)$$

Using relation (2.140), the left-hand side term of Equation (2.139) can be developed as

$$\frac{d}{dt}\sum_{\mu=1}^{n}\tilde{p}_{\mu}\delta\pi_{\mu} = \sum_{\mu=1}^{n}\frac{d}{dt}\left(\frac{\partial\tilde{T}}{\partial\omega_{\mu}}\right)\delta\pi_{\mu} + \sum_{\mu=1}^{n}\tilde{p}_{\mu}(\delta\pi_{\mu})^{\cdot}. \tag{2.141}$$

The first, third, and fourth terms on the right-hand side of Equation (2.139) can be transformed as

$$\delta\tilde{T} = \sum_{\mu=1}^{n}\frac{\partial\tilde{T}}{\partial q_{\mu}}\delta q_{\mu} + \sum_{\mu=1}^{n}\frac{\partial\tilde{T}}{\partial\omega_{\mu}}\delta\omega_{\mu} = \sum_{\alpha=1}^{n}\left(\frac{\partial\tilde{T}}{\partial q_{\mu}}\frac{\partial\dot{q}_{\mu}}{\partial\omega_{\alpha}}\right)\delta\pi_{\alpha} + \sum_{\mu=1}^{n}\frac{\partial\tilde{T}}{\partial\omega_{\mu}}\delta\omega_{\mu}$$

and due to formula (2.17),

$$\delta\tilde{T} = \sum_{\mu=1}^{n}\frac{\partial\tilde{T}}{\partial\pi_{\mu}}\delta\pi_{\mu} + \sum_{\mu=1}^{n}\frac{\partial\tilde{T}}{\partial\omega_{\mu}}\delta\omega_{\mu} = \sum_{\mu=1}^{n}\left(\frac{\partial\tilde{T}}{\partial\pi_{\mu}}\delta\pi_{\mu} + \frac{\partial\tilde{T}}{\partial\omega_{\mu}}\delta\omega_{\mu}\right), \tag{2.142}$$

where the subscript α is replaced by μ in the first sum on the right-hand side. Next,

$$\sum_{r=1}^{n}\tilde{p}_{r}\left[(\delta\pi_{r})^{\cdot} - \delta\omega_{r} - \sum_{\mu=1}^{n}W_{\mu}^{r}\delta\pi_{\mu}\right] = \sum_{r=1}^{n}\tilde{p}_{r}(\delta\pi_{r})^{\cdot} - \sum_{r=1}^{n}\frac{\partial\tilde{T}}{\partial\omega_{r}}\delta\omega_{r} - \sum_{r=1}^{n}\sum_{\mu=1}^{n}\frac{\partial\tilde{T}}{\partial\omega_{r}}W_{\mu}^{r}\delta\pi_{\mu}.$$

Replacing the subscript r by μ in the first and second sum on the right-hand side of the last relation, we obtain

$$\sum_{r=1}^{n}\tilde{p}_{r}\left[(\delta\pi_{r})^{\cdot} - \delta\omega_{r} - \sum_{\mu=1}^{n}W_{\mu}^{r}\delta\pi_{\mu}\right] = \sum_{\mu=1}^{n}\tilde{p}_{\mu}(\delta\pi_{\mu})^{\cdot} - \sum_{\mu=1}^{n}\frac{\partial\tilde{T}}{\partial\omega_{\mu}}\delta\omega_{\mu} - \sum_{\mu=1}^{n}\sum_{r=1}^{n}\frac{\partial\tilde{T}}{\partial\omega_{r}}W_{\mu}^{r}\delta\pi_{\mu}.$$
$$\tag{2.143}$$

Relations (2.141) through (2.143) inserted into Equation (2.139) and after terms rearrangement yield

$$\sum_{\mu=1}^{n}\left[\frac{d}{dt}\left(\frac{\partial\tilde{T}}{\partial\omega_{\mu}}\right) - \frac{\partial\tilde{T}}{\partial\pi_{\mu}} + \sum_{r=1}^{n}\frac{\partial\tilde{T}}{\partial\omega_{r}}W_{\mu}^{r} - \tilde{Q}_{\mu}\right]\delta\pi_{\mu} = 0. \tag{2.144}$$

For a holonomic system, variations $\delta\pi_\mu, \mu = 1,...,n$, are independent and equations of motion are

$$\frac{d}{dt}\left(\frac{\partial\tilde{T}}{\partial\omega_\mu}\right) - \frac{\partial\tilde{T}}{\partial\pi_\mu} + \sum_{r=1}^{n}\frac{\partial\tilde{T}}{\partial\omega_r}W_\mu^r = \tilde{Q}_\mu. \qquad \mu = 1,...,n \qquad (2.145)$$

Equations (2.145) are the generalized Boltzmann–Hamel equations for the holonomic system. Relations between the quasi-velocities and generalized velocities may be nonlinear as in (2.10) and (2.11). When they are linear, we obtain the Boltzmann–Hamel equations derived, for example, in Nejmark and Fufaev (1972). It can be verified that when quasi-coordinates are equivalent to generalized coordinates, i.e., $\pi_r = q_r$, and quasi-velocities are equivalent to generalized velocities, i.e., $\omega_r = \dot{q}_r, r = 1,...,n$, then Equations (2.145) are Lagrange's equations with $W_\mu^r = 0, \alpha, \mu, r = 1,...,n$.

Consider now a system subjected to material or programmed nonholonomic constraints. Assume that the constraints are of the first order and have the form of Equation (2.10):

$$\omega_\beta = \omega_\beta(t, q_\sigma, \dot{q}_\sigma) = 0. \qquad \beta = 1,...,b \qquad (2.146)$$

Based on (2.14), the relations

$$\delta\pi_\beta = \sum_{\sigma=1}^{n}\frac{\partial\omega_\beta}{\partial\dot{q}_\sigma}\delta q_\sigma = 0, \qquad \beta = 1,...,b \qquad (2.147)$$

hold for all ω_β; hence, Equations (2.144) can be written as

$$\sum_{\mu=b+1}^{n}\left[\frac{d}{dt}\left(\frac{\partial\tilde{T}}{\partial\omega_\mu}\right) - \frac{\partial\tilde{T}}{\partial\pi_\mu} + \sum_{r=1}^{n}\frac{\partial\tilde{T}}{\partial\omega_r}W_\mu^r - \tilde{Q}_\mu\right]\delta\pi_\mu = 0. \qquad (2.148)$$

The system possesses $(n-b)$ degrees of freedom and the variations $\delta\pi_{b+1},...,\delta\pi_n$ are independent. We obtain then $(n-b)$ equations of motion in the form

$$\frac{d}{dt}\left(\frac{\partial\tilde{T}}{\partial\omega_\mu}\right) - \frac{\partial\tilde{T}}{\partial\pi_\mu} + \sum_{r=1}^{n}\frac{\partial\tilde{T}}{\partial\omega_r}W_\mu^r = \tilde{Q}_\mu, \qquad \mu = b+1,...,n \qquad (2.149)$$

to which n kinematic relations

$$\dot{q}_\lambda = \dot{q}_\lambda(t, q_\sigma, \omega_r), \qquad \sigma, \lambda = 1, \dots, n, \quad r = b+1, \dots, n \qquad (2.150)$$

have to be added. Equations (2.149) are the generalized Boltzmann–Hamel equations for a nonholonomic system. The set of Equations (2.149) and (2.150) consists of $(2n\text{-}b)$ equations for n unknown q's and $(n\text{-}b)$ ω's, and they specify the system motion completely. Notice that b of ω's are satisfied based on the constraint equations (2.146). The rest of quasi-velocities are selected arbitrarily by a designer or a control engineer. We may say that the first order constraints are "swallowed" by these b ω's and the rest of $(n\text{-}b)$ ω's are selected arbitrarily. These are the main advantages of the introduction of the quasi-velocities. Equations (2.149) and (2.150) may be presented as first order differential equations in ω's

$$M(q)\dot{\omega} + C(q,\omega) + D(q) = \tilde{Q},$$

$$B(q,\omega) = 0. \qquad (2.151)$$

Numerical simulation advantages, mostly no need to numerically stabilize the constraint equations, are demonstrated in the example below.

Example 2.10

In Example 1.3, we presented a nonholonomic toy, a roller-racer whose fundamental means of propulsion is the pivoting of the steering handlebar around the joint axis and the nonholonomic constraints. The purely kinematic analysis of the roller-racer is not allowed. Kinematics must be complemented with the system dynamics. All studies that concern constrained dynamics and control are based on Lagrange's equations with multipliers followed by the reduction procedure. Let us derive equations of motion of the roller-racer using Equations (2.149) (Jarzębowska and Lewandowski 2006).

Motion of the roller-racer is described by $q = (x, y, \theta, \psi) \in \Omega$, $\Omega = SE(2) \times S^1$, where (x, y, θ) describe a position, and ψ is a shape variable. A distance from the center of mass of the rear platform to the rotary joint is l_1, and the length of the second platform is l_2, masses of platforms are equal to m_1 and m_2 (Figure 1.4). Both masses are allowed for constrained dynamics and control. Moments of inertia of platforms are J_1 and J_2. We assume that a viscous friction develops between wheels and a ground. Equations of nonholonomic constraints are as in (1.3):

$$\dot{x}\sin\theta - \dot{y}\cos\theta = 0,$$

$$-\dot{x}\sin\psi + \dot{y}\cos\psi + \dot{\theta}l_1\cos(\theta - \psi) + l_2\dot{\psi} = 0. \qquad (2.152)$$

The kinetic energy for the roller-racer is

$$T = \frac{1}{2}m_1(\dot{x}^2 + \dot{y}^2) + \frac{1}{2}J_1\dot{\theta}^2 + \frac{1}{2}J_2\dot{\psi}^2 + \frac{1}{2}m_2[(\dot{x} - l_1\dot{\theta}\sin\theta)^2 + (\dot{y} + l_1\dot{\theta}\cos\theta)^2].$$

A (4×1) vector of quasi-velocities, in which the Boltzmann–Hamel equations are derived, is

$$\omega_1 = v = \dot{x}\cos\theta + \dot{y}\sin\theta, \quad \omega_2 = \dot{\theta},$$

$$\omega_3 = \dot{x}\sin\theta - \dot{y}\cos\theta = 0, \quad \omega_4 = -\dot{x}\sin\psi + \dot{y}\cos\psi + l_1\dot{\theta}\cos(\psi - \theta) + l_2\dot{\psi} = 0. \tag{2.153}$$

This selection requires that quasi-velocities, in a number equal to the number of constraint equations, are identically equal to them. The rest of quasi-velocities are due to a designer. We are motivated by a control goal, i.e., tracking and a subsequent controller design. The quasi-velocity ω_1 is the roller-racer forward velocity, and ω_2 is the rate of change of its orientation with respect to the world coordinates. The kinetic energy for the roller-racer written in quasi-velocities is

$$T^* = \frac{1}{2}(m_1 + m_2)[\omega_1^2 + \omega_3^2] + \frac{1}{2}J_1\omega_2^2 + + m_2 l_1 \omega_2 \omega_3 + \frac{1}{2}\frac{J_2}{l_2^2}[\omega_1^2 \sin^2(\psi - \theta)$$

$$- \omega_1(\omega_2 l_1 + \omega_3)\sin 2(\psi - \theta) + (\omega_2 l_1 + \omega_3)^2 \cos^2(\psi - \theta)$$

$$- 2\omega_4(\omega_2 l_1 + \omega_3)\cos(\psi - \theta) + \omega_4^2 + 2\omega_1 \omega_4 \sin(\psi - \theta)].$$

The generalized forces determined in "directions" of quasi-velocities are $Q_j^* = \overset{4}{\underset{\sigma=1}{\Sigma}} Q_\sigma b_{\sigma j}, j = 1,\ldots,4,$ and Q_{jf}^*- friction forces. The Boltzmann–Hamel equations require computations of the Boltzmann symbols, which all can be computerized. We use MATLAB^R to derive the constrained dynamics. The Boltzmann–Hamel equations that govern the roller-racer dynamics are

$$M(\quad)\dot{\omega} + C(\quad,\quad,\omega)\omega + D\omega = E(\quad)\tau,$$

$$\dot{q} = B(q)\omega, \tag{2.154}$$

$$\dot{\quad} = \omega_1 e_1 + \omega_2(e_2 - 1).$$

where now $\omega = (\omega_1, \omega_2)$ is a (2×1) vector, and ω_3, ω_4 are set equal to zero because they satisfy the constraints as indicated in Equations (2.146). Additionally, we introduced a relative orientation angle $= \psi - \theta$. We do not need it in the development of our constrained dynamics; however, with the aid of this angle, we can easily show how to obtain the reduced-state dynamics equivalent to the one that results from the reduction procedure applied to Lagrange's equations with multipliers (Tsakiris and Krishnaprasad 1998). To this end, it is enough to take the time derivative $\ddot{\quad}$ and eliminate $\dot{\omega}_1, \dot{\omega}_2$.

Matrices that furnish Equations (2.154) are

$$
M = \begin{bmatrix} m + \dfrac{J_2}{l_2^2}\sin^2 & -\dfrac{J_2 l_1}{2 l_2^2}\sin 2 \\[2ex] -\dfrac{J_2 l_1}{2 l_2^2}\sin 2 & J_1 + J_2\left(\dfrac{l_1}{l_2}\right)^2 \cos^2 \end{bmatrix}, \quad
B = \begin{bmatrix} \cos\theta & 0 \\ \sin\theta & 0 \\ 0 & 1 \\ \dfrac{\sin}{l_2} & -\dfrac{l_1}{l_2}\cos \end{bmatrix}
$$

$$
C_{11} = \frac{J_2}{l_2^2}\left[\sin 2 \;\dot{} + \omega_2\left(\cos + \frac{l_1}{l_2}\right)\sin\right],\quad
C_{12} = \frac{-J_2 l_1}{l_2^2}\left[\cos 2 \;\dot{} + \omega_2\left(\cos + \frac{l_1}{l_2}\right)\cos\right] - m_2\omega_2 l_1,
$$

$$
C_{21} = \frac{-J_2}{l_2^2}\left[l_1\cos 2 \;\dot{} + \omega_1\left(\cos + \frac{l_1}{l_2}\right)\sin\right] + m_2\omega_2 l_1,
$$

$$
C_{22} = \frac{-J_2 l_1}{l_2^2}\left[l_1\sin 2 \;\dot{} - \omega_1\left(\cos + \frac{l_1}{l_2}\right)\cos\right],
$$

$$
E = [e_1 \;\; e_2]^T, \text{ with } e_1 = \frac{\sin}{l_2}, e_2 = -\frac{l_1}{l_2}\cos, \; D = diag(d_1 \;\; d_2), \; m = m_1 + m_2.
$$

To demonstrate the roller-racer uncontrolled motion, we also present an influence of mass m_2 on a total motion of the system. In many studies, small mass m_2 is neglected even if J_2 is not (see, e.g., Krishnaprasad and Tsakiris 2001). In all simulations, the following parameters, in SI units, are selected: $m_1 = 33$, $m_2 = 2$, $J_1 = 0.7$, $J_2 = 0.006$, $l_1 = 0.8$, $l_2 = 0.15$. An application of an external oscillatory torque $\tau = 0.1\sin(3t)$ that mimics propulsion of the roller-racer results in a typical pattern of its undulatory motion. Initial motion conditions are $q_0 = (0,0,0,\pi)$, $\omega_0 = (1,0)$. The influence of m_2 is shown in Figures 2.6 and 2.7, where one can

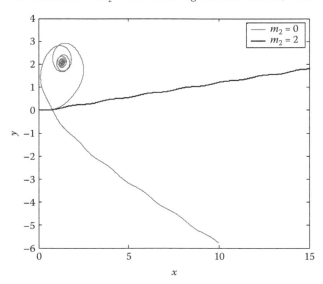

FIGURE 2.6 (SEE COLOR INSERT.)
Motion patterns of the uncontrolled motion of the roller-racer.

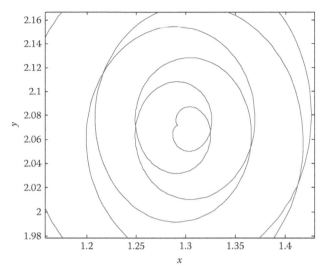

FIGURE 2.7 (SEE COLOR INSERT.)
Magnification of the roller-racer motion with $m_2 = 0$.

see that the roller-racer with $m_2 = 0$ turns and starts moving backward, which is physically incorrect. The roller-racer dynamics (Equation 2.154) is control oriented, and we take advantage of it in Chapter 5.

2.4 Equations of Motion for Systems with High Order Constraints

2.4.1 An Extended Concept of Constraints—Programmed Constraints

Nonlinear control uses dynamic models based mostly upon Lagrange's equations with multipliers. They may be presented in the matrix form as

$$M(q)\ddot{q} + C(q,\dot{q}) + D(q) = J^T(q)\lambda + Q(q,\dot{q}),$$

$$J(q)\dot{q} = 0,$$

(2.155)

where the $(n \times n)$ matrix function $M(q)$ is assumed to be symmetric and positive definite, $C(q,\dot{q})$ is a n-vector function, $D(q)$ is a vector of gravitational forces, $Q(q,\dot{q})$ is a vector of external forces, and $J(q)$ denotes a $(b \times n)$ matrix function, which is assumed to have full rank. All these functions are assumed to be smooth and defined on an appropriate open subset of the $(q,\dot{q})$ phase space. The scope of applications of these equations is limited, because only first order constraints, material or generated by conservation laws, are merged into this dynamic model. Requirements for motion that are often specified by equations for controlled systems are not merged into Equations (2.155). First order constraints that are adjoined to the equations of motion via the introduction of Lagrange multipliers can also be embedded through the reduction procedure, which avoids the addition of auxiliary variables.

This book concerns modeling for control, and another approach to mechanics must be discussed. This is geometric mechanics, which gives attention to the geometric structure of mechanical systems (see, e.g., Bloch 2003; Isidori 1989; Marsden 1992). Geometric mechanics enables writing equations of motion for mechanical systems in a coordinate-free fashion. The key idea is to regard the kinetic energy of a system as a Riemannian metric and to write the Euler-Lagrange equations in terms of the associated Riemannian connection. Geometric mechanics, developed mostly since 1970, and nonlinear geometric control theory provide a new insight into dynamics and control of mechanical systems. These efforts cast the ideas of nonholonomic mechanics in a more mathematical setting to make it consistent with the treatment received by the unconstrained mechanics. A coupling between geometric mechanics and control theory is interesting in this development, e.g., many recent results in nonholonomic systems with symmetry were motivated by recent techniques in nonlinear control. The geometric approach, by exploring the geometry of nonholonomic systems, helps understand the structure of equations of motion in a way that aids analysis and isolation of important geometric objects that govern the system motion. One example of results where the geometry associated with nonholonomic mechanics is useful is in analyzing controllability properties. Results in stabilization and adaptive control of nonlinear systems using geometric methods are reported in Bloch (2003), Caundar de Vit and Sordalen (1992), and Samson (1990, 1995). Another example is a snake-board presented in Example 1.2, which is a modern version of a skateboard (Lewis and Murray 1997). Its locomotion is achieved using coupling between nonholonomic constraints and symmetry properties. For this system, a traditional analysis of its dynamics does not readily explain the mechanism of locomotion. Using the momentum equation, which plays an important role in control problems, the interaction between the constraints and the symmetry becomes clear, and the basic mechanics underlying locomotion can be explained (Ostrowski et al. 1997). There are more interesting "nonholonomic toys" and devices, e.g., a roller-racer presented in Example 1.3, or a snake-board or a hopping robot, which motivate research on dynamics, control, and geometric mechanics by their astonishing properties. The geometric approach does not generate "new equations" but uses Lagrangian mechanics viewed from the geometric perspective, which enables new insight into control problems and proves to be useful in engineering applications.

The geometric mechanics approach to constraints is that of the traditional setup for constraint reaction forces, i.e., in this framework they do no work, (Bloch 2003; Marsden 1992). Nonholonomic systems come in two varieties in the geometric setting. First, there are those with dynamic nonholonomic constraints, that is, constraints preserved by the Euler-Lagrange or Hamilton equations, such as angular momentum conservation. These constraints are viewed as not externally imposed on the system but rather are the consequences of the equations of motion, and sometimes they are treated as conservation laws rather than constraints per se. They are represented by the differential equations of the first order in coordinates. Unanchored robots in space, the motion of astronauts on space walks, gymnasts, and springboard divers are examples of systems that undergo the conservation law. Constraints from the second group are kinematic and are imposed by conditions of rolling without slipping.

The conclusion is then that the scope of applications of Lagrange's equations (2.155) to constrained systems modeling, in both geometric mechanics and control, is limited.

In Section 2.3.5, we derived Nielsen's equations starting from Lagrange's equations. They may be presented in a different form (Jarzębowska 2007b):

$$\left(\frac{\partial \dot{T}}{\partial \dot{q}_\mu} - 2 \frac{\partial T}{\partial q_\mu} \right) = Q_\mu. \qquad \mu = 1, ..., k. \tag{*}$$

Tzenoff derived new equations from the Gauss principle. They are referred to as Tzenoff's equations of the second kind and have the following form (Tzenoff 1924):

$$\frac{1}{2} \left(\frac{\partial \ddot{T}}{\partial \ddot{q}_\mu} - 3 \frac{\partial T}{\partial q_\mu} \right) = Q_\mu. \tag{**}$$

They are applied to systems with second order constraints. Other equations by Tzenoff that are referred to as Tzenoff's equations of the third kind were derived from the postulated variational principle in the following form:

$$\sum_{v=1}^{N} (F_v - m_v \ddot{r}_v) \delta \ddot{r}_v = 0,$$

where $\delta t = 0, \delta r_v = \delta \dot{r}_v = \delta \ddot{r}_v = 0, \delta \dddot{r}_v \neq 0$ and F_v are components of a N-dimensional vector of external forces, m_v denotes mass of v-th particle, and r_v are components of a N-dimensional position vector such that $\delta \dddot{r}_v$ are defined as

$$\delta \dddot{r}_v = \sum_{\sigma=1}^{n} \frac{\partial \dddot{r}_v}{\partial \dddot{q}_\sigma} \delta \dddot{q}_\sigma = \sum_{\sigma=1}^{n} \frac{\partial r_v}{\partial q_\sigma} \delta \dddot{q}_\sigma. \qquad v = 1, ..., N.$$

Tzenoff's equations of the third kind have the form

$$\frac{1}{3}\left(\frac{\partial \dddot{T}}{\partial \dddot{q}_\mu} - 4\frac{\partial T}{\partial q_\mu} \right) = Q_\mu. \qquad \mu = 1,\ldots,k. \qquad (***)$$

They can be applied to systems with third order constraints. The author does not know any continuation of this research on the derivation of equations of motion for systems with constraints of an order higher than three by Tzenoff or anyone else until the 1960s.

A completely new stage in approaching constrained systems started with an extension of a constraint order. Constraints may be of p-th order:

$$G_\beta(t, q_\sigma, \dot{q}_\sigma, \ldots, q_\sigma^{(p)}) = 0. \qquad \beta = 1,\ldots,b \qquad (2.156)$$

They are extensions of the constraints specified by Equations (2.22) or (2.30). They are extensions of the ideas by Appell, Mieszczerski, or Beghin. The new approach needed an introduction of a generalized virtual displacement concept, which would be an extension of the virtual displacement in the Appell–Chetaev meaning to ideal p-th-order constraints. Mangeron and Deleanu (1962) postulated the generalized virtual displacement that suits the constraints (2.156) as

$$\sum_{\sigma=1}^{n} \frac{\partial G_\beta}{\partial q_\sigma^{(p)}} \delta q_\sigma = 0. \qquad (2.157)$$

The existence of three variational principles in classical mechanics, i.e., the d'Alembert, Jourdain, and Gauss principle, raises the question about their equivalence from the point of view of the generation of equations of motion. It can be shown that for holonomic systems, these three principles are equivalent. For nonholonomic systems, they are equivalent for systems for which the constraint equations are linear in velocities (see, e.g., Nejmark and Fufaev 1972). When the nonholonomic constraint equations are nonlinear in velocities, the problem of equivalence should be examined separately. It must be stated clearly what is the virtual displacement for nonlinear nonholonomic constraints and whether these constraints are ideal. Systems for which virtual displacements are related to first order nonholonomic constraints by (2.64) are referred to as Appell–Chetaev systems. The d'Alembert, Jourdain, and Gauss principles are equivalent for Appell–Chetaev systems. In general, for systems that are not Appell–Chetaev, these principles are not equivalent. We consider Appell–Chetaev systems that satisfy the generalized virtual displacement definition (2.157). A more detailed discussion about variational principles can be found in Nejmark and Fufaev (1972) and Papastavridis (1998, 2002).

A generalized variational principle for ideal constraints was postulated by Mangeron and Deleanu in the form

$$\sum_{v=1}^{N}(F_v - m_v\ddot{r}_v)\delta r_v^{(p)} = 0. \qquad \delta t = 0,\ \delta r_v = \delta\dot{r}_v = \cdots = \delta r_v^{(p-1)} = 0,\ \delta r_v^{(p)} \neq 0 \quad (2.158)$$

For Appell–Chetaev systems, for $p = 1,2,3$, the principle (2.158) coincides with the classical mechanics principles and with the principle postulated by Tzenoff. The proof and comments on the principle can be found in Mangeron and Deleanu (1962), as well as in Jarzębowska (2007b).

Remark 2.1: Works (Rong 1996a, 1996b, 1997; Tianling 1992) consider systems with high order constraints, but the equations derived there are in vector forms, and no examples or applications to high order constraints are presented. They do not generate, in the author's opinion, any new insight, from the application point of view, into modeling systems with high order constraints.

Remark 2.2: In Tianling (1992), the principle (2.158) is referred to as the universal d'Alembert principle. We refer to it as the generalized variational principle, following the original work (Mangeron and Deleanu 1962).

To the best of the author's knowledge, the principle (2.158) and equations obtained from it in Mangeron and Deleanu (1962) were the limits of analytical mechanics resulting in the derivation of equations of motion for constrained systems. Equations (*) through (***) were not used for control applications. The latest results are presented in this section following Jarzębowska (2005, 2006a, 2006b, 2007b, 2008a).

Let us start the derivation of equations of motion for systems subjected to high order constraints from the specification of these constraints as

$$B_\beta(t,q,\dot{q},...,q^{(p)}) = 0, \qquad \beta = 1,...,k,\ k < n, \qquad (2.159)$$

where p is a constraint order, and B_β is a k-dimensional vector. Equations (2.159) can be nonlinear in $q^{(p)}$. Differentiation of (2.159) with respect to time, until the highest derivative of a coordinate is linear, results in constraint equations linear with respect to this highest coordinate derivative. From now on, without loss of generality, we assume that "p" stands for the highest order derivative of a coordinate that appears linearly in a constraint equation. It is possible that the constraints are linear with respect to only one p-th order coordinate derivative. For simplicity of the development presented herein, we assume that they are linear in all p-th order derivatives of coordinates. Based on the assumptions, we rewrite Equations (2.159) as

$$B(t,q,\dot{q},...,q^{(p-1)})q^{(p)} + s(t,q,\dot{q},...,q^{(p-1)}) = 0, \qquad (2.160)$$

where B is a $(k \times n)$-dimensional full rank matrix, $n > k$, and s is a $(k \times 1)$ vector.

According to Definitions 2.1 and 2.2 from Section 2.2.3, and Equations (2.30), driving and task constraints, performance goals, or other requirements on a system motion to obtain its specified performance may be included into the "programmed constraints" class whose generalized specification may be presented by Equations (2.160). They can get the unified name as they play the same role—they program the motion. There are many possible formulations of programmed constraints. It has to be verified whether a programmed constraint formulated for a system is reachable for it. It can be done by inspection of solutions of equations of a programmed motion; we discuss these problems in Chapters 5 and 7. For the constraints (2.160), we formulate the following definition.

Definition 2.3: The constraint equations (2.160) are completely nonholonomic if they cannot be integrated with respect to time, i.e., we cannot obtain constraint equations of a lower order. If we can integrate Equations (2.160) $(p - 1)$ or less times, i.e., we can obtain nonholonomic constraints of first orders or orders lower than p we say that (2.160) are partially integrable. If Equations (2.160) can be integrated completely, i.e., we obtain constraints of the form $A_\beta(t, q) = 0$, they are holonomic.

Remark 2.3: We assume that Equations (2.160) are completely nonholonomic. Then they do not restrict positions $q(t)$ and their derivatives up to $(p-1)$-th order. Definition 2.3 is an extension of the definition of completely nonholonomic first order constraints (Bloch 2003) and completely nonholonomic second order constraints (Reyhanoglu et al. 1996). Necessary and sufficient integrability conditions for the differential equations of arbitrary order are formulated in Tarn, Zhang, and Serrani (2003).

Remark 2.4: Throughout this book, we consider programs referred to as partly specified where the number of programmed and material constraint equations is smaller than the number of system degrees of freedom. An exception is made in Chapter 7, where we consider an example of a program referred to as fully specified.

The examples that follow illustrate formulations of the programmed constraints (2.160). They are related to examples of constrained systems presented in Chapter 1 and given in the literature (e.g., Fukao et al. 2000; Lewis et al. 2004; Oriolo et al. 2002; Scheuer and Laugier 1998; Zotov 2003). They are simple, yet instructive, and they demonstrate how one can formulate programmed constraint equations based on motion specifications. Most of these motion specifications are formulated in forms of constraint equations for the first time. The examples are presented throughout the book, and they illustrate modeling and control theoretical results. The first two examples concern wheeled systems. They are a unicycle and a wheeled mobile robot of type (2,0). We selected them because they illustrate many theoretical results in both mechanics and control. Also, the most common for single-body robots are differential drive and synchro-drive, which are both kinematically equivalent to a unicycle. The third example, already presented in Chapter 1, is a two-link planar manipulator model, which can be easily extended to an n-link manipulator.

Example 2.11

In Koh and Cho (1999), Oriolo, de Luca, and Vendittelli (2002), Scheuer and Laugier (1998), and Shekl and Lumelsky (1998) for trajectory planning or tracking purposes, to preserve continuity of the trajectory curvature, constraints are put upon the curvature and its derivative. For a real car-like robot, the orientation between directing wheels and its main axis is bounded, which implies that the turning radius is lower bounded or that the curvature is bounded $|\kappa| \le \kappa_{max}$. The orientation of the directing wheels may change with a limited speed and the derivative of the curvature has to be bounded $|\dot{\kappa}| \le \dot{\kappa}_{max}$. Within these bounds, the curvature or its derivative may be specified functions. Also, bounds are put on linear and angular velocities to preserve the curvature radius corresponding to nominal velocities. These bounds are called control constraints. As discussed earlier, these constraints are not merged into kinematic or dynamic control models of robots. They are taken into account at a controller design stage. We may pose the problem as follows (Jarzębowska 2007b): Find a trajectory $(x(t), y(t))$, for which the curvature profile is a given function $\Phi(t)$. This specifies a programmed constraint which, for planar motion, has the form

$$\Phi(t) = \frac{\begin{vmatrix} \dot{x} & \dot{y} \\ \ddot{x} & \ddot{y} \end{vmatrix}}{(\dot{x}^2 + \dot{y}^2)^{3/2}}. \tag{2.161}$$

Equation (2.161) is a nonlinear nonholonomic constraint of the second order.

To specify a constraint on the rate of change of the curvature profile $\dot{\Phi}(t)$, Equation (2.161) has to be differentiated. We obtain a linear nonholonomic constraint equation of the third order, which solved for one of derivatives of the Cartesian coordinates yields

$$\dddot{x} = \frac{-\Phi(\dot{x}^2 + \dot{y}^2)^2[\dot{\Phi}(\dot{x}^2 + \dot{y}^2) + 3\Phi(\dot{x}\ddot{x} + \dot{y}\ddot{y})]}{\dot{y}(\dot{x}\ddot{y} - \ddot{x}\dot{y})} + \dddot{y}\frac{\dot{x}}{\dot{y}}. \tag{2.162}$$

We can reuse (2.161) in (2.162) to obtain

$$\dddot{x} = \frac{-\dot{\Phi}(\dot{x}^2 + \dot{y}^2)^{3/2}}{\dot{y}} - \frac{3(\dot{x}\ddot{x} + \dot{y}\ddot{y})(\dot{x}\ddot{y} - \dot{y}\ddot{x})}{\dot{y}(\dot{x}^2 + \dot{y}^2)} + \dddot{y}\frac{\dot{x}}{\dot{y}}. \tag{2.162a}$$

Introducing $\Phi_1 = \dot{\Phi}$ and

$$F_1 = -\frac{3(\dot{x}\ddot{x} + \dot{y}\ddot{y})(\dot{x}\ddot{y} - \dot{y}\ddot{x})}{\dot{y}(\dot{x}^2 + \dot{y}^2)},$$

Equation (2.162a) can be written as

$$\dddot{x}\dot{y} - \dddot{y}\dot{x} = -\Phi_1(\dot{x}^2 + \dot{y}^2)^{3/2} + F_1\dot{y}$$

or in the form (2.160):

$$[\dot{y} \quad -\dot{x}]\begin{bmatrix}\ddot{x}\\\ddot{y}\end{bmatrix} + [\Phi_1(\dot{x}^2 + \dot{y}^2)^{3/2} - F_1\dot{y}] = 0. \tag{2.162b}$$

If a vehicle is the two-wheeled platform from Example 1.1, (x,y) are the coordinates of its mass center, equations of material constraints (1.1) have to be added to (2.162b).

Example 2.12

Take a unicycle whose wheel radius is r and it rolls without slipping on a plane surface as shown in Figure 2.8. The wheel admits a nonholonomic behavior typical for mobile robots, so it is representative for the constraint and dynamic modeling, and for control purposes. Let φ denote the heading angle of the wheel, measured from the axis x, θ is the rotation angle due to rolling, measured from a fixed reference, and (x,y) are the coordinates of the wheel contact point with the ground. Nonholonomic material constraints for the wheel are given by

$$\dot{x} = r\dot{\theta}\cos\varphi \quad , \quad \dot{y} = r\dot{\theta}\sin\varphi \quad . \tag{2.163}$$

We want the unicycle to follow a trajectory specified by a programmed constraint equation:

$$\varphi(t) = \cos(0.5\pi - t). \tag{2.164}$$

This desired trajectory is similar to the one in Fukao, Nakagava, and Adachi (2000) and Oriolo, De Luca, and Vendittelli (2002). However, we specify it by the programmed constraint equation. Similar reference trajectories are often specified for mobile robots, but they are not in a constraint form and are not merged into system dynamics in the process of designing a tracking controller (You and Chen 1993, Yun and Sarkar 1998).

Two wheels can model a two-wheeled mobile robot as presented in Figure 1.2. Material constraint equations for the mobile robot have the form of Equation (1.1).

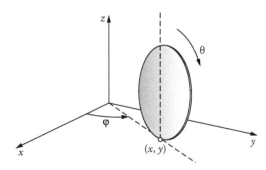

FIGURE 2.8
Model of a vertical wheel rolling on the plane.

Additionally, we want the robot to move along a trajectory specified by a pro-grammed constraint equation in the form

$$x_C^2 + y_C^2 = R^2(t). \tag{2.165}$$

Both constraint equations (1.1) and (2.165) have to be satisfied when a mobile robot is controlled. In Chapter 7, we develop a tracking strategy to control the robot subjected to the programmed constraints like (2.162b) or (2.165).

Example 2.13

Take a space two-link manipulator model as shown in Figure 1.7. The law of con-servation of the angular momentum implies that moving the manipulator links causes the base body to rotate. The conservation law is viewed as a nonholo-nomic constraint on the system, and it has the form as in Equation (1.8).

Additionally, we would like to add a position programmed constraint that speci-fies motion along a straight line for the end of the second link. This constraint has the form

$$l_1[\sin(\theta_1 + \phi) - \alpha\cos(\theta_1 + \phi)] + l_2[\sin(\theta_1 + \theta_2 + \phi) - \alpha\cos(\theta_1 + \theta_2 + \phi)] - \beta = 0, \tag{2.166}$$

where α and β specify the line position. The constraints on the system are (1.8) and (2.166). They both can be transformed to the form (2.160) as follows. We write (1.8) in the form (2.160):

$$B_1{}^{\cdot} + B_{1\theta1}\dot{\theta}_1 + B_{1\theta2}\dot{\theta}_2 = 0,$$

where $B_1 = J + (m_1 + m_2)l_1^2 + m_2 l_2^2 + 2m_2 l_1 l_2 \cos\theta_2$, $B_{1\theta1} = (m_1 + m_2)l_1^2 + m_2 l_2^2 + 2m_2 l_1 l_2 \cos\theta_2$, and $B_{1\theta2} = m_2 l_2^2 + m_2 l_1 l_2 \cos\theta_2$.

Differentiating Equation (2.166) with respect to time, we obtain

$$A_1{}^{\cdot} + A_{1\theta1}\dot{\theta}_1 + A_{1\theta2}\dot{\theta}_2 + \sigma A_1 = 0,$$

where $A_1 = l_1[\sin(\theta_1 +) - \alpha\cos(\theta_1 +)] + l_2[\sin(\theta_1 + \theta_2 +) - \alpha\cos(\theta_1 + \theta_2 +)] - \beta = 0$,

$A_1 = l_1\cos(\theta_1 +) + l_1\alpha\sin(\theta_1 +) + l_2\cos(\theta_1 + \theta_2 +) + l_2\alpha\sin(\theta_1 + \theta_2 +)$,

$A_{1\theta1} = l_1\cos(\theta_1 +) + l_1\alpha\sin(\theta_1 +) + l_2\cos(\theta_1 + \theta_2 +) + l_2\alpha\sin(\theta_1 + \theta_2 +)$,

$A_{1\theta2} = l_2\cos(\theta_1 + \theta_2 +) + l_2\alpha\sin(\theta_1 + \theta_2 +)$, and σ is a positive constant selected to stabilize the constraint in numerical computations.

All equations of the constraints for the space manipulator are

$$\begin{bmatrix} A_1 & A_{1\theta1} & A_{1\theta2} \\ B_1 & B_{1\theta1} & B_{1\theta2} \end{bmatrix} \begin{bmatrix} \dot{\theta}_1 \\ \dot{\theta}_2 \end{bmatrix} = \begin{bmatrix} -\sigma A_1 \\ 0 \end{bmatrix}. \tag{2.167}$$

They specify both the property of free-floating and the program imposed on the second link.

2.4.2 Generalized Programmed Motion Equations Specified in Generalized Coordinates

This section details a derivation of equations of motion for constrained systems. These are generalized programmed motion equations (GPME) (Jarzębowska 2002; Jarzębowska 2007a, 2007b, 2008a). They provide one unified theoretical framework that enables obtaining equations of motion for systems with arbitrary order nonholonomic constraints. Equations of motion from classical mechanics arise as their peculiar cases. The background for the derivation of equations of motion for systems with arbitrary order nonholonomic constraints was developed in Mangeron and Deleanu (1962). The generalized variational principle (Equation 2.158) was the basis there to derive these equations in a vector form. This vector form of equations can be compared with one of "principal forms" of equations of motion in classical mechanics. Mangeron and Deleanu did not derive equations of motion for any specific system with high order nonholonomic constraints. They applied their equations to systems with first order constraints and investigated their relation to classical methods of mechanics. The GPME are not derived from the principle (2.158). In our derivation, we do not assume any specific form of the kinetic energy—for example, that it is a quadratic function of velocities. Such form of the kinetic energy is assumed in Mangeron and Deleanu (1962). The derivation of equations of motion without any assumption about the form of the kinetic energy is a significant improvement, because they can be applied to systems for which the kinetic energy is not a quadratic function of velocities. The GPME are developed in an analytical form, and they are an extension of the work by Mangeron and Deleanu.

The general form of the nonholonomic constraints of p-th order is as in Equations (2.160). To develop the GPME, we formulate the following lemma.

Lemma 2.1

Assume that a function $F = F(t, q_1, ..., q_n, \dot{q}_1, ..., \dot{q}_n)$ is regular enough, i.e., all derivatives up to the certain order p exist. Then, the following identity holds for F:

$$\frac{d}{dt}\left(\frac{\partial F}{\partial \dot{q}_\sigma} \right) \equiv \frac{1}{p}\left(\frac{\partial F^{(p)}}{\partial q_\sigma^{(p)}} - \frac{\partial F}{\partial q_\sigma} \right). \qquad \sigma = 1, ..., n, p = 1, 2, 3, ... \qquad (2.168)$$

Proof: The proof can be completed by mathematical induction. To this end, we compute total time derivatives of F of orders $p = 1, 2, 3, ...$, i.e.,

$$\frac{dF}{dt}, \frac{d^2 F}{dt^2}, ..., \frac{d^{(p)} F}{dt^{(p)}}.$$

Skipping time-consuming calculations, we can verify that $\frac{d^{(p)}F}{dt^{(p)}} \equiv F^{(p)}$ has the form

$$
F^{(p)} = p\left(\sum_{\sigma=1}^{n} \frac{\partial^2 F}{\partial \dot{q}_\sigma \partial t} q_\sigma^{(p)} + \sum_{\sigma=1}^{n}\sum_{\alpha=1}^{n} \frac{\partial^2 F}{\partial \dot{q}_\sigma \partial q_\alpha} \dot{q}_\alpha q_\sigma^{(p)} + \sum_{\sigma=1}^{n}\sum_{\alpha=1}^{n} \frac{\partial^2 F}{\partial \dot{q}_\sigma \partial \dot{q}_\alpha} \ddot{q}_\alpha q_\sigma^{(p)} \right)
$$

$$
+ \sum_{\sigma=1}^{n} \frac{\partial F}{\partial q_\sigma} q_\sigma^{(p)} + \sum_{\sigma=1}^{n} \frac{\partial F}{\partial \dot{q}_\sigma} q_\sigma^{(p+1)} + \Gamma, \qquad\qquad p > 2
$$

(2.169)

where the function Γ does not contain derivatives $q_\sigma^{(p)}$ and $q_\sigma^{(p+1)}$, $\sigma = 1,\ldots,n$. Hence, we obtain

$$
\frac{\partial F^{(p)}}{\partial q_\sigma^{(p)}} = p\left(\frac{\partial^2 F}{\partial \dot{q}_\sigma \partial t} + \sum_{\alpha=1}^{n} \frac{\partial^2 F}{\partial \dot{q}_\sigma \partial q_\alpha} \dot{q}_\alpha + \sum_{\alpha=1}^{n} \frac{\partial^2 F}{\partial \dot{q}_\sigma \partial \dot{q}_\alpha} \ddot{q}_\alpha \right) + \frac{\partial F}{\partial q_\sigma}. \qquad (2.170)
$$

Relation (2.170) holds for any $p = 1, 2, \ldots$ and $\sigma = 1, \ldots, n$. We can verify that for $p = 1$, we have

$$
\dot{F} = \frac{\partial F}{\partial t} + \sum_{\alpha=1}^{n} \frac{\partial F}{\partial q_\alpha} \dot{q}_\alpha + \sum_{\alpha=1}^{n} \frac{\partial F}{\partial \dot{q}_\alpha} \ddot{q}_\alpha \qquad (2.171a)
$$

and hence,

$$
\frac{\partial \dot{F}}{\partial \dot{q}_\sigma} = \frac{\partial^2 F}{\partial t \partial \dot{q}_\sigma} + \sum_{\alpha=1}^{n} \frac{\partial^2 F}{\partial q_\alpha \partial \dot{q}_\sigma} \dot{q}_\alpha + \sum_{\alpha=1}^{n} \frac{\partial^2 F}{\partial \dot{q}_\alpha \partial \dot{q}_\sigma} \ddot{q}_\alpha + \frac{\partial F}{\partial q_\sigma}. \qquad \sigma = 1,\ldots,n \quad (2.171b)
$$

In the same way, we can demonstrate that relation (2.170) holds for $p = 2$. Indeed,

$$
\ddot{F} = 2\sum_{\sigma=1}^{n} \frac{\partial^2 F}{\partial \dot{q}_\sigma \partial t} \ddot{q}_\sigma + 2\sum_{\sigma=1}^{n}\sum_{\alpha=1}^{n} \frac{\partial^2 F}{\partial q_\sigma \partial \dot{q}_\alpha} \dot{q}_\sigma \ddot{q}_\alpha + \sum_{\sigma=1}^{n}\sum_{\alpha=1}^{n} \frac{\partial^2 F}{\partial \dot{q}_\sigma \partial \dot{q}_\alpha} \ddot{q}_\sigma \ddot{q}_\alpha
$$

$$
+ \sum_{\sigma=1}^{n} \frac{\partial F}{\partial q_\sigma} \ddot{q}_\sigma + \sum_{\sigma=1}^{n} \frac{\partial F}{\partial \dot{q}_\sigma} \dddot{q}_\sigma + \Lambda,
$$

(2.171c)

where $\Lambda = 2\displaystyle\sum_{\sigma=1}^{n} \frac{\partial^2 F}{\partial q_\sigma \partial t} \dot{q}_\sigma + \sum_{\sigma=1}^{n}\sum_{\alpha=1}^{n} \frac{\partial^2 F}{\partial q_\sigma \partial q_\alpha} \dot{q}_\sigma \dot{q}_\alpha + \frac{\partial^2 F}{\partial t^2}.$

Interchanging subscripts σ and α in the second term in (2.171c), we obtain

$$\frac{\partial \ddot{F}}{\partial \ddot{q}_\sigma} = 2\left(\frac{\partial^2 F}{\partial \dot{q}_\sigma \partial t} + \sum_{\alpha=1}^{n} \frac{\partial^2 F}{\partial \dot{q}_\sigma \partial q_\alpha} \dot{q}_\alpha + \sum_{\alpha=1}^{n} \frac{\partial^2 F}{\partial \dot{q}_\sigma \partial \dot{q}_\alpha} \ddot{q}_\alpha \right) + \frac{\partial F}{\partial q_\sigma}. \qquad (2.171d)$$

Next, we calculate

$$\frac{d}{dt}\left(\frac{\partial F}{\partial \dot{q}_\sigma} \right) = \frac{\partial^2 F}{\partial t \partial \dot{q}_\sigma} + \sum_{\alpha=1}^{n} \frac{\partial^2 F}{\partial q_\alpha \partial \dot{q}_\sigma} \dot{q}_\alpha + \sum_{\alpha=1}^{n} \frac{\partial^2 F}{\partial \dot{q}_\alpha \partial \dot{q}_\sigma} \ddot{q}_\alpha. \qquad \sigma = 1,...,n \quad (2.172)$$

From Equations (2.170) and (2.172), we get the desired result:

$$\frac{d}{dt}\left(\frac{\partial F}{\partial \dot{q}_\sigma} \right) = \frac{1}{p}\left(\frac{\partial F^{(p)}}{\partial q_\sigma^{(p)}} - \frac{\partial F}{\partial q_\sigma} \right). \qquad p = 1,2,3,... \qquad (2.173)$$

q.e.d.

If to replace F with the kinetic energy function $T = T(t, q_1,...,q_n, \dot{q}_1,...,\dot{q}_n)$ in (2.173) and insert it into Lagrange's equations, we obtain

$$\frac{1}{p}\left[\frac{\partial T^{(p)}}{\partial q_\sigma^{(p)}} - (p+1)\frac{\partial T}{\partial q_\sigma} \right] = Q_\sigma, \qquad \sigma = 1,...,n \qquad (2.174)$$

where all potential forces are in Q_σ. Equations (2.174) are the generalized programmed motion equations (GPME). They become Nielsen's equations (*) for $p = 1$, Tzenoff's equations of the second kind (**) for $p = 2$, and Tzenoff's equations of the third kind (***) for $p = 3$.

In what follows, we present a detailed examination of Equations (2.174). It clarifies how to apply them for systems with constraints of a given order. For a system subjected to first order constraints, Nielsen's equations were examined in Section 2.3.5.

Now, examine the case of a system subjected to high order nonholonomic constraints:

$$G_\beta(t, q_1,...,q_n, \dot{q}_1,...,\dot{q}_n,...,q_1^{(p)},...,q_n^{(p)}) = 0 \qquad (2.175)$$

Assuming that the first k out of p-th order derivatives of generalized coordinates in Equations (2.175) are dependent, a vector of p-th order derivatives can be partitioned at least locally in the same fashion as in Equations (2.100), and the constraint equations can be transformed to the form

$$q_\beta^{(p)} = g_\beta^{(p)}(t, q_1,...,q_n, \dot{q}_1,...,\dot{q}_n,...,q_{k+1}^{(p)},...,q_n^{(p)}). \qquad \beta = 1,...,k \qquad (2.176)$$

By differentiating Equations (2.176) with respect to time, we obtain

$$q_\beta^{(p+1)} = \gamma_\beta^{(p+1)}(t, q_1, ..., q_n, \dot{q}_1, ..., \dot{q}_n, ..., q_{k+1}^{(p)}, ..., q_n^{(p)}, q_{k+1}^{(p+1)}, ..., q_n^{(p+1)}). \quad (2.177)$$

Replacing the term $\frac{d}{dt}\left(\frac{\partial T}{\partial \dot{q}_\sigma}\right)$ in Lagrange–d'Alembert equations with Equations (2.173) results in

$$\sum_{\sigma=1}^{n} \left[\frac{1}{p}\left(\frac{\partial T^{(p)}}{\partial q_\sigma^{(p)}} \right) - (p+1)\frac{\partial T}{\partial q_\sigma} \right) - Q_\sigma \right] \delta q_\sigma = 0 \quad (2.178)$$

or, equivalently,

$$\sum_{\beta=1}^{k} \left[\frac{1}{p}\left(\frac{\partial T^{(p)}}{\partial q_\beta^{(p)}} - (p+1)\frac{\partial T}{\partial q_\beta} \right) - Q_\beta \right] \delta q_\beta + \sum_{\mu=k+1}^{n} \left[\frac{1}{p}\left(\frac{\partial T^{(p)}}{\partial q_\mu^{(p)}} - (p+1)\frac{\partial T}{\partial q_\mu} \right) - Q_\mu \right] \delta q_\mu = 0.$$
$$(2.179)$$

Based on the definition of the generalized virtual displacement relations (2.157) and on relations (2.176), we have

$$\delta q_\beta = \sum_{\mu=k+1}^{n} \frac{\partial g_\beta^{(p)}}{\partial q_\mu^{(p)}} \delta q_\mu, \qquad \beta = 1, ..., k \quad (2.180)$$

so Equation (2.179) takes the form

$$\sum_{\mu=k+1}^{n} \left\{ \frac{1}{p}\left(\frac{\partial T^{(p)}}{\partial q_\mu^{(p)}} - (p+1)\frac{\partial T}{\partial q_\mu} \right) - Q_\mu + \sum_{\beta=1}^{k} \left[\frac{1}{p}\left(\frac{\partial T^{(p)}}{\partial q_\beta^{(p)}} - (p+1)\frac{\partial T}{\partial q_\beta} \right) - Q_\beta \right] \frac{\partial g_\beta^{(p)}}{\partial q_\mu^{(p)}} \right\} \delta q_\mu = 0.$$
$$(2.181)$$

Variations $\delta q_\mu, \mu = k+1, ..., n$, are independent, so we obtain equations of motion in the form

$$\frac{1}{p}\left[\frac{\partial T^{(p)}}{\partial q_\mu^{(p)}} - (p+1)\frac{\partial T}{\partial q_\mu} \right] - Q_\mu + \sum_{\beta=1}^{k} \left\{ \frac{1}{p}\left[\frac{\partial T^{(p)}}{\partial q_\beta^{(p)}} - (p+1)\frac{\partial T}{\partial q_\beta} \right] - Q_\beta \right\} \frac{\partial g_\beta^{(p)}}{\partial q_\mu^{(p)}} = 0. \quad (2.182)$$
$$\mu = k+1, ..., n$$

These are the GPME equivalent to Equations (2.174), and together with (2.175), they are a set of n equations for n unknown $q'_\sigma s$. They become Nielsen's equations in Maggi's form for $p = 1$, so they hold for any p. We could stop our analysis of Equations (2.182) at this statement. However, Equations (2.182) do not hold for $p = 1$ when the right-hand sides of (2.176) and (2.177) replace $q_\beta^{(p)}$ and $q_\beta^{(p+1)}$ in the expression for p-th order derivatives of the kinetic energy. To demonstrate this, denote by $T_{(c)}^{(p)}$ the p-th order derivative of the kinetic energy and replace $q_\beta^{(p)}$ and $q_\beta^{(p+1)}$ with relations (2.176) and (2.177) in it. This operation yields

$$T_{(c)}^{(p)} = T^{(p)}\left(t, q_\sigma, \dot{q}_\sigma, \ddot{q}_\sigma, ..., g_\beta^{(p)}(t, q_\sigma, \dot{q}_\sigma, \ddot{q}_\sigma, ..., q_\mu^{(p)}), q_\mu^{(p)},\right.$$

$$\left.\gamma_\beta^{(p+1)}(t, q_\sigma, \dot{q}_\sigma, \ddot{q}_\sigma, ..., q_\mu^{(p)}, q_\mu^{(p+1)}), q_\mu^{(p+1)}\right).$$

(2.183)

From (2.183), we obtain

$$\frac{\partial T_{(c)}^{(p)}}{\partial q_\mu^{(p)}} = \frac{\partial T^{(p)}}{\partial q_\mu^{(p)}} + \sum_{\beta=1}^{k} \frac{\partial T^{(p)}}{\partial q_\beta^{(p)}} \frac{\partial g_\beta^{(p)}}{\partial q_\mu^{(p)}} + \sum_{\beta=1}^{k} \frac{\partial T^{(p)}}{\partial q_\beta^{(p+1)}} \frac{\partial \gamma_\beta^{(p+1)}}{\partial q_\mu^{(p)}}. \qquad \mu = k+1,...,n \quad (2.184)$$

It can be verified that for $p = 1$, relations (2.184) are equivalent to (2.123b), but (2.123a) differs from an analogous relation, which we can obtain from (2.184) when constraints are of order $p > 1$, that is, the derivatives of the kinetic energy with respect to coordinates are

$$\frac{\partial T_{(c)}}{\partial q_\mu} = \frac{\partial T}{\partial q_\mu}. \qquad \mu = k+1,...,n \quad (2.185)$$

The reason is that the function of the kinetic energy depends on coordinates q_μ and their first order derivatives. Constraint equations for $p > 1$, do not affect them. Also, relations (2.184) result in

$$\frac{\partial T^{(p)}}{\partial q_\mu^{(p)}} = \frac{\partial T_{(c)}^{(p)}}{\partial q_\mu^{(p)}} - \sum_{\beta=1}^{k} \frac{\partial T^{(p)}}{\partial q_\beta^{(p)}} \frac{\partial g_\beta^{(p)}}{\partial q_\mu^{(p)}} - \sum_{\beta=1}^{k} \frac{\partial T^{(p)}}{\partial q_\beta^{(p+1)}} \frac{\partial \gamma_\beta^{(p+1)}}{\partial q_\mu^{(p)}}. \quad (2.186)$$

From (2.169) taken for T, it follows that

$$\frac{\partial T^{(p)}}{\partial q_\sigma^{(p+1)}} = \frac{\partial T}{\partial \dot{q}_\sigma}. \qquad \sigma = 1,...,n \quad (2.187)$$

Substitution of (2.187) into the third term of the right-hand side of (2.186) results in

$$\frac{\partial T^{(p)}}{\partial q_\mu^{(p)}} = \frac{\partial T_{(c)}^{(p)}}{\partial q_\mu^{(p)}} - \sum_{\beta=1}^{k} \frac{\partial T^{(p)}}{\partial q_\beta^{(p)}} \frac{\partial g_\beta^{(p)}}{\partial q_\mu^{(p)}} - \sum_{\beta=1}^{k} \frac{\partial T}{\partial \dot{q}_\beta} \frac{\partial \gamma_\beta^{(p+1)}}{\partial q_\mu^{(p)}}. \quad (2.188)$$

Finally, substitution of (2.185) and (2.188) into Equations (2.182) yields

$$\frac{1}{p}\left[\frac{\partial T_{(c)}^{(p)}}{\partial q_\mu^{(p)}} - (p+1)\frac{\partial T_{(c)}}{\partial q_\mu}\right] - \frac{1}{p}\sum_{\beta=1}^{k}\frac{\partial T}{\partial \dot{q}_\beta}\frac{\partial \gamma_\beta^{(p+1)}}{\partial q_\mu^{(p)}} - \frac{p+1}{p}\sum_{\beta=1}^{k}\frac{\partial T}{\partial q_\beta}\frac{\partial g_\beta^{(p)}}{\partial q_\mu^{(p)}} = Q_{(c)\mu}, \quad (2.189)$$

$$\mu = k+1,...,n$$

where

$$Q_{(c)\mu} = Q_\mu + \sum_{\beta=1}^{k}Q_\beta\frac{\partial g_\beta^{(p)}}{\partial q_\mu^{(p)}}.$$

Equations (2.189) are equivalent to Equations (2.182). They do not become Nielsen's equations in Maggi's form for $p = 1$, because of the reasons explained earlier. They hold for $p > 1$.

The GPME in the forms of Equations (2.127), (2.182), and (2.189) are not convenient for derivations of equations of motion. We may use the development that resulted in Equations (2.127) and (2.189) to design two algorithms (Jarzębowska 2007b).

Algorithm 2.1

To derive equations of Nielsen's type (2.127) for a system subjected to first order constraint equations (2.100), we proceed as follows:

1. Construct a function P_1 such that

$$P_1 = \dot{T} - 2\dot{T}_0, \tag{2.190}$$

where $\dot{T}_0$ for $p = 1$ is computed from the relation

$$T_0^{(p)} = \sum_{\sigma=1}^{n}\frac{\partial T}{\partial q_\sigma}q_\sigma^{(p)}. \tag{2.191}$$

2. Construct a function R_1 such that

$$R_1 = P_1 - \sum_{\sigma=1}^{n}\dot{q}_\sigma Q_\sigma = R_1(t,q_1,...,q_n,\dot{q}_1,...,\dot{q}_n,\ddot{q}_1,...,\ddot{q}_n) = R_1(t,q_\sigma,\dot{q}_\mu,\dot{q}_\beta,\ddot{q}_\sigma). \tag{2.192}$$

3. Construct a function R_1^* in which $\dot{q}_\beta$ from (2.192) are replaced by relations (2.100):

$$R_1^* = R_1(t, q_\sigma, \dot{q}_\mu, g_\beta^{(1)}(t, q_\sigma, \dot{q}_\mu), \ddot{q}_\sigma) = R_1^*(t, q_\sigma, \dot{q}_\mu, \ddot{q}_\sigma). \qquad (2.193)$$

4. The desired equations of the programmed motion have the form

$$\frac{\partial R_1^*}{\partial \dot{q}_\mu} = \frac{\partial R_1}{\partial \dot{q}_\mu} + \sum_{\beta=1}^{k} \frac{\partial R_1}{\partial \dot{q}_\beta} \frac{\partial g_\beta^{(1)}}{\partial \dot{q}_\mu} = 0. \qquad \mu = k+1, \ldots, n \qquad (2.194)$$

Algorithm 2.1 provides a computationally efficient way for deriving equations of motion for a system with the constraints (2.100). Equations (2.194) together with (2.100) are a set of $\mu + \beta = (n - k) + k = n$ equations for n unknown q_σ's. It can be verified by direct calculations based on Equations (2.190) through (2.192) that

$$\frac{\partial R_1^*}{\partial \dot{q}_\mu} = \frac{\partial \dot{T}}{\partial \dot{q}_\mu} - 2\frac{\partial}{\partial \dot{q}_\mu}\left(\sum_{\sigma=1}^{n} \frac{\partial T}{\partial q_\sigma}\dot{q}_\sigma\right) - \frac{\partial}{\partial \dot{q}_\mu}\left(\sum_{\sigma=1}^{n} Q_\sigma \dot{q}_\sigma\right) + \sum_{\beta=1}^{k} \frac{\partial \dot{T}}{\partial \dot{q}_\beta} \frac{\partial g_\beta^{(1)}}{\partial \dot{q}_\mu}$$

$$- 2\sum_{\beta=1}^{k} \frac{\partial}{\partial \dot{q}_\beta}\left(\sum_{\sigma=1}^{n} \frac{\partial T}{\partial q_\sigma}\dot{q}_\sigma\right)\frac{\partial g_\beta^{(1)}}{\partial \dot{q}_\mu} - \sum_{\beta=1}^{k} \frac{\partial}{\partial \dot{q}_\beta}\left(\sum_{\sigma=1}^{n} Q_\sigma \dot{q}_\sigma\right)\frac{\partial g_\beta^{(1)}}{\partial \dot{q}_\mu}$$

$$= \frac{\partial \dot{T}}{\partial \dot{q}_\mu} - 2\frac{\partial T}{\partial q_\mu} - Q_\mu + \sum_{\beta=1}^{k}\left(\frac{\partial \dot{T}}{\partial \dot{q}_\beta} - 2\frac{\partial T}{\partial q_\beta} - Q_\beta\right)\frac{\partial g_\beta^{(1)}}{\partial \dot{q}_\mu} = 0,$$

which is equivalent to Nielsen's equations in Maggi's form (2.119). Substitution of (2.125) and (2.126) into it results in Equations (2.127).

Remark 2.5: In the derivation above, it is assumed that $\partial Q_\sigma / \partial \dot{q}_\sigma = 0$ for all $\dot{q}_\sigma$ to obtain a simpler form of final equations of motion. This assumption can be relaxed, and if $Q = Q(t, q_\sigma, \dot{q}_\sigma)$, we have to use relations (2.102) and present Q as $Q = Q(t, q_\sigma, \dot{q}_\mu)$.

Algorithm 2.2

To derive the GPME (2.189) for a system subjected to p-th order constraints, proceed as follows:

Assume that the constraints can be solved at least locally with respect to $q_\beta^{(p)}$, that is, relations (2.176) hold.

1. Construct a function P_p such that

$$P_p = \frac{1}{p}\left[T^{(p)} - (p+1)T_0^{(p)}\right] \tag{2.195}$$

and $T_0^{(p)}$ is defined by (2.191).

2. Construct a function R_p such that

$$R_p = P_p - \sum_{\sigma=1}^{n} q_\sigma^{(p)}Q_\sigma = R_p(t, q_\sigma, \dot{q}_\sigma, ..., q_\mu^{(p)}, q_\beta^{(p)}, q_\sigma^{(p+1)}). \tag{2.196}$$

3. Construct a function R_p^*, in which $q_\beta^{(p)}$ from (2.196) are replaced with (2.176):

$$R_p^* = R_p^*(t, q_\sigma, \dot{q}_\sigma, ..., q_\mu^{(p)}, g_\beta^{(p)}(t, q_\sigma, ..., q_\mu^{(p)}), q_\sigma^{(p+1)}) = R_p^*(t, q_\sigma, \dot{q}_\sigma, ..., q_\mu^{(p)}, q_\sigma^{(p+1)}). \tag{2.197}$$

4. Assuming that

$$\frac{\partial Q_\sigma}{\partial q_\sigma^{(p)}} = 0,$$

the desired GPME for a system with p-th order constraints (2.176) have the form

$$\frac{\partial R_p^*}{\partial q_\mu^{(p)}} = \frac{\partial R_p}{\partial q_\mu^{(p)}} + \sum_{\beta=1}^{k} \frac{\partial R_p}{\partial q_\beta^{(p)}} \frac{\partial g_\beta^{(p)}}{\partial q_\mu^{(p)}} = 0. \qquad \mu = k+1, ..., n \tag{2.198}$$

We can demonstrate that Equations (2.198) are $(n-k)$ second order equations of motion. To this end, we formulate a theorem.

Theorem 2.1

The GPME (2.198) are $(n-k)$ second order differential equations of motion of a system subjected to high order constraints specified by k equations (2.176). The set of n equations has the form

$$M(q)\ddot{q} + V(q, \dot{q}) + D(q) = Q(t, q, \dot{q}),$$

$$B(t, q, \dot{q}, ..., q^{(p-1)})q^{(p)} + s(t, q, \dot{q}, ..., q^{(p-1)}) = 0, \tag{2.199}$$

where M is a $(n-k) \times n$ inertia matrix; V is a $(n-k)$-dimensional velocity dependent vector; D is a $(n-k)$-dimensional vector of gravity forces, and Q is a $(n-k)$-dimensional vector of external forces.

Proof: First, we demonstrate by direct calculations based on relations (2.195) and (2.196) that Equations (2.198) are equivalent to Equations (2.189).

Then we use Lemma 2.1 to show that Equations (2.198) are second order differential equations in the form (2.199). Functions (2.195) and (2.197) substituted into Equations (2.198) yield

$$\frac{\partial R_p^*}{\partial q_\mu^{(p)}} = \frac{1}{p}\frac{\partial T^{(p)}}{\partial q_\mu^{(p)}} - \frac{p+1}{p}\frac{\partial}{\partial q_\mu^{(p)}}\left(\sum_{\sigma=1}^{n}\frac{\partial T}{\partial q_\sigma}q_\sigma^{(p)}\right) - \frac{\partial}{\partial q_\mu^{(p)}}\left(\sum_{\sigma=1}^{n}Q_\sigma q_\sigma^{(p)}\right) + \sum_{\beta=1}^{k}\frac{1}{p}\frac{\partial T^{(p)}}{\partial q_\beta^{(p)}}\frac{\partial g_\beta^{(p)}}{\partial q_\mu^{(p)}}$$

$$-\frac{p+1}{p}\sum_{\beta=1}^{k}\frac{\partial}{\partial q_\beta^{(p)}}\left(\sum_{\sigma=1}^{n}\frac{\partial T}{\partial q_\sigma}q_\sigma^{(p)}\right)\frac{\partial g_\beta^{(p)}}{\partial q_\mu^{(p)}} - \sum_{\beta=1}^{k}\frac{\partial}{\partial q_\beta^{(p)}}\left(\sum_{\sigma=1}^{n}Q_\sigma q_\sigma^{(p)}\right)\frac{\partial g_\beta^{(p)}}{\partial q_\mu^{(p)}} = 0.$$

Performing all differentiations in the expression above, we obtain

$$\frac{1}{p}\left[\frac{\partial T^{(p)}}{\partial q_\mu^{(p)}} - (p+1)\frac{\partial T}{\partial q_\mu}\right] - Q_\mu + \sum_{\beta=1}^{k}\left\{\frac{1}{p}\left[\frac{\partial T^{(p)}}{\partial q_\beta^{(p)}} - (p+1)\frac{\partial T}{\partial q_\beta}\right] - Q_\beta\right\}\frac{\partial g_\beta^{(p)}}{\partial q_\mu^{(p)}} = 0,$$

which are exactly Equations (2.182). They can be transformed to Equations (2.189) using relations (2.183) through (2.188). Next, using Lemma 2.1, specifically (2.170), we have that

$$\frac{\partial T^{(p)}}{\partial q_\mu^{(p)}} = p\left(\frac{\partial^2 T}{\partial \dot{q}_\mu \partial t} + \sum_{\alpha=k+1}^{n}\frac{\partial^2 T}{\partial \dot{q}_\mu \partial q_\alpha}\dot{q}_\alpha + \sum_{\alpha=k+1}^{n}\frac{\partial^2 T}{\partial \dot{q}_\mu \partial \dot{q}_\alpha}\ddot{q}_\alpha\right) + \frac{\partial T}{\partial q_\mu}.$$

These derivatives produce functions that depend on derivatives of coordinates up to second orders. The same holds for derivatives of $T^{(p)}$ with respect to $q_\beta^{(p)}$. Then, Equations (2.198) are second order differential equations that can be written in the form (2.199), and this completes the proof.

The most important property of Equations (2.199) is that they are free of constraint reaction forces. It makes them suitable for direct control applications.

Equations (2.199) are referred to as a unified dynamic model for a constrained system. Dynamic models presented in, e.g., Yun and Sarkar (1998); Udwadia and Kalaba (1996); You and Chen (1993), are peculiar cases of Equations (2.199) for $p = 1$.

The GPME are used in Chapter 7 to design a tracking control strategy.

Example 2.14

Develop a constrained dynamic model for a unicycle from Example 2.12 using Algorithm 2.1. Its wheel radius is r, mass is m, and I_θ, I are moments of inertia.

The nonholonomic constraint equations are given by Equations (2.163), and they are the only constraints on the unicycle. Following Algorithm 2.1, we obtain

$$T = \frac{1}{2}m(\dot{x}^2 + \dot{y}^2) + \frac{1}{2}I_\theta\dot{\theta}^2 + \frac{1}{2}I_\phi\dot{\phi}^2, \quad \dot{T} = m\ddot{x}\dot{x} + m\ddot{y}\dot{y} + I_\theta\dot{\theta}\ddot{\theta} + I_\phi\dot{\phi}\ddot{\phi}, \quad (2.200)$$

$$\dot{T}_o = \sum_{i=1}^{4}\frac{\partial T}{\partial q_\sigma}\dot{q}_\sigma = 0, \quad P_1 = \dot{T} - 2\dot{T}_o = \dot{T}. \quad (2.201)$$

From the coordinate vector $q = (x, y, \theta, \phi)$, we select dependent coordinates (x,y) and independent (θ, ϕ). Then, $\dot{x} = r\dot{\theta}\cos\phi$, $\dot{y} = r\dot{\theta}\sin\phi$, and we can generate R_1^*, which is

$$R_1^* = mr\dot{\theta}\ddot{x}\cos\phi + mr\dot{\theta}\ddot{y}\sin\phi + I_\theta\dot{\theta}\ddot{\theta} + I_\phi\dot{\phi}\ddot{\phi}. \quad (2.202)$$

Final equations are

$$\frac{\partial R_1^*}{\partial \dot{q}_{1i}} = 0,$$

where q_{1i} stand for the independent coordinates θ, ϕ, i.e.,

$$mr\ddot{x}\cos\phi + mr\ddot{y}\sin\phi + I_\theta\ddot{\theta} = \tau_1,$$

$$I_\phi\ddot{\phi} = \tau_2,$$

$$\dot{x} = r\dot{\theta}\cos\phi, \quad (2.203)$$

$$\dot{y} = r\dot{\theta}\sin\phi.$$

After reusing the nonholonomic constraint equations, the constrained dynamics become

$$(mr^2 + I_\theta)\ddot{\theta} = \tau_1,$$

$$I_\phi\ddot{\phi} = \tau_2,$$

$$\dot{x} = r\dot{\theta}\cos\phi, \quad (2.204)$$

$$\dot{y} = r\dot{\theta}\sin\phi,$$

where τ_1 and τ_2 are control torques needed to control the rolling and heading angles, respectively. The reader may compare the result (Equation 2.204) with the equations of the unicycle motion obtained after Lagrange multipliers elimination in Murray, Li, and Sastry (1994), example 6.2, pages 272–274.

The constrained dynamics (2.204) is decoupled into two sets: one that depends on θ and ϕ, and the second that depends on x and y. However, for most systems, the constrained dynamics has the general form as in Equations (2.199). Also, observe that the constrained dynamics (2.204) is already the reduced-state form; compare with equations (6.6.10) in Bloch (2003) which are derived by the Lagrange approach, and they are

$$\ddot{x} = \lambda_1,$$
$$\ddot{y} = \lambda_2,$$
$$\ddot{\theta} = -\lambda_1 \cos\ - \lambda_2 \sin\ + u_1,$$
$$\ddot{} = u_2,$$

(2.205)

and it is assumed there that mass, radius, and moments of inertia are all equal to one. u_1 denotes the control torque about the rolling axis (it is τ_1 in Equation (2.204)), and u_2 is the control torque about the vertical axis through the point of contact (it is τ_2 in Equation (2.204)). Then, multipliers λ_1 and λ_2 have to be eliminated, and for new variables $z_1 = \theta$, $z_2 = \ $, $z_3 = x$, $z_4 = y$, $z_5 = \dot{\theta}$, $z_6 = \dot{\ }$, the reduced equations (6.6.12) in Bloch (2003) are

$$\dot{z}_1 = z_5, \quad \dot{z}_2 = z_6, \quad \dot{z}_3 = z_5 \cos z_2,$$
$$\dot{z}_4 = z_5 \sin z_2, \quad \dot{z}_5 = u_1/2, \quad \dot{z}_6 = u_2.$$

(2.206a)

They are the same as Equations (2.204) when to define $z_1 = \theta$, $z_2 = \ $, $z_3 = x$, $z_4 = y$, $z_5 = \dot{\theta}$, $z_6 = \dot{\ }$ to obtain

$$\dot{z}_1 = z_5, \quad \dot{z}_2 = z_6, \quad \dot{z}_3 = z_5 r \cos z_2,$$
$$\dot{z}_4 = z_5 r \sin z_2, \quad \dot{z}_5 = \tau_1/(mr^2 + I_\theta), \quad \dot{z}_6 = \tau_2/I\ .$$

(2.206b)

It can be seen that the derivation of constrained dynamics using the GPME is faster and straightforward. The dynamic equations, which we obtain, are ready to design a model-based controller. We demonstrate this in Chapter 7.

Example 2.15

Consider the same unicycle model as in Example 2.14 and specify a programmed constraint for it. The nonholonomic material constraints are given by Equations (2.163) and the programmed constraint by Equation (2.164). All the constraints on the unicycle are then

$$\dot{x} = r\dot{\theta}\cos\ , \quad \dot{y} = r\dot{\theta}\sin\ , \quad (t) = \cos(0.5\pi - t).$$

(2.207)

To derive a unified dynamic model (2.199) for the unicycle, we use Algorithm 2.1 again. First, the three constraint equations (2.207) are presented in the form (2.160):

$$\dot{x} = r\dot{\theta}\cos\ , \quad \dot{y} = r\dot{\theta}\sin\ = 0, \quad \dot{\ } = \sin(0.5\pi - t).$$

(2.208)

Next, Algorithm 2.1 yields that

$$\frac{\partial \dot{T}}{\partial \dot{q}_\sigma} - 2\frac{\partial T}{\partial q_\sigma} = 0,$$

the kinetic energy T, its time derivative $\dot{T}_o$, and P_1 are given in Example 2.14. Selecting x, y, and φ as dependent coordinates, we generate R_1^*, which is

$$R_1^* = m(r\dot{\theta}\cos\ )\ddot{x} + m(r\dot{\theta}\sin\ )\ddot{y} + I_\theta\ddot{\theta}\dot{\theta} + I\ \sin(0.5\pi - t)\dddot{}. \qquad (2.209)$$

The unified dynamic model (2.199) for the unicycle is as follows:

$$mr\ddot{x}\cos\ + mr\ddot{y}\sin\ + I_\theta\ddot{\theta} = 0,$$
$$\dot{x} = r\dot{\theta}\cos\ ,$$
$$\dot{y} = r\dot{\theta}\sin\ , \qquad (2.210)$$
$$\dot{}\ = \sin(0.5\pi - t).$$

Solutions of these equations provide position time histories $x_p(t), y_p(t)$ and their time derivatives along the programmed motion.

Simulation results in programmed motion tracking according to constraints (2.207) are presented in Chapter 7.

Example 2.16

Consider a model of the planar two-link manipulator with two degrees of freedom described by Θ_1, Θ_2. Its geometry and inertia properties are given in Example 1.5 and Figure 1.6. Select the programmed constraint for the end-effector given by Equation (1.7), i.e.,

$$\ddot{\Theta}_2 = F_1 - F_2\ddot{\Theta}_1, \qquad (2.211)$$

where F_1, F_2, and A_ϕ, A_1, A_2 are defined in Example 1.5.

To obtain the unified dynamic model for the manipulator constrained by the programmed constraint, we use Algorithm 2.2 for $p = 3$. The kinetic energy T of the manipulator is equal to

$$T = \frac{1}{2}(\alpha + 2\beta\cos\Theta_2)\dot{\Theta}_1^2 + \frac{1}{2}\delta\dot{\Theta}_2^2 + (\delta + \beta\cos\Theta_2)\dot{\Theta}_1\dot{\Theta}_2,$$

where $\alpha = I_{z1} + I_{z2} + m_1 r_1^2 + m_2(l_1^2 + r_2^2), \beta = m_2 l_1 r_2, \delta = I_{z2} + m_2 r_2^2, r_i = l_i/2, i = 1,2.$

The third time derivative of T is

$$\dddot{T} = \dddot{\Theta}_1 A + \dddot{\Theta}_2 B + W, \qquad (2.212)$$

and

$$A = 3(\alpha + 2\beta\cos\Theta_2)\ddot{\Theta}_1 + 3(\delta + \beta\cos\Theta_2)\ddot{\Theta}_2 - 3\beta\dot{\Theta}_2^2\sin\Theta_2 - 6\beta\dot{\Theta}_1\dot{\Theta}_2\sin\Theta_2,$$

$$B = 3(\delta + \beta\cos\Theta_2)\ddot{\Theta}_1 + 3\delta\ddot{\Theta}_2 - 4\beta\dot{\Theta}_1\dot{\Theta}_2\sin\Theta_2.$$

The term W does not contain terms with third order time derivatives of coordinates and can be omitted. Next, we generate

$$P_3 = \frac{1}{3}\dddot{T} - \frac{4}{3}\sum_{n=1}^{2}\frac{\partial T}{\partial q_n}\dddot{q}_n = \frac{1}{3}A\dddot{\Theta}_1 + \frac{1}{3}B\dddot{\Theta}_2 + \frac{4}{3}\beta\dot{\Theta}_1\sin\Theta_2(\dot{\Theta}_1+\dot{\Theta}_2)\ddot{\Theta}_2. \quad (2.213)$$

Substituting the third time derivative of a dependent coordinate $\dddot{\Theta}_2$ extracted from the constraint equation to P_3, we obtain

$$R_3^* = \frac{1}{3}A\dddot{\Theta}_1 + \frac{1}{3}B(F_1 - F_2\dddot{\Theta}_1) + \frac{4}{3}\beta\dot{\Theta}_1\sin\Theta_2(\dot{\Theta}_1+\dot{\Theta}_2)(F_1 - F_2\dddot{\Theta}_1). \quad (2.214)$$

The equation of the programmed motion is

$$\frac{\partial R_3^*}{\partial \dddot{\Theta}_1} = 0.$$

Adding the constraint equation, we obtain the unified dynamic control model for the constrained manipulator in the form

$$(b_1 - b_2 F_2)\ddot{\Theta}_1 + (b_2 - \delta F_2)\ddot{\Theta}_2 + c = 0,$$
$$\ddot{\Theta}_2 = F_1 - F_2\ddot{\Theta}_1 \quad (2.215)$$

where $b_1 = \alpha + 2\beta\cos\Theta_2$, $b_2 = \delta + \beta\cos\Theta_2$, $c = -\beta\dot{\Theta}_2(\dot{\Theta}_2 + 2\dot{\Theta}_1)\sin\Theta_2 - 4/3\beta\dot{\Theta}_1^2 F_2 \sin\Theta_2$.

2.4.3 Generalized Programmed Motion Equations Specified in Quasi-Coordinates

Equations (2.149) do not apply to systems with nonholonomic constraints of an order higher than one. To extend them on systems with high order constraints, let us prove the following.

Lemma 2.2 (Jarzębowska 2008b)

Assume that there exists a function $\tilde{F}$ of the form

$$\tilde{F} = \tilde{F}(t, q_\sigma, \omega_r), \qquad \sigma, r = 1, ..., n \quad (2.216)$$

where generalized coordinates q_σ and quasi-velocities ω_r are related by (2.128). Assume also that $\tilde{F}$ is regular enough, i.e., its time derivatives up to some order p exist. Then, the following identity holds:

$$\frac{d}{dt}\left(\frac{\partial\tilde{F}}{\partial\omega_\sigma}\right) = \frac{1}{p}\left(\frac{\partial\tilde{F}^{(p)}}{\partial\omega_\sigma^{(p-1)}} - \frac{\partial\tilde{F}}{\partial\pi_\sigma}\right). \qquad \sigma = 1, ..., n, \quad p = 1, 2, 3, ... \quad (2.217)$$

Proof: The proof can be completed by mathematical induction. To this end, we compute total time derivatives of $\tilde{F}$ of orders $p = 1, 2, 3, \ldots$, that is,

$$\frac{d\tilde{F}}{dt}, \frac{d^2\tilde{F}}{dt^2}, \ldots, \frac{d^{(p)}\tilde{F}}{dt^{(p)}}.$$

Skipping time-consuming calculations, we can verify that

$$\frac{d^{(p)}\tilde{F}}{dt^{(p)}} \equiv \tilde{F}^{(p)}$$

has the form

$$\tilde{F}^{(p)} = p\left(\sum_{\sigma=1}^{n} \frac{\partial^2 \tilde{F}}{\partial \omega_\sigma \partial t} \omega_\sigma^{(p-1)} + \sum_{\sigma=1}^{n} \sum_{\alpha=1}^{n} \frac{\partial^2 \tilde{F}}{\partial \omega_\sigma \partial q_\alpha} \dot{q}_\alpha \omega_\sigma^{(p-1)} + \sum_{\sigma=1}^{n} \sum_{\alpha=1}^{n} \frac{\partial^2 \tilde{F}}{\partial \omega_\sigma \partial \omega_\alpha} \dot{\omega}_\alpha \omega_\sigma^{(p-1)} \right)$$

$$+ \sum_{\sigma=1}^{n} \frac{\partial \tilde{F}}{\partial \pi_\sigma} \omega_\sigma^{(p-1)} + \tilde{\Gamma}, \qquad\qquad p > 2 \qquad (2.218)$$

where the function $\tilde{\Gamma}$ does not contain derivatives $\omega_\sigma^{(p-1)}$, $\sigma = 1, \ldots, n$. Hence, we obtain

$$\frac{\partial \tilde{F}^{(p)}}{\partial \omega_\sigma^{(p-1)}} = p\left(\frac{\partial^2 \tilde{F}}{\partial \omega_\sigma \partial t} + \sum_{\alpha-1}^{n} \frac{\partial^2 \tilde{F}}{\partial \omega_\sigma \partial q_\alpha} \dot{q}_\alpha + \sum_{\alpha=1}^{n} \frac{\partial^2 \tilde{F}}{\partial \omega_\alpha \partial \omega_\sigma} \dot{\omega}_\alpha \right) + \frac{\partial \tilde{F}}{\partial \pi_\sigma}. \qquad (2.219)$$

Relation (2.219) holds for any $p = 1, 2, \ldots$ and $\sigma = 1, \ldots, n$. We can verify that for $p = 1$, we have

$$\dot{\tilde{F}} = \frac{\partial \tilde{F}}{\partial t} + \sum_{\alpha=1}^{n} \frac{\partial \tilde{F}}{\partial q_\alpha} \dot{q}_\alpha + \sum_{\alpha=1}^{n} \frac{\partial \tilde{F}}{\partial \omega_\alpha} \dot{\omega}_\alpha, \qquad\qquad (2.220a)$$

$$\frac{\partial \dot{\tilde{F}}}{\partial \omega_\sigma} = \frac{\partial^2 \tilde{F}}{\partial t \partial \omega_\sigma} + \sum_{\alpha=1}^{n} \frac{\partial^2 \tilde{F}}{\partial \omega_\sigma \partial q_\alpha} \dot{q}_\alpha + \sum_{\alpha=1}^{n} \frac{\partial^2 \tilde{F}}{\partial \omega_\alpha \partial \omega_\sigma} \dot{\omega}_\alpha + \frac{\partial \tilde{F}}{\partial \pi_\sigma}. \qquad \sigma = 1, \ldots, n \quad (2.220b)$$

In the same way, we can demonstrate that relation (2.219) holds for $p = 2$. Next, we calculate

$$\frac{d}{dt}\left(\frac{\partial \tilde{F}}{\partial \omega_\sigma} \right) = \frac{\partial^2 \tilde{F}}{\partial t \partial \omega_\sigma} + \sum_{\alpha=1}^{n} \frac{\partial^2 \tilde{F}}{\partial q_\alpha \partial \omega_\sigma} \dot{q}_\alpha + \sum_{\alpha=1}^{n} \frac{\partial^2 \tilde{F}}{\partial \omega_\alpha \partial \omega_\sigma} \dot{\omega}_\alpha. \qquad \sigma = 1, \ldots, n \quad (2.221)$$

From (2.219) and (2.221), we get the desired result:

$$\frac{d}{dt}\left(\frac{\partial \tilde{F}}{\partial \omega_\sigma} \right) = \frac{1}{p}\left(\frac{\partial \tilde{F}^{(p)}}{\partial \omega_\sigma^{(p-1)}} - \frac{\partial \tilde{F}}{\partial \pi_\sigma} \right). \qquad \sigma = 1, \ldots, n, \quad p = 1, 2, 3, \ldots \quad (2.222)$$

q.e.d.

If we replace $\tilde{F}$ with the kinetic energy function $\tilde{T} = \tilde{T}(t, q_\sigma, \omega_\sigma), \sigma = 1, ..., n,$ in (2.222) and insert it into the generalized Boltzmann–Hamel Equations (2.145), we obtain

$$\frac{1}{p}\left[\frac{\partial \tilde{T}^{(p)}}{\partial \omega_\mu^{(p-1)}} - (p+1)\frac{\partial \tilde{T}}{\partial \pi_\mu}\right] + \sum_{r=1}^{n} \frac{\partial \tilde{T}}{\partial \omega_r} W_\mu^r = \tilde{Q}_\mu. \qquad \mu = 1, ..., n, \, p = 1, 2, 3, \quad (2.223)$$

Equations (2.223) are the extended form of the Boltzmann–Hamel equations. They can be modified for applications to systems with nonholonomic constraints of high order. These high order constraints specified in quasi-coordinates have the form

$$\tilde{G}_\beta\left(t, q_\sigma, \omega_r, \dot{\omega}_r, ..., \omega_r^{(p-1)}\right) = 0. \qquad \beta = 1, ..., b, \, \sigma, r = 1, ..., n, \, b < n \quad (2.224)$$

Based on relations (2.128), (2.14) and the definition of the generalized virtual displacement (2.157),

$$\delta G_\beta = \sum_{\sigma=1}^{n} \frac{\partial G_\beta}{\partial q_\sigma^{(p)}} \delta q_\sigma = 0, \quad (2.225)$$

we obtain that

$$\delta \tilde{G}_\beta = \sum_{r=1}^{n} \frac{\partial \tilde{G}_\beta}{\partial \omega_r^{(p-1)}} \delta \pi_r = 0. \quad (2.226)$$

In the constraint equation (2.224), we may partition the vector $\omega^{(p-1)}$ of $(p-1)$ order derivatives as $\omega^{(p-1)} = \left(\omega_\beta^{(p-1)} \quad \omega_\mu^{(p-1)}\right)$ with

$$\omega_\beta^{(p-1)} = \Omega_\beta^{(p-1)}\left(t, q_\sigma, \omega_\sigma, \dot{\omega}_\sigma, ..., \omega_\mu^{(p-1)}\right). \quad \sigma = 1, ..., n, \beta = 1, ..., b, \mu = b+1, ..., n$$

$$(2.227)$$

The underlying assumption is that the above partition is possible, i.e., the Jacobian of the transformation is nonsingular. By differentiating Equation (2.227) with respect to time, we obtain

$$\omega^p = \Omega_\beta^p\left(t, q_\sigma, \omega_\sigma, \dot{\omega}_\sigma, ..., \omega_\mu^{(p-1)}, \omega_\mu^p\right). \quad (2.228)$$

Now, using (2.222), we rewrite Equations (2.144) in the form

$$
\sum_{\beta=1}^{b} \left\{ \frac{1}{p} \left[\frac{\partial \tilde{T}^{(p)}}{\partial \omega_\beta^{(p-1)}} - (p+1) \frac{\partial \tilde{T}}{\partial \pi_\beta} \right] + \sum_{r=1}^{n} \frac{\partial \tilde{T}}{\partial \omega_r} W_\beta^r - \tilde{Q}_\beta \right\} \delta \pi_\beta
$$

$$
+ \sum_{\mu=b+1}^{n} \left\{ \frac{1}{p} \left[\frac{\partial \tilde{T}^{(p)}}{\partial \omega_\mu^{(p-1)}} - (p+1) \frac{\partial \tilde{T}}{\partial \pi_\mu} \right] + \sum_{r=1}^{n} \frac{\partial \tilde{T}}{\partial \omega_r} W_\mu^r - \tilde{Q}_\mu \right\} \delta \pi_\mu = 0.
$$

(2.229)

Based on (2.226),

$$
\delta \pi_\beta = \sum_{\mu=b+1}^{n} \frac{\partial \Omega_\beta^{(p-1)}}{\partial \omega_\mu^{(p-1)}} \delta \pi_\mu. \qquad \beta = 1, \dots, b
$$

(2.230)

Then Equation (2.229) takes the form

$$
\frac{1}{p} \left[\frac{\partial \tilde{T}^{(p)}}{\partial \omega_\mu^{(p-1)}} - (p+1) \frac{\partial \tilde{T}}{\partial \pi_\mu} \right] + \sum_{r=1}^{n} \frac{\partial \tilde{T}}{\partial \omega_r} W_\mu^r - \tilde{Q}_\mu + \sum_{\beta=1}^{b} \left\{ \frac{1}{p} \left[\frac{\partial \tilde{T}^{(p)}}{\partial \omega_\beta^{(p-1)}} - (p+1) \frac{\partial \tilde{T}}{\partial \pi_\beta} \right] \right.
$$

$$
\left. + \sum_{r=1}^{n} \frac{\partial \tilde{T}}{\partial \omega_r} W_\beta^r - \tilde{Q}_\beta \right\} \frac{\partial \Omega_\beta^{(p-1)}}{\partial \omega_\mu^{(p-1)}} = 0. \qquad \mu = b+1, \dots, n
$$

(2.231)

We refer to Equations (2.231) as the GPME in quasi-coordinates. For $p = 1$, they become Equations (2.149). To automate the derivation of Equations (2.231), algorithms similar to Algorithms 2.1 and 2.2 can be developed.

Example 2.17

To illustrate the application of the modeling framework developed in quasi-coordinates, select a unicycle model as in Example 2.14. Select the programmed constraint to be

$$
x^2 + y^2 - \phi(t) = 0 = \Phi(t)
$$

(2.232)

with $\phi(t) = 0.2t + 1$. Material constraints for the robot are given by Equations (2.163).

Quasi-velocities for the unified dynamic model of the constrained unicycle are selected as

$$
\omega_1 = \dot{x} - r\dot{\theta}\cos = 0, \quad \omega_2 = \dot{y} - r\dot{\theta}\sin = 0,
$$

$$
\omega_3 = \dot{\Phi}(t) + \alpha\Phi(t) = 0, \quad \omega_4 = \cdot,
$$

(2.233)

where for simulation, only the programmed constraint has to be numerically stabilized, i.e., the relation $\dot{\Phi}(t) + \alpha\Phi(t) = 0$ is used. The GPME in quasi-coordinates (2.151) for the unicycle are as follows:

$$I\,\dot{\omega}_4 = \frac{-\dot{\eta}^2(t)(mr^2 + I_\theta)(x\sin\eta + y\cos\eta)}{4r^2(x\cos\eta + y\sin\eta)^3},$$

$$\dot{x} = r\dot{\theta}\cos\eta,$$

$$\dot{y} = r\dot{\theta}\sin\eta,$$

$$\dot{\theta} = \frac{\dot{\eta}(t)}{2r(x\cos\eta + y\sin\eta)}, \tag{2.234}$$

$$\dot{\eta} = \omega_4.$$

We can see that a programmed constraint is treated in the same way as other constraints on a system, and according to the concept of the quasi-velocity, it is "swallowed" by it. Equations (2.234) are free of constraint reaction forces; they are ready for control applications. The unicycle motion according to constraint (2.232) is presented in Figure 2.9.

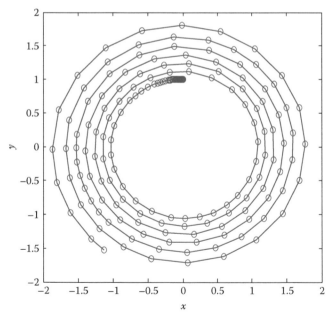

FIGURE 2.9
Programmed motion of a unicycle.

Example 2.18

Consider the two-link planar manipulator model from Example 2.4.6 and derive the GPME in quasi-coordinates for it. Select the same programmed constraint (1.7). Quasi-velocities may be selected as

$$\omega_1 = \dot{\Theta}_1 l_1, \quad \omega_2 = (\dot{\Theta}_1 + \dot{\Theta}_2) l_2. \tag{2.235}$$

With this selection of ω_1, ω_2, the end-effector velocity can also be controlled i.e., this velocity is determined as $v^2 = \omega_1^2 + \omega_2^2 + 2\omega_1\omega_2 \cos\Theta_2$. The programmed constraint (1.7) specified in quasi-velocities has the form

$$\ddot{\omega}_2 - (1 - F_2)\frac{l_2}{l_1}\ddot{\omega}_1 - F_1 l_2 = 0, \tag{2.236}$$

where

$$F_1 = \frac{A_\phi - A_1 - A_2 a_0}{a_2 + a_4 a_0} \text{ and } F_2 = \frac{a_1 + a_2 + a_0(a_3 + a_4)}{a_2 + a_4 a_0}$$

are specified in quasi-velocities as

$$A_\phi = \frac{-\Phi(a_5^2 + a_6^2)^2 \left[\dot{\Phi}(a_5^2 + a_6^2) + 3\Phi(a_5 a_7 + a_6 a_8)\right]}{a_6(a_5 a_8 - a_7 a_6)},$$

$$A_1 = 3a_3 \frac{\omega_1 \dot{\omega}_1}{l_1^2} + 3a_4 \frac{\omega_2 \dot{\omega}_2}{l_2^2} - a_1 \frac{\omega_1^3}{l_1^3} - a_2 \frac{\omega_2^3}{l_2^3}, \quad A_2 = 3a_3 \frac{\omega_1 \dot{\omega}_1}{l_1^2} + 3a_2 \frac{\omega_2 \dot{\omega}_2}{l_2^2} + a_3 \frac{\omega_1^3}{l_1^3} + a_4 \frac{\omega_2^3}{l_2^3},$$

$$a_1 = -l_1 \sin\Theta_1, \ a_2 = -l_2 \sin(\Theta_1 + \Theta_2), \ a_0 = a_5/a_6,$$

$$a_3 = -l_1 \cos\Theta_1, \ a_4 = -l_2 \cos(\Theta_1 + \Theta_2), \ a_5 = a_1 \frac{\omega_1}{l_1} + a_2 \frac{\omega_2}{l_2}, \ a_6 = -a_3 \frac{\omega_1}{l_1} - a_4 \frac{\omega_2}{l_2},$$

$$a_7 = a_1 \frac{\omega_1}{l_1} + a_3 \frac{\omega_1^2}{l_1^2} + a_2 \frac{\dot{\omega}_2}{l_2} + a_4 \frac{\omega_2^2}{l_2^2}, \ a_8 = -a_3 \frac{\dot{\omega}_1}{l_1} + a_1 \frac{\omega_1^2}{l_1^2} - a_4 \frac{\dot{\omega}_2}{l_2} + a_2 \frac{\omega_2^2}{l_2^2}.$$

The unified dynamic model of the manipulator (2.231) has the form

$$\frac{b_1 - b_2 - F_2(b_2 - \delta)}{l_1} \dot{\omega}_1 + \frac{b_2 - F_2 \delta}{l_2} \dot{\omega}_2 + c = 0,$$

$$\ddot{\omega}_2 - \frac{(1 - F_2)l_2}{l_1} \ddot{\omega}_1 - F_1 l_2 = 0. \tag{2.237}$$

where $b_1 = \alpha + 2\beta\cos\Theta_2, b_2 = \delta + \beta\cos\Theta_2, c = -\beta\sin\Theta_2\left[(1 - 4/3F_2)\frac{\omega_1^2}{l_1^2} - \frac{\omega_2^2}{l_2^2}\right]$,

$\alpha = l_{z1} + l_{z2} + m_1 r_1^2 + m_2(l_1^2 + r_2^2), \beta = m_2 l_1 r_2, \delta = l_{z2} + m_2 r_2^2$.

The control dynamics for the manipulator will be derived in Chapter 7.

We can see that control oriented modeling in quasi-coordinates results in constrained dynamic models, which are first order ordinary differential equations and numerical simulations are faster for them; for more details see Chapter 7.

Problems

1. Derive the two formulas given by (2.24) for the nonholonomic constraint equations for the rolling falling disk from Example 2.3.
2. Derive the constraint equation (1.8) for the space manipulator.
3. Derive Lagrange's equations for the space manipulator presented in Figure 1.7. Assume that the only constraint is (1.8) that comes from the angular momentum conservation.
4. Derive the Boltzmann–Hamel equations for the rolling ball presented in Figure 2.5.
5. Derive the Maggi equations for the rolling ball presented in Figure 2.5.
6. Derive the GPME for the space manipulator presented in Figure 1.7. Assume that the constraint on it is given by Equation (1.8), and it comes from the angular momentum conservation.

References

Appell, P. 1911. Exemple de mouvement d'un point assujeti a une liaison exprimee par une relation non lineaire entre les composantes de la vitesse. *Comptes Renduss* 48–50.

Appell, P. *Traite de mecanique rationnell.* Vol. 1, 1941; Vol. 2, 1953. Paris: Gauthier-Villars.

Beghin, H. 1947. *Course de mecanique*. Paris.

Bloch, A.M. 2003. *Nonholonomic mechanics and control*. New York: Springer-Verlag.

Boltzmann, L. 1902. Sitzungsber. Akad. Wiss, Wien, Vol. III.

Bushnell, L., D. Tilbury, and S. Sastry. 1995. Steering three-input nonholonomic systems: The fire truck example. *Int. J. Robot. Res.* 14:366–381.

Campion, G., B. d'Andrea-Novel, and G. Bastin. 1991. Controllability and state feedback stabilizability of nonholonomic nechanical systems. In *Advanced Robot Control*, ed. C. Canudas-de-Wit, LNCIS 162, 106–124. New York: Springer.

Canudas-de-Wit, C., and O.J. Sordalen. 1992. Exponential stabilization of mobile robots with nonholonomic constraints. *IEEE Trans. Automat. Contr.* 37(1):1791–1797.

Chang Y.-Ch., and B.-S. Chen. 1996. Adaptive tracking control design of nonholonomic mechanical systems. In *Proc. 35th Conf. Decision Contr.* 4739–4744.

Chaplygin, S.A. 1897. About a motion of a heavy body on a horizontal plane. In *Izbrannye Trudy Klassiki Nauki*, chap. 3, 363–375. Moscow: Nauka (in Russian).

de Jalon, J.G., and E. Bayo. 1994. *Kinematic and dynamic simulation of multibody systems*. Mech. Eng. Series, Berlin: Springer-Verlag.

Dobronravov, V.V. 1970. *Foundations of mechanics of non-holonomic systems*. Moscow: Nauka (in Russian).

Fukao, T., H. Nakagawa, and N. Adachi. 2000. Adaptive tracking control of a nonholonomic mobile robot, *IEEE Trans. Robot. Automat.* 16(5):609–615.

Gutowski, R. 1971. *Analytical mechanics*. Warsaw: PWN (in Polish).

Gutowski, R., and B. Radziszewski. 1969. The behavior of the solutions of equations of motion of the mechanical system with program constraints. *Bull. Acad. Polon. Sci.*, Serie sci. tech. 17(2):17–25.

Hamel G. 1904. *Die Lagrange-Eulerischen Gleichungen der Mechanik. Z. Math. Phys.* 50:1–57.

Hamel, G. 1967. *Theoretische mechanik.* Berlin: Springer-Verlag.

Hogan, N. 1987. Stable execution of contact tasks using impedance control. In *Proc. IEEE Int. Conf. Robot. Autom.*, 595–601, Raleigh, NC.

Isidori, A. 1989. *Nonlinear control systems.* Berlin: Springer.

Jarzębowska, E. 2002. On derivation of motion equations for systems with nonholonomic high-order program constraints. *Multibody Syst. Dyn.* 7(3):307–329.

Jarzębowska, E. 2005. Dynamics modeling of nonholonomic mechanical systems: Theory and applications. *Nonlin. Anal.* 63(5–7):185–197.

Jarzębowska, E. 2006a. Control oriented dynamic formulation of robotic systems with program constraints. *Robotica* 24(1):61–73.

Jarzębowska, E. 2006b. Tracking control design for underactuated constrained systems. *Robotica* 24(1):591–593.

Jarzębowska, E. 2007a. Stabilizability and motion tracking conditions for nonholonomic control systems. *Math. Prob. Eng.* Cairo: Hindawi.

Jarzębowska, E. 2007b. *Model-based tracking control strategies for constrained mechanical systems.* Palermo: International Society for Advanced Research.

Jarzębowska, E. 2008a. Advanced programmed motion tracking control of nonholonomic mechanical systems. *IEEE Trans. Robot.* 24(6):1315–1328.

Jarzębowska, E. 2008b. Quasi-coordinates based dynamics modeling and control design for nonholonomic systems. *Nonlin. Anal.* 16(16):1741–1754.

Jarzębowska, E., and N.H. McClamroch. 2000. On nonlinear control of the Ishlinsky problem as an example of a nonholonomic non-Chaplygin system. In *Proc. Am. Contr. Conf.* 3249–3253.

Jarzębowska, E., and R. Lewandowski. 2006. Modeling and control design using the Boltzmann–Hamel equations: A roller-racer example. In *Proc. 8th IFAC Symp. Robot Contr.* SYROCO, Bologna, Italy.

Kamman, J.W., and R.L. Huston. 1984. Dynamics of constrained multibody systems. *J. Appl. Mech.* 51:899–903.

Kane, T.R., and D.L. Levinson. 1985. *Dynamics—Theory and applications.* New York: McGraw-Hill.

Koh, K.C., and H.S. Cho. 1999. A smooth path tracking algorithm for wheeled mobile robots with dynamic constraints. *J. Intell. Robot. Syst.* 24:367–385.

Krishnaprasad, P.S., and D.P. Tsakiris. 2001. Oscillations, SE(2)-snakes and motion control: A study of the roller-racer. *Dyn. Syst.* 16(4):347–397.

Kwatny, H.G., and G.L. Blankenship. 2000. *Nonlinear control and analytical mechanics. A computational approach.* Boston: Birkhauser.

Lancos, C. 1986. *The variational principles of mechanics.* New York: Dover.

Layton, R.A. 1998. *Principles of analytical system dynamics.* New York: Springer-Verlag.

Lewis, F.L., C.T. Abdallah, and D.M. Dawson. 2004. *Control of robot manipulators.* New York: Marcel Dekker, 2nd ed.

Lewis, A.D., and R.M. Murray. 1997. Controllability of simple mechanical control systems. *SIAM J. Contr. Optim.* 35(3):766–790.

Mangeron, D., and S. Deleanu. 1962. Sur une classe d'equations de la mecanique analytique ou sens de Tzenoff. *Comptes Rendus de l'Academie Bulgare des Sciences* 15:1–9.

Marsden, J.E. 1992. *Lectures on mechanics*. London Math. Soc. Lect. Note Ser., 174, Cambridge University Press.

Moon, F.C. 1998. *Applied dynamics*. New York: John Wiley & Sons.

Murray, R.M., Z.X. Li, and S.S. Sastry. 1994. *A mathematical introduction to robotic manipulation*. Boca Raton, FL: CRC Press.

Nejmark, J.I., and N.A. Fufaev. 1972. *Dynamics of nonholonomic systems*. Providence, RI: American Mathematical Society.

Nielsen, J. 1935. *Vorlesungen uber elementare mechanik*. Berlin: Verlag von J. Springer.

Nijmeijer, H., and A. van der Schaft. 1990. *Nonlinear dynamical control systems*. New York: Springer-Verlag.

Nikravesh, P. 1988. *Computer aided analysis of mechanical systems*. Englewood Cliffs, NY: Prentice Hall.

Oriolo, G., and Y. Nakamura. 1991. Control of mechanical systems with second order nonholonomic constraints: Underactuated manipulators. In *Proc. 30th Conf. Decision Control*. 2398–2403.

Oriolo, G., A. De Luca, and M. Vendittelli. 2002. WMR control via dynamic feedback linearization: Design, implementation, and experimental validation. *IEEE Trans. Cont. Syst. Techn.* 10(6):835–852.

Ostrowski, J., J.P. Desai, and V. Kumar. 1997. Optimal gait selection for nonholonomic locomotion systems. In *Proc. IEEE Int. Conf. Robot. Automat.* 786–791.

Papastavridis, J.G. 1988. On the nonlinear Appell's equations and the determination of generalized reaction forces. *Int. J. Eng. Sci.* 26(6):609–625.

Papastavridis, J.G. 1994. On the Boltzmann–Hamel equations of motion: A vectorial treatement. *J. Appl. Mech.* 61:453–459.

Papastavridis, J.G. 1998. A panoramic overview of the principles and equations of motion of advanced engineering dynamics. *Appl. Mech. Rev.* 51(4):239–265.

Papastavridis, J.G. 2002. *Analytical mechanics. A comprehensive treatise on the dynamics of constrained systems; For engineers, physicians, and mathematicians*. New York: Oxford University Press.

Pars, L.A. 1965. *Treatise of analytical dynamics*. London: W. Heinemann, Ltd.

Reyhanoglu, M., A. van der Schaft, N.H. McClamroch, and I. Kolmanovsky. 1996. Nonlinear control of a class of underactuated systems. In *Proc. 35th Conf. Decision Contr.* 1682–1687. Japan.

Rong, Q. 1996a. The equations of motion for the system of the variable mass in the nonlinear nonholonomic space. *Appl. Math. Mech.* (English Edition) 17(4):379–384.

Rong, Q. 1996b. The method of derivation of MacMillans equations for the nonlinear nonholonomic systems. *Appl. Math. Mech.* (English Edition) 17(6):585–587.

Rong, Q. 1997. The fundamental equations of dynamics using representation of quasi-coordinates in the space of nonlinear nonholonomic constraints. *Appl. Math. Mech.* (English Edition) 18(11):1105–1113.

Samson, C. 1990. Velocity and torque feedback control of a nonholonomic cart. In *Advanced robot control*, C. Canudas de Wit, ed. Lecture Notes in Control and Information Sciences, New York: Springer-Verlag, 1990.

Samson, C. 1995. Control of chained systems: Application to path following and time-varying point stabilization of mobile robots. *IEEE Trans. Automat. Contr.* 40:64–77.

Scheuer, A., and Ch. Laugier. 1998. Planning sub-optimal and continuous-curvature paths for car-like robots. In *Proc. IEEE/RSJ Intl. Conf. Intell. Robot. Syst.* 25–31.

Shekl, A.M., and V.L. Lumelsky. 1998. Motion planning for nonholonomic robots in a limited workspace. In *Proc. 1998 IEEE/RSJ Intl. Conf. Intell. Robot. Syst.* 1473–1478.

Slotine, J.J., and W. Li. 1991. *Applied nonlinear control*. Englewood Cliffs, NJ: Prentice Hall.

Soltakhanov, Sh., M.P. Yushkov, and S.A. Zegzhda. 2009. *Mechanics of nonholonomic systems*. Berlin: Springer.

Tarn, T-J., M. Zhang, and A. Serrani. 2003. New integrability conditions for differential constraints. *Syst. Contr. Lett.* 49:335–345.

Tianling, L. 1992. Universal d'Alembert principle and equations of motion for any order nonholonomic systems in derivative space. *Chinese Sci. Bull.* 24.

Tsakiris, D.P. and P.S. Krishnaprasad. 1998. Oscillations, SE(2)-snakes and motion control: a study of the roller-racer. *Technical Report, CDSS*, University of Maryland.

Tzenoff, J. 1924. Sur les equations du mouvement des systemes mecanique nonholonomes. *Mat. Ann.* 91.

Udwadia, F., and R. Kalaba. 1996. *Analytical dynamics—A new approach*. London: Cambridge University Press.

Vafa, Z. 1991. Space manipulator motion with no satellite attitude disturbances. In *Proc. IEEE Int. Conf. on Robot. Automat.* 1770–1775.

Voronets, P.V. 1901. On the equations of motion for nonholonomic systems. *Matematicheskii Sbornik.* 22(4) (in Russian).

Vranceanu, G. 1929. Studio geometrico dei sistemi annolonomi. *Bull. Mat.* 30.

You, L.-S., and B.-S. Chen. 1993. Tracking control designs for both holonomic and nonholonomic constrained mechanical systems: A unified viewpoint. *Int. J. Contr.* 58(3): 587–612.

Yun, X., and N. Sarkar. 1998. Unified formulation of robotic systems with holonomic and nonholonomic constraints. *IEEE Trans. Robot. Automat.* 14(4):640–650.

Zotov, Yu. K., and A.V. Tomofeyev. 1992. Controllability and stabilization of programmed motions of reversible mechanical and electromechanical systems. *J. Appl. Math. Mech.* (trans. from Russian) 56(6):873–880.

Zotov, Yu. K. 2003. Controllability and stabilization of programmed motions of an automobile-type transport robot. *J. Appl. Math. Mech.* (trans. from Russian) 67(3):303–327.

3

Introduction to Nonlinear Control Theory

3.1 Stability Properties of Nonlinear Systems

Most real system models are nonlinear, and many of them cannot be linearized and presented by linear state-variable equations. The nonlinear system models we are interested in are models of ground, mobile, and space manipulators, ground car-like vehicles, underwater vehicles, biomechanical or biomimetic systems, and many others. Their dynamics properties are governed by nonlinear ordinary differential equations. For some of these system models, kinematics relations may be used as a basis for design of control algorithms. These kinematics relations are discussed in Section 3.4. For some of them, only dynamics models may be used to design control algorithms; such systems are discussed in Section 3.5.

3.1.1 State-Space Representation of Nonlinear Systems

A nonlinear dynamic system can be presented in a state-space form as

$$\dot{q}(t) = f(q(t), t), \tag{3.1}$$

where f is a nonlinear vector function, q is a state variable vector, and $q(t) \in R^n$. By definition, a state vector is not unique. If q is a state vector, which is sufficient to completely determine the behavior of a system for all $t > t_0$, then any $\bar{q}(t) = Tq(t)$, where T is a $(n \times n)$ invertible matrix is also a state vector.

Dynamic models of systems as presented in Chapter 2 may be transformed to the form of Equation (3.1).

For control systems, an input vector is introduced into the state-space form and

$$\dot{q}(t) = f(q(t), u(t), t) \tag{3.2}$$

is referred to as a state-space control formulation. The input vector u is assumed to appear linearly in Equation (3.2). Usually, $u = u(q, \dot{q}, t)$.

The state-space system is said to be nonautonomous when time t appears explicitly in Equation (3.2). If f does not depend upon time explicitly, the system is said to be autonomous. The nonautonomous systems (3.2) are also referred to as nonlinear time-varying systems. Autonomous systems are also referred to as time invariant.

Example 3.1

Consider the system of two disks whose equations of motion are derived in Example 2.8. This is an example of a holonomic system model. Let us present it in a state-space formulation as follows. Introduce two-dimensional vectors $w = (x, \ )$ and $q = (w, \dot{w})$. Motion equations of the two disks may be then presented as

$$M\ddot{w} + C(w, \dot{w}) + D(w) = \tau, \qquad (3.3)$$

where an external force vector is added, and the matrices are as follows:

$$M = \begin{bmatrix} M + 3m/2 & MR\cos \\ MR\cos & 3MR/2 \end{bmatrix}, \quad C = \begin{bmatrix} 0 & MR\dot{\ }\sin \\ 0 & 0 \end{bmatrix}, \quad D = \begin{bmatrix} cx \\ MgR\sin \end{bmatrix}.$$

For rigid bodies M is invertible, so the equations may be written as

$$\dot{q} = \begin{bmatrix} \dot{w} \\ \ddot{w} \end{bmatrix} = \begin{bmatrix} \dot{w} \\ -M^{-1}\big[C(q)q + D(q) \big] \end{bmatrix} + \begin{bmatrix} 0 \\ M^{-1}\tau \end{bmatrix} \qquad (3.4)$$

or $\dot{q}(t) = f(q) + g(q)\tau$. For the state-space control form as above, one may design some control algorithm τ based upon a desired control goal.

The state-space control form (3.2) may also result from a kinematic model. Such models are derived based on constraint equations for nonholonomic systems. It is interesting to note that for holonomically constrained systems, a kinematic state-space control form is trivial, but the dynamics-based control problem is not, as presented in Example 3.1.

Example 3.2

Consider the unicycle model presented in Figure 2.8. The nonholonomic constraint equations (2.163) may be presented in the form (3.2) as follows. Select the forward velocity of the unicycle $r\dot{\theta}$ and the rate of change of the heading angle $\dot{\ }$ as two control inputs. Then, a state-space control form, which is a kinematic control model for the unicycle, yields

$$\begin{aligned} \dot{x} &= u_1 \cos \ . \\ \dot{y} &= u_1 \sin \ , \\ u_1 &= r\dot{\theta}, \\ u_2 &= \dot{\ }. \end{aligned} \qquad (3.5)$$

The next step is to design control laws u_1 and u_2 that can steer the unicycle to a desired location.

A detailed discussion of kinematic control models like (3.5) is presented in Section 3.4.

3.1.2 Stability Theorems of the Lyapunov Direct Method

Lyapunov stability theory is a powerful tool to investigate stability of an equilibrium point of a differential equation without solving it. The Lyapunov method, sometimes referred to as direct, provides us with qualitative results to the stability problems, which may be used in designing stabilizing controllers for nonlinear dynamical systems. Our goal in this section is to present the necessary portion of the Lyapunov stability theory to enable a stability analysis of controllers that we are going to design in the next chapters. This material is actually standard in textbooks on nonlinear systems. More details can be found in Isidori (1989), Khalil (2001), LaSalle and Liefschetz (1961), Lewis et al. (2004), Sontag (1990), Tsinias (1989, 1991), and Vidyasagar (1992).

The Lyapunov method of investigating stability concerns the behavior of unforced nonlinear systems described by the differential equations of the form of Equation (3.1):

$$\dot{q}(t) = f(q(t), t). \qquad t > 0, \quad q(t) \in R^n \tag{3.6}$$

Without loss of generality, we may assume that the equilibrium point of (3.6) is the origin. Also, we will assume that Equation (3.6) has a unique solution. It corresponds to $q(t_0) = q_0$, and it is $q(q_0, t_0, t)$. We refer to it as $q(t)$.

Before we introduce Lyapunov theorems, let us recall definitions of certain classes of functions and their properties.

Definition 3.1: A continuous function $g : R \to R$ belongs to class K if

1. $g(0) = 0$,
2. $g(q) > 0$ for all $q > 0$,
3. g is nondecreasing, that is, $g(q_1) \geq g(q_2)$ for $q_1 > q_2$.

Definition 3.2: A continuous function $V : R^+ \times R^n \to R$, where $R^+ = [0, \infty)$ is

1. Locally positive definite, if there exist a class K function $g(\cdot)$ and a neighborhood N of the origin of R^n such that $V(t, q) \geq g(\|q\|)$ for all $t > 0$ and all $q \in N$.
2. Positive definite if $N = R^n$.
3. Locally negative definite if $-V$ is locally positive definite.
4. Negative definite if $-V$ is positive definite.

Example 3.3

The functions $V_1 = q_1^2 + \cos^2(q_2)$ and $V_3 = (1 - \cos(q_2)) + \frac{q_2^2}{2}$ are locally positive definite, and $V_2 = (1+t)(q_1^2 + q_2^2)$ is positive definite.

Definition 3.3: A continuous function $V : R^+ \times R^n \to R$, $R^+ = [0, \infty)$, is said to be

1. Locally decrescent if there exist a class K function $g(\cdot)$ and a neighborhood N of the origin of R^n such that $V(t,q) \leq g(\|q\|)$ for all $t > 0$ and all $q \in N$.
2. Decrescent (globally decrescent) if $N = R^n$.

Definition 3.4: For a system of differential equations (3.6), a function $V : R^+ \times R^n \to R$ may be selected and the derivative of V along (3.6) is defined as a function $\dot{V} : R^+ \times R^n \to R$, given by

$$\dot{V}(t,q) \equiv \frac{dV(t,q)}{dt} = \frac{\partial V(t,q)}{\partial t} + \frac{\partial V(t,q)}{\partial q} f(t,q).$$

This function is often referred to as a Lyapunov function.

Example 3.4

Consider a pendulum that may rotate in the (x,y) plane about a fixed point O. Its dynamics is governed by

$$ml^2 \ddot{\varphi} + b \dot{\varphi} + mgl \sin \varphi = 0, \tag{3.7}$$

where φ is the angle measured from the downward vertical position, m is the pendulum mass, l is its length, b is a friction coefficient for friction at the joint at O. The pendulum dynamics can be transformed to the state-space form (3.6) as follows. Let us introduce a two-dimensional vector $q = (q_1, q_2) = (\varphi, \dot{\varphi})$. Then,

$$\dot{q}_1 = q_2,$$

$$\dot{q}_2 = -\frac{b}{ml^2} q_2 - \frac{g}{l} \sin q_1. \tag{3.8}$$

Select a function V as

$$V = mgl(1 - \cos \varphi) + \frac{ml^2}{2} \dot{\varphi}^2.$$

Equilibrium points for the pendulum are those for which $q_2 = 0$ and $\sin q_1 = 0$, that is, $(0,0)$ and $(\pi,0)$. The function V is locally positive definite. Its derivative along the pendulum dynamics is $\dot{V} = mgl \dot{\varphi} \sin \varphi + ml^2 \dot{\varphi} \ddot{\varphi} = -b \dot{\varphi}^2 \leq 0$.

Theorem 3.1: Stability in the sense of Lyapunov

The nonlinear system $\dot{q}(t) = f(q(t),t)$, $q \in R^n$, $t \geq 0$, $q(0) = q_0$ with its equilibrium point at the origin, i.e., $f(0,t) = 0$, is given, and N is a neighborhood of the origin of size ε, that is, $N = \{q; \|q\| \leq \varepsilon\}$. The following hold:

1. The origin is stable if for $q \in N$ there exists a scalar function $V(t, q)$ with continuous first order partial derivatives such that
   a. $V(t, q)$ is positive definite
   b. $\dot{V}(t, q)$ is negative semi-definite
2. The origin is uniformly stable, if conditions (1a) and (1b) are satisfied and $V(t, q)$ is decrescent for $q \in N$.
3. The origin is asymptotically stable if $V(t, q)$ satisfies (1a) and $\dot{V}(t, q)$ is negative definite for $q \in N$.
4. The origin is globally asymptotically stable if $V(t, q)$ satisfies (1a) and $\dot{V}(t, q)$ is negative definite for all $q \in R^n$.
5. The origin is uniformly asymptotically stable if $V(t, q)$ satisfies (1a), $V(t, q)$ is decrescent, and $\dot{V}(t, q)$ is negative definite for $q \in N$.
6. The origin is globally uniformly asymptotically stable if $N = R^n$, $V(t, q)$ satisfies (1a), $V(t, q)$ is decrescent, $\dot{V}(t, q)$ is negative definite, and $V(t, q)$ is radially unbounded, i.e., it goes to infinity uniformly in time as $\|q\| \to \infty$.
7. The origin is exponentially stable if there exist positive constants α, β, γ such that $\alpha \|q\|^2 \leq V(t, q) \leq \beta \|q\|^2$ and $\dot{V}(t, q) \leq -\gamma \|q\|^2$ $\forall q \in N$.
8. The origin is globally exponentially stable if it is exponentially stable for all $q \in R^n$.

Note that Theorem 3.1 provides sufficient conditions for the stability of the origin. If we cannot provide any Lyapunov function for a system, it does not mean that it is unstable.

Example 3.5

For the pendulum considered in Example 3.4, it can be concluded that if $\dot{V} = -b \cdot^2 \leq 0$, the origin is stable, actually uniformly stable. It may be justified physically, i.e., the presence of friction in the system is responsible for the energy absorption. The function $\dot{V}(t, q)$ indicates the energy loss in the system. $V(t, q)$ has been selected as the sum of the potential and kinetic energies of the pendulum. Quite often, a Lyapunov function is selected to be the energy or part of the system energy.

3.1.3 Special Formulations of Stability Theorems

Lyapunov's direct method theorems are very useful in designing tracking or stabilizing controllers. By choosing a control signal to make $\dot{V}(q, t)$ negative definite, the stability of the closed-loop system is guaranteed. However, it is not always easy due to the fact that $\dot{V}(q, t)$ may be shown negative but not necessarily negative definite. The other problem, even more significant than the latter one, is how to find a Lyapunov function candidate because Lyapunov theorems do not provide the receipts.

Consider then some cases of the system (3.6) for which the Lyapunov theory simplifies. Consider first an autonomous system, i.e., the right-hand side of Equation (3.6) does not depend upon time explicitly. For such a system, a Lyapunov function does not depend upon time, and the following may be formulated.

Definition 3.5: A set D is said to be an invariant set of a dynamical system if every trajectory that starts in D remains in D.

Lemma 3.1

A time-invariant continuous function $V(q)$ is positive definite if $V(0) = 0$ and $V(q) > 0$ for $q \neq 0$. $V(q)$ is locally positive definite if the above holds in some neighborhood of the origin.

With the assumption about the independence of time, the Lyapunov results hold except that there is no distinction between the stability and the uniform stability results.

In control problems, asymptotic stability of the equilibrium is of interest. It may be verified with the LaSalle theorem.

Theorem 3.2 (LaSalle theorem)

For an autonomous nonlinear system $\dot{q}(t) = f(q(t))$, $q(0) = q_0$, with the origin being an equilibrium point, select a Lyapunov function $V(q)$ such that for $q \in N \quad R^n$, $V(q) > 0$ and $\dot{V}(q) \leq 0$. Then, the origin is asymptotically stable if and only if $\dot{V}(q) = 0$ only at $q = 0$.

The origin is globally asymptotically stable if $N = R^n$ and $V(q)$ is radially unbounded.

Generally, the LaSalle theorem can be used to prove the stability of sets rather than equilibrium points. The basic idea is that because $V(q)$ is lower bounded, i.e., $V(q) > V_l$, then the derivative $\dot{V}(q, \dot{q})$ has to gradually vanish, and the trajectory is confined to the set where $\dot{V}(q, \dot{q}) = 0$.

There are some special cases when a Lyapunov function can be easily obtained. It works for systems that may be treated by the Krasovskii theorem.

Theorem 3.3 (Krasovskii theorem)

For an autonomous nonlinear system $\dot{q}(t) = f(q(t))$, $q(0) = q_0$, with the equilibrium point at the origin, let $A(q) = \partial f / \partial q$. Then, a sufficient condition for the origin to be asymptotically stable is that there exist two symmetric positive definite matrices, P and Q, such that for all $q \neq 0$, the matrix $F(q) = A(q)^T P + PA(q) + Q \geq 0$ in some ball B about the origin.

A Lyapunov function candidate for a system is then $V(q) = f(q)^T Pf(q)$. If $B = R^n$ and if $V(q)$ is radially unbounded, i.e., $V(q) \to \infty$ for $\|q\| \to \infty$, then the system is globally asymptotically stable.

There is one problem with the application of the Krasovskii theorem; it may be difficult to verify whether $F(q)$ is negative semi-definite.

In many applications, trajectories are time varying, and open-loop systems are not autonomous. More advanced stability results are needed to prove global asymptotic stability. The extension of the LaSalle result on nonautonomous systems is governed by the LaSalle–Yoshizawa theorem.

Definition 3.6: A continuous class K function g is said to be class K_∞ function if $g(q) \to \infty$ for $q \to \infty$.

Theorem 3.4 (LaSalle–Yoshizawa theorem)

For a nonautonomous nonlinear system $\dot{q}(t) = f(q(t), t)$ with the origin being an equilibrium point, select a C^1 class function $V : R^n \times R \to R$ such that for all $t > 0$, $q(t) \in R^n$

$$g_l(\|q\|) \leq V(q, t) \leq g_u(\|q\|),$$

$$\dot{V}(q, t) = \frac{\partial V}{\partial t} + \frac{\partial V}{\partial q} f(q, t) \leq -W(q) \leq 0,$$

where g_l and g_u are class K_∞ functions, and $W(q)$ is a continuous function. Then, all trajectories $q(t)$ of a system $\dot{q}(t) = f(q(t), t)$ are globally bounded and satisfy $\lim_{t\to\infty} W(q(t)) = 0$. Additionally, if $W(q)$ is positive definite, then $q = 0$ is a globally uniformly asymptotically stable equilibrium point.

Another lemma that facilitates investigating stability of nonautonomous systems is the Barbalat lemma that leads to results similar to those of LaSalle's theorem.

Lemma 3.2 (Barbalat lemma)

Let $f(t)$ be a differentiable function of t. The two equivalent formulations hold:

1. If $\dot{f}(t) = df/dt$ is uniformly continuous and $\lim_{t\to\infty} f(t) = k < \infty$, then $\lim_{t\to\infty} \dot{f}(t) = 0$.
2. If $f(t) \geq 0$, $\dot{f}(t) \leq 0$ and $\ddot{f}(t)$ is bounded, then $\lim_{t\to\infty} \dot{f}(t) = 0$.

In order to use Barbalat's lemma, one has to assert uniform continuity of a function. It is hard to do directly from the definition. It can be easier accomplished by examining the function's derivative. A simple sufficient condition for a differentiable function to be uniformly continuous is that its derivative be bounded.

We may apply Barbalat's lemma to examine stability of dynamical systems. To this end, we use the lemma that looks like an invariant set theorem in Lyapunov analysis for autonomous systems.

Lemma 3.3 (Lyapunov-like lemma)

If a scalar function $V(q,t)$ satisfies the following conditions:

- $V(q,t)$ is lower bounded,
- $\dot{V}(q,t)$ is negative semi-definite,
- $\dot{V}(q,t)$ is uniformly continuous in time,

then $\dot{V}(q,t)$ tends to 0 as t tends to ∞.

An analysis based on Barbalat's lemma is sometimes referred to as a Lyapunov-like analysis. It differs from the Lyapunov analysis in two properties. The one is that the function V can be a lower-bounded function of q and t instead of a positive definite function. The second is that the derivative of V must be shown to be uniformly continuous in addition to being negative or zero. This can be shown by proving that $\ddot{V}$ is bounded. Using either way for stability analysis, the same difficulty arises at the very beginning—this is the selection of the scalar function V.

To get a broader view of the use of the Lyapunov analysis, consider an example that demonstrates how to prove stability of closed-loop dynamics.

Example 3.6

The control law known as the repetitive control law (RCL), originally designed for trajectory tracking for holonomic systems, was modified for model-based tracking for nonholonomic systems in Jarzębowska (2007). We discuss its application in Chapter 7.

The system control dynamics in the reduced state form is given as

$$\bar{M}_c(q)\ddot{q}_1 + \bar{C}_c(q,\dot{q}_1)\dot{q}_1 + \bar{D}_c(q) = \tau_p,$$

$$\dot{q} = G(q)\dot{q}_1,$$

$$(3.9)$$

where $\bar{M}_c(q) = M(q)G(q)$, $\bar{C}_c(q,\dot{q}_1) = M(q)\dot{G}(q) + C(q,\dot{q})G(q)$, $\bar{D}_c(q) = D(q)$.

The modified RCL has the form

$$\tau_p = \hat{u}_p(t) + k_v r_1 + k_s e_1 + k_a \|e_1\|^2 r_1,$$

$$(3.10)$$

where $\hat{u}_p(t)$ is a n-dimensional vector of a learning term that compensates for $u_p(t)$, and it is updated by the learning update rule

$$\hat{u}_p(t) = \hat{u}_p(t - T) + k_L r_1,$$

$$(3.11)$$

with k_L being a positive scalar control gain. Other quantities in (3.10) are as follows: r_1 is a filtered tracking error defined as $r_1 = e_1 + \dot{e}_1$ with e_1 being a tracking error $e_1 = q_{1p} - q_1$, q_{1p} are desired values of states, and k_v, k_s, k_a are scalar, constant, control gains. The learning update rule (3.11) can be written in terms of the learning error that is defined as

$$\tilde{u}_p(t) = u_p(t) - \hat{u}_p(t). \qquad (3.12)$$

Relation (3.11) can be presented in the form

$$u_p(t) - \hat{u}_p(t) = u_p(t) - \hat{u}_p(t - T) - k_L r_1,$$

and utilizing the assumption about the repeatability of motion, the learning error update rule is

$$\tilde{u}_p(t) = u_p(t - T) - \hat{u}_p(t - T) - k_L r_1 = \tilde{u}_p(t - T) - k_L r_1. \qquad (3.13)$$

To verify the type of stability for the closed-loop dynamics (3.9) with the controller (3.10), we formulate the following theorem.

Theorem: The control dynamics (3.9) with the RCL (3.10) is the closed-loop dynamics that provides asymptotic stability of the position and velocity tracking errors in programmed, i.e., desired motion tracking.

Proof: Based on properties of the inertia matrix in Equation (3.9), we conclude that it is lower bounded. We assume that q_p are periodic trajectories, and $\dot{q}_p$ and $\ddot{q}_p$ are bounded. The second of equations in (3.9) is the constraint equation, and it is enough to control q_1 to have q_2 satisfied by the constraints. Let us write the first equation in (3.9) in terms of r_1. It yields

$$\bar{M}_c(q)\dot{r}_1 = -\bar{C}_c(q, \dot{q}_1)r_1 + u_a(t) - \tau_p, \qquad (3.14)$$

where $u_a(t)$ is a $(n\text{-}k)$-dimensional vector referred to as the actual system dynamics given by

$$u_a(t) = \bar{M}_c(q)(\ddot{q}_{1p} + \dot{e}_1) + \bar{C}_c(q, \dot{q}_1)(\dot{q}_{1p} + e_1) + \bar{D}_c(q) + z(\dot{q}_1). \qquad (3.15)$$

The error dynamics (3.14) can be transformed to the form

$$\bar{M}_c(q)\dot{r}_1 = -\bar{C}_c(q, \dot{q}_1)r_1 + u_p(t) + \tilde{U} - \tau_p, \qquad (3.16)$$

where $\tilde{U} = u_a(t) - u_p(t)$ is the difference between the actual system dynamics and its repeatable programmed dynamics. $\tilde{U}$ can be quantified as follows:

$$\|\tilde{U}\| \leq a_1 \|e_1\| + a_2 \|e_1\|^2 + a_3 \|r_1\| + a_4 \|r_1\|\|e_1\|, \qquad (3.17)$$

where a_1, a_2, a_3, and a_4 are positive constants that depend on a predefined motion that is to be controlled and properties of a system. Substituting (3.10) into (3.16), we obtain

$$\bar{M}_c(q)\dot{r}_1 = -\bar{C}_c(q,\dot{q}_1)r_1 - k_v r_1 - k_s e_1 - k_a \|e_1\|^2 \, r_1 + \tilde{u}_p(t) + \tilde{U}. \qquad (3.18)$$

We assume that the learning term $\hat{u}_p(t)$ is prevented from growing. To analyze the stability of the error dynamics (3.18), the Lyapunov function candidate is selected to be

$$V = \frac{1}{2}r_1^T \bar{M}_c(q)r_1 + \frac{1}{2}k_s e_1^T e_1 + \frac{1}{2k_L}\int_{t-T}^{t} \tilde{u}_p^T(\eta)\tilde{u}_p(\eta)\,d\eta. \qquad (3.19)$$

The Lyapunov function time derivative is equal to

$$\dot{V} = \frac{1}{2}r_1^T \dot{\bar{M}}_c(q)r_1 + r_1^T \bar{M}_c(q)\dot{r}_1 + k_s e_1^T \dot{e}_1 + \frac{1}{2k_L}\left[\tilde{u}_p^T(t)\tilde{u}_p(t) - \tilde{u}_p^T(t-T)\tilde{u}_p(t-T)\right]. \quad (3.20)$$

Substituting (3.18) into (3.20), we obtain

$$\begin{aligned} \dot{V} &= k_s e_1^T \dot{e}_1 - k_v r_1^T r_1 - k_s r_1^T e_1 - k_a \|e_1\|^2 \, r_1^T r_1 + r_1^T \tilde{U} \\ &\quad + \frac{1}{2}r_1^T[\dot{\bar{M}}_c(q) - 2\bar{C}_c(q,\dot{q}_1)]\,r_1 + r_1^T \tilde{u}_p(t) \\ &\quad + \frac{1}{2k_L}\left[\tilde{u}_p^T(t)\tilde{u}_p(t) - \tilde{u}_p^T(t-T)\tilde{u}_p(t-T)\right]. \end{aligned} \qquad (3.21)$$

By a direct inspection of Equation (3.21), we can see that utilizing the skew-symmetry property and the update rule (3.13), the second and third line terms reduce to $-\frac{1}{2}k_L r_1^T r_1$. Hence, Equation (3.21) reduces to

$$\dot{V} = -k_s e_1^T e_1 - \left(k_v + \frac{1}{2}k_L\right)r_1^T r_1 - k_a \|e_1\|^2 \, r_1^T r_1 + r_1^T \tilde{U}. \qquad (3.22)$$

Now, we can place an upper bound on $\dot{V}$ as

$$\dot{V} \leq -k_s \|e_1\|^2 - \left(k_v + \frac{1}{2}k_L\right)\|r_1\|^2 - k_a \|e_1\|^2\|r_1\|^2 + \|r_1\|\|\tilde{U}\|. \qquad (3.23)$$

The signal $\tilde{U}$ is estimated by relation (3.17), so $\dot{V}$ can be upper bounded as

$$\dot{V} \leq -k_s \|e_1\|^2 - \left(k_v + \frac{1}{2}k_L\right)\|r_1\|^2 - k_a \|e_1\|^2\|r_1\|^2 + \|r_1\|\left(a_1 \|e_1\| + a_2 \|e_1\|^2 + a_3 \|r_1\| + a_4 \|e_1\|\|r_1\|\right). \qquad (3.24)$$

Rearrangement of terms in (3.24) yields

$$\dot{V} \leq -\|e_1\|^2 \left(k_s - \frac{1}{4}a_2\right) - \|e_1\|^2\|r_1\|^2 (k_a - a_4 - a_2) - \|r_1\|^2 \left(k_v + \frac{1}{2}k_L - a_3 - \frac{1}{4}a_4\right) + a_1\|r_1\|\|e_1\|$$

$$- a_2\|e_1\|^2 \left(\frac{1}{2} - \|r_1\|\right)^2 - a_4\|r_1\|^2 \left(\frac{1}{2} - \|e_1\|\right)^2. \tag{3.25}$$

If the controller gain k_a is selected to be $k_a > a_2 + a_4$, then the second and the last two terms in relation (3.25) are all negative. The derivative $\dot{V}$ can get a new upper bound

$$\dot{V} \leq -\|e_1\|^2 \left(k_s - \frac{1}{4}a_2\right) - \|r_1\|^2 \left(k_v + \frac{1}{2}k_L - a_3 - \frac{1}{4}a_4\right) + a_1\|r_1\|\|e_1\|. \tag{3.26}$$

Introducing a new variable $z = (\|e_1\|, \|r_1\|)^T$ and a matrix N,

$$N = \begin{bmatrix} k_s - \dfrac{1}{4}a_2 & -\dfrac{1}{2}a_1 \\ -\dfrac{1}{2}a_1 & k_v + \dfrac{1}{2}k_L - a_3 - \dfrac{1}{4}a_4 \end{bmatrix} \tag{3.27}$$

relation (3.26) takes the form

$$\dot{V} \leq -z^T N z. \tag{3.28}$$

With regard to (3.28), we need N to be positive definite. This condition can be met if the gains k_v, k_s, and k_L are selected as

$$k_s > \frac{1}{2}a_1 + \frac{1}{4}a_2, \quad k_v + \frac{1}{2}k_L > \frac{1}{2}a_1 + a_3 + \frac{1}{4}a_4. \tag{3.29}$$

Conditions (3.28) and (3.29) preserve that $\dot{V}$ is negative semi-definite. Because $\dot{V}$ is negative semi-definite, we can state that V is upper bounded. If V is upper bounded, we state, in turn, that $e_1, \dot{e}_1, r_1, \tilde{u}(t)$ are bounded. Because they are bounded, we can use Equation (3.18) to show that $\dot{r}_1, \ddot{q}_1$ and, hence, $\ddot{V}$ are bounded. Also, it follows that $q_2, \dot{q}_2, \ddot{q}_2$ are bounded. Because $\bar{M}_c(q)$ is lower bounded, we can state that V is lower bounded. Concluding, if V is lower bounded, $\dot{V}$ is negative semi-definite, and $\ddot{V}$ is bounded. Based on Barbalat's lemma, we obtain that

$$\lim_{t \to \infty} \dot{V} = 0. \tag{3.30}$$

From the argument above and from relation (3.28), we have that

$$\lim_{t\to\infty}(e_1,r_1)^T = 0. \tag{3.31}$$

Then, we can conclude that the position tracking error e_1 is asymptotically stable so is different meaning than $e_1 \to e_2$. Because r_1 is defined to be a stable first order differential equation in e_1, we can write that

$$\lim_{t\to\infty}\dot{e}_1 = 0. \tag{3.32}$$

We conclude that the velocity tracking errors $\dot{e}_1$ and $\dot{e}_2$ are asymptotically stable.

q.e.d.

In the proof, we assumed that the learning term $\hat{u}_p(t)$ might be "artificially" kept from growing. This let us prove that the velocity tracking error $\dot{e}$ is asymptotically stable. If to relax this assumption, all we can say about the velocity tracking error is that it is bounded.

We are interested in the stability of control systems, and in this context, a relation between input and output signals should be examined—specifically, whether a bounded input results in a bounded output of the system. The Lyapunov theory does not answer this question.

Example 3.7

Consider a control system $\dot{y}(t) + \frac{y(t)}{t} = u(t)$. It is asymptotically stable with an equilibrium point $y = 0$. A unit step input, starting at $t = 0$, may be applied to the system and it is bounded. However, it results in an output $y(t) = t/2$, which is unbounded when t increases.

Consider a control system

$$\dot{q}(t) = f(q(t), u(t), t) \quad \text{with} \quad q(t_0) = q_0, \tag{3.33}$$

$$z(t) = g(q(t), u(t), t),$$

where $z(t)$ is the system output.

Definition 3.7: The dynamical system (3.33) is bounded-input–bounded-output (BIBO) stable if for any $\|u(t)\| \le M < \infty$ there exists finite $\varepsilon > 0$ and b such that $\|z(t)\| \le \varepsilon M + b$.

BIBO stability implicates the uniform boundedness of all equilibrium states.

A local version of BIBO stability, with small inputs, is a total stability concept. The total stability theory addresses stability of a system that undergoes disturbances $d(t)$. The $d(t)$ term may represent unmodeled friction, parameter variations, disturbances from motors, or an external environment. In the total stability theory, a system dynamics is presented as

$$\dot{q} = f(q, t) + d(q, t), \tag{3.34}$$

where $d(q, t)$ is a perturbation term. The unperturbed system is

$$\dot{q} = f(q, t). \tag{3.34a}$$

The idea of total stability relies upon the ability of a system to withstand small persistent disturbances. It is defined as follows.

Definition 3.8: The equilibrium point $q = 0$ for the unperturbed system of Equation (3.34a) is said to be totally stable if for every $\varepsilon \geq 0$, two numbers γ_1 and γ_1 exist such that $\|q(t_0)\| < \gamma_1$ and $\|d(q, t)\| < \gamma_2$ imply that every solution $q(t)$ of the perturbed system (3.34) satisfies the condition $\|q(t)\| < \varepsilon$.

Definition 3.8 states that the equilibrium $q = 0$ is totally stable if the state of the perturbed system can be kept arbitrarily close to zero by restricting the initial state and the perturbation to be sufficiently small. To assert the total stability of a nonlinear system, the following theorem applies.

Theorem 3.5

If the equilibrium point of Equation (3.34a) is uniformly asymptotically stable, then it is totally stable.

This theorem states that uniformly asymptotically stable systems can withstand small disturbances. Uniform asymptotic stability can be verified by the Lyapunov theorem 3.1 so that total stability of a system may be established by the same theorem.

Also, based on the Lyapunov theory, asymptotic stability is not sufficient to guarantee total stability of a nonlinear system. Exponentially stable systems are always totally stable because exponential stability implies uniform asymptotic stability.

If the range of disturbances that act upon a system is extended, the global uniform stability cannot guarantee the boundedness of the state in the presence of large bounded disturbances (see, e.g., Slotine and Li 1996).

For systems with disturbances, the local notion of total stability was replaced with a more global concept of input-to-state stability (ISS) (for details, see, e.g., Sontag 2000; Tanner 2003).

3.2 Classification of Control Problems

An approach to control problems for nonlinear systems is strongly related to a system in question, i.e., whether it is holonomic or nonholonomic. As stressed in Chapter 1, the two classes of nonlinear control systems vary, and different types of control methods and controllers have to be pursued for them. For holonomic systems, control algorithms have to be designed at the dynamic level for both fully actuated and underactuated systems. The latter are usually included in nonholonomic systems, and we follow this classification. For nonholonomic systems, control is possible at the kinematic or dynamic level.

From the control perspective, the difference between the two classes of nonlinear control systems can be summarized as follows. Nonholonomic control systems, specifically systems with material constraints, are usually completely controllable in their configuration space, but they cannot be stabilized to a desired configuration by using smooth state feedback control (Brockett 1983). Holonomic control systems can be stabilized by using smooth state feedback control (Byrnes and Isidori 1989; Isidori 1989).

Another aspect of control design for nonlinear systems is that the relative difficulty depends not only on the nonholonomic nature of a system but also on a control objective. For some of them, classical nonlinear control approaches like feedback linearization and dynamic inversion are effective. The control objective of this kind is stabilization to a manifold that contains the equilibria manifold (Bloch and McClamroch 1989; Bloch 2003), stabilization to certain trajectories (Walsh et al. 1994), dynamic path following (Sarkar et al. 1993), or output tracking (van Nieuwstadt 1997).

However, there are control objectives that cannot be solved using the standard nonlinear control methods. They are motion planning or stabilization to an equilibrium state. Much research on motion planning and trajectory generation for driftless nonholonomic control systems resulted in numerous interesting solutions to the steering problem (e.g., Bloch 2003; Kolmanovsky and McClamroch 1995; Murray et al. 1994). These are beyond the scope of this book.

In this section, we review control problems formulated for nonlinear systems from the point of view of the nonholonomicity property.

3.2.1 Stabilization

A stabilization problem consists of designing a control law, which guarantees that equilibrium of a closed-loop system is asymptotically stable or at least locally asymptotically stable.

Generally, stabilization for a holonomic nonlinear system can be formulated as follows (Khalil 2001; Slotine and Li 1996).

Definition 3.9 (Asymptotic Stability): Given a nonlinear system whose dynamics is described by Equation (3.2), find a control law u such that, starting from anywhere in a region in Ω, the state $q \to 0$ as $t \to \infty$.

If the control law u depends upon measurement signals directly, it is referred to as a static control law. If it depends upon derivatives of the states, it is a dynamic control law.

When a control objective is to stabilize the state to some nonzero set-point q_d, we can transform the problem into a zero-point stabilization by taking $q - q_d$ as a new state.

Example 3.8

Consider the pendulum dynamics, similar to that in Example 3.4, but take the inverted pendulum and without the friction term. Its dynamics can be written as

$$J\ddot{\varphi} - mgl\sin\varphi = \tau, \tag{3.35}$$

where φ is the angle measured from the upright vertical position, J and m are inertia properties, and τ is a stabilizing torque. The stabilizing control goal is to bring the pendulum to the vertical upright position. The stabilizing torque τ can be selected as

$$\tau = -k_d\dot{\varphi} - k_p\varphi - mgl\sin\varphi, \tag{3.36}$$

where k_d and k_p are positive constants. The stabilizing torque τ is a proportional derivative (PD) plus a feed-forward part for gravity compensation. Its application leads to a globally stable closed-loop dynamics:

$$J\ddot{\varphi} + k_d\dot{\varphi} + k_p\varphi = 0. \tag{3.37}$$

Another stabilizing torque τ that leads to a stable closed-loop dynamics may be of the form

$$\tau = -k_d\dot{\varphi} - 2mgl\sin\varphi, \tag{3.38}$$

and the closed-loop dynamics yields

$$J\ddot{\varphi} + k_d\dot{\varphi} + mgl\sin\varphi = 0. \tag{3.39}$$

Many stabilization problems for nonlinear systems are not easy to solve. If we take the pendulum from Example 3.8, put it on a cart, let the cart move, and make its velocity a control input that is to stabilize the pendulum in its upright vertical position, the problem becomes a stabilization of an under-actuated system. There are two degrees of freedom of the pendulum–cart system and one control input. A different control law has to be designed for such a system.

A detailed discussion of local stabilization results using the Lyapunov direct methods and the center manifold theory is presented in Nijmeijer and van der Shaft (1990). It is conducted for affine autonomous nonlinear holonomic systems whose control state-space representation is

$$\dot{q} = f(q) + \sum_{i=1}^{m} g_i(q)u_i, \quad f(q_0) = 0. \tag{3.40}$$

where f is a drift vector field, and g_i, $i = 1,\ldots,m$, are control vector fields, both smooth on M.

For a nonholonomic nonlinear system, a control problem looks different. Although controllability implies that any configuration can theoretically be reached in finite time from any other configuration by applying adequate control inputs, no smooth control law depending solely on configuration error variables, and thus independent of initial conditions, is able to ensure asymptotic convergence of the nonholonomic system to a predefined configuration.

This may be formalized together with the problem of selection of a control objective. Let us make a connection between the control model (Equation 3.40) for systems with material and non-material nonholonomic constraints, and control objectives stated for them. We assume for Equation (3.40) that the constraint set Σ contains an open neighborhood of the origin in R^m.

If we select a control objective, which is stabilization (local asymptotic stabilization) by a continuous static state feedback strategy, the following has to be satisfied.

Definition 3.10: The system of Equation (3.40) is said to be locally asymptotically stabilizable (LAS) if there exists a feedback $u(x)$ defined in a neighborhood of 0 such that $0 \in M$ is an asymptotically stable equilibrium of the closed-loop system.

A feedback controller $u(x)$ is said to be a static state feedback when it is a continuous map $u : M \to U : x \to u(x)$, $u(0) = 0$, such that the closed-loop system (3.40) has a unique solution $x(t, x(0))$, $t \geq 0$, for sufficiently small initial state $x(0)$.

The asymptotic stabilizability of the target equilibrium holds only if the dimension of the equilibria set including the target is equal to the number of control inputs. This result is equivalent to Brockett's necessary condition for feedback stabilization (Brockett 1983). Based on Brockett's condition, control models of nonholonomic systems are not asymptotically stabilizable, even locally. However, we can still formulate control objectives for some control problems that make them LAS; we demonstrate this in Chapter 4.

The property above motivated us to look for alternative feedback control laws, e.g., discontinuous (Astolfi 1996; Bloch et al. 1992). Another solution is a feedback control law that depends on the time variable. Time-varying feedback stabilization techniques are quite established now, e.g., see Samson 1990, where such a technique is used to stabilize a cart equipped with two independently actuated rear wheels and a self-aligning front wheel. The method used there to derive a set of globally stabilizing time-varying smooth feedback control laws is based on finding an adequate time-varying Lyapunov-like function, and the convergence proof is based upon the LaSalle theorem. In Samson (1993), a set of smooth time-varying feedbacks for a class of nonlinear systems encompassing simple car models as well as sufficient conditions for asymptotic convergence of the state vector to zero are derived. It is shown in

Coron and Pomet (1992) that it is possible to stabilize a nonholonomic system using smooth time-varying state feedback. Samson and Pomet present constructive approaches to stabilization (Pomet 1992; Samson 1995; Samson and Ait-Abderrahim 1991a), and other results on stabilization are in (Kokotovic and Sussmann 1989; M'Closkey and Murray 1994; Morin et al. 1998; Morin and Samson 2003). In Astolfi (1996); Canudas de Wit and Sordalen (1992); Kolmanovsky et al. (1994), among many others, discontinuous feedback laws for specific systems are proposed. In Bloch and Crough (1995), Bloch and McClamroch (1989, 1990) and Bloch et al. (1992, 1994), problems related to stabilization to a manifold instead of to a point are studied. Finally, hybrid stabilization methods are discussed in Kolmanovsky and McClamroch (1995) and references therein. For a detailed survey of stabilization methods for nonholonomic systems, see Bloch (2003) and Kolmanovsky and McClamroch (1995).

Another possible approach to the control of nonholonomic systems is the study of controllability along a reference trajectory. If we are provided a desired state trajectory, we can construct a controller that stabilizes the system to this trajectory. Methods in this spirit are presented in Walsh et al. (1994) and references therein.

Time-varying feedback controllers have good properties of being smooth and globally stabilizing. However, they produce slow convergence of system states and cannot be exponentially stabilizing. Exponential convergence is desirable in practice, and results in obtaining faster convergence by using nonsmooth feedback are reported (M'Closkey and Murray 1997; Sordalen and Egeland 1995) as are results by using hybrid feedback (Hespanha and Morse 1999). Works on exponentially stabilizing time-varying controllers for nonholonomic systems are also reported (Laiou and Astolfi 1999; Tian and Li 1999). An interesting result is presented in Morin and Samson (2003), where the authors develop a practical stabilization with exponential convergence for driftless nonholonomic systems. In Tian and Li (2002), exponential stabilization is achieved for a dynamic control model of a nonholonomic system by smooth time-varying control. This dynamic control model has two inputs and can represent a large class of nonholonomic systems, such as systems in chained or power forms, or some underactuated systems. The results reported above are obtained for nonholonomic systems transformable to chained or power forms, or to their extensions. Problems of stabilization and motion planning for systems with first order nonholonomic constraints that cannot be transformed to chained or power forms can be solved by other approaches, for example, those that use the notion of Liouvillian systems (Chelouah 1997; Sira-Ramirez et al. 2000).

3.2.2 Trajectory and Motion Tracking

Stabilization is difficult and challenging, but tracking is practically important. Motion tracking and trajectory tracking as a peculiar case are the main focus of this book, so we review tracking problems from the perspective of control research. Tracking for fully actuated systems is usually specified by a control objective to track a trajectory with the exponential convergence rate in order to guarantee performance and robustness (Yazdapanah and Khosrowshahi 2003). The tracking problem for systems like manipulators and robots is addressed in the literature (Bayard and Wen 1988; Slotine and Li 1989; Whitcomb et al. 1991), where asymptotic, exponential, and adaptive tracking are achieved via nonlinear analysis. These results are now standard in text-books on control (Isidori 1989; Nijmeijer and van der Schaft 1990) and robotics (Murray et al. 1994; Slotine and Li 1996). Similar techniques are applied to the attitude control problem for satellites (Wen and Kreutz-Delgado 1991), and to the attitude and position control of underwater vehicles (Fossen 1994).

Trajectory tracking for nonholonomic systems is achieved based on two kinds of models. One of them considers velocities of a system as control inputs and uses a system kinematic model. It addresses tracking at a kinematic con-trol level and ignores the system dynamics (see, e.g., Bloch 2003; Bushnell et al. 1995; Canudas-de-Wit and Sordalen 1992; De Luca et al. 2002; Oriolo et al. 2002; Samson 1995; Walsh et al. 1994). These models assume perfect velocity track-ing to generate actual robot control inputs. In practice, it is not easy to realize perfect velocity tracking. The second research stream uses the system dynam-ics, where control forces and torques as well as velocities can be control inputs (Isidori 1989; Kwatny and Blankenship 2000). For a nonholonomic system, a dynamic control model is usually integrated with a kinematic control model (Chen et al. 1996; Cheng and Tsai 2003; M'Closkey and Murray 1997; Yang and Kim 1999; Yun and Sarkar 1998). A control system developed in such a way has two-level architecture. The lower control level operates within a kine-matic model of a nonholonomic system to stabilize its motion to a desired tra-jectory. The upper control level uses a dynamic model and stabilizes feedback obtained on the lower control level. This allows steering commands, which are velocities at the kinematic level, to be converted to forces and torques. The control system developed within this two-level architecture assumes a per-fect knowledge of the system dynamics in most cases. Some works propose two-level control architecture in the case of model uncertainties (Chang and Chen 1996; Fukao et al. 2000; Kim et al. 2000).

A controller based on a dynamic model only can also achieve trajectory tracking for nonholonomic systems. This is a model in the reduced state form (Papadopoulos and Chasparis 2002; Sarkar et al. 1994; Yun and Sarkar 1998). It is obtained by the elimination of constraint reaction forces from the constrained dynamic model. In this way, control dynamics is decoupled from the constraint forces that can be determined separately when needed.

There are systems like a snake-board or a roller-racer that cannot be con-trolled at the kinematic level. The reason is that there are fewer control inputs than degrees of freedom, and kinematic equations are insufficient for control. They require both kinematic and dynamic models for control (Jarzębowska and Lewandowski 2006; Lewis et al. 2004).

In classical nonlinear control, tracking usually means trajectory tracking, and it is defined as follows.

Definition 3.11 (Asymptotic Trajectory Tracking): Given a nonlinear system whose dynamics is described by Equation (3.33) and a desired output $z_d(t)$, find a control law u such that, starting from anywhere in a region in Ω, the tracking errors $e = z(t) - z_d(t)$ go to zero and the whole state remains bounded.

Other tracking convergence, e.g., exponential can be defined similarly. Usually, the desired output $z_d(t)$ and its derivatives up to a sufficiently high order are required to be continuous and bounded. Also, it is assumed that they are available for online control law computation. This latter assumption is satisfied because the desired trajectory is often preplanned, and its time derivatives can be easily obtained. In some control problems, this assumption is not satisfied, for instance in tracking aircraft by radar.

Example 3.9

In tracking aircraft by radar, the only available signal is an aircraft position $z_A(t)$. The so called reference model can be used to provide derivatives of signals. The control dynamics is then generated as

$$\ddot{z}_d(t) + k_1 \dot{z}_d(t) + k_2 z(t) = k_2 z_A(t), \tag{3.41}$$

where k_1 and k_2 are positive constants. The tracking problem is formulated as tracking the output of the reference model.

Tracking control for holonomic systems is based on model-based controllers as explained earlier. It is a well-established area of control (see, e.g., Lewis et al. 2004). Model-based tracking for nonholonomic systems is discussed in Section 3.7 and in Chapters 5 and 7.

The statement that "tracking problems are more difficult to solve than stabilization problems, because in tracking problems the controller should not only keep the whole state stabilized but also drive the system output toward the desired output" should, in the author's opinion, rather refer to holonomic nonlinear systems. It is rather opposite in control of nonholonomic systems, where tracking is usually considered simpler than stabilization.

In control problems for both holonomic and nonholonomic systems, stabilization and trajectory tracking are sometimes considered related. Tracking is regarded as stabilization to some predefined trajectory. Then, stabilization may be formulated for the nonautonomous dynamics, that is, the error dynamics, $\ddot{e} + f(e, \dot{e}, u, z_d, \dot{z}_d, \ddot{z}_d) = 0$ with $e = z(t) - z_d(t)$.

Both stabilization and tracking can be considered globally or locally.

3.2.3 Path Following

In the tracking problem, a system is to track a desired motion, typically a trajectory, which means that it is expected to reach a sequence of desired positions in specified time instants. It means that the reference is time parameterized. In following, typically path following, a system is required to reach and follow some path without any temporal specification. The basic property of this control task is that a controller acts on a system orientation to drive it to the specified path. The path following control problem can be formulated then as the stabilization to zero of a scalar path error function, which is a distance to the path. We may say that path following can be treated either as a subproblem of trajectory tracking or stabilization. In the first instance, the separation of the geometric and timing information along the trajectory is needed. In the second, the controlled output, which is a distance to the path, is to be made zero.

The path following problem can be formulated globally or locally. The local formulation means that a controller works properly provided that we start sufficiently close to the path. For the tracking formulation, "sufficiently close" is evaluated with respect to a current position of the reference position, often specified by a reference robot. For trajectory tracking and path following, a kinematically feasible path should be provided by a path planner. It means that for a nonholonomic vehicle, the path should comply with the nonholonomic constraints of the vehicle. Only for omnidirectional vehicles is any path feasible. More research on path following can be found in the literature (Canudas-de-Wit et al. 1993; Encarnacao and Pascoal 2000; Jiang and Nijmeijer 1999; De Luca et al. 1998; Micaelli and Samson 1993; Samson and Ait-Abderrahim 1991b; del Rio et al. 2002) and references therein. Path following strategies are discussed briefly in Chapter 6.

3.3 Control Properties of Nonlinear Systems

Many papers and monographs have been published about control of holonomic systems. This control problem may be considered a solved problem, at least theoretically (Arimoto 1990; Berghuis 1993; Krstic et al. 1995; Lewis et al. 2004; Spong and Vidyasagar 1989). In this book, the emphasis is on control of constrained systems, so the rest of this chapter favors the constrained systems.

3.3.1 Classification of Constrained Control Systems

As mentioned in Chapter 1, geometric mechanics provides powerful tools to specify many control properties of nonlinear systems. Specifically, Lie algebra concepts provide quantitative tools useful in control design. We will use some of them in this chapter, and in what follows, we recall basic concepts and definitions only. Readers interested in advanced geometric control theory may consult, for example, Bloch (2003), Isidori (1989), Kwatny and Blankenship (2000), Sontag (1990), and references therein.

Holonomic and nonholonomic system models exhibit different control properties, so it is important to distinguish the two types of systems. Suppose that a set of m constraint equations on a system may be presented as

$$A(q)\dot{q} = 0, \tag{3.42}$$

where $A(q) \in R^{m \times n}$ is a full rank matrix, and q is a n-dimensional coordinate vector. Suppose also that there are a holonomic and $(m-a)$ nonholonomic constraint equations. The number of degrees of freedom of the system is $(n-m)$ then. Let s_i, $i = 1,...,n-m$, be a set of continuously differentiable, i.e., smooth, and linearly independent vector fields in the null space of $A(q)$, $\mathcal{N}(q)$, and then

$$A(q)s_i(q) = 0. \qquad i = 1,...,n-m \tag{3.43}$$

Vectors s_i form a full rank matrix $S(q)$. The derivatives $\dot{q}$ are generalized velocities. The constrained velocities of the system are always in the null space of $A(q)$, so we can define a $(n-m)$ velocity vector $v(t) = (v_1, v_2, ..., v_{n-m})$ such that for all t the constraint equations can be written as

$$\dot{q} = S(q)v(t). \tag{3.44}$$

Equation (3.44) presents the common form of the constraint equations. Physically, $v(t)$ are control inputs, and they are selected based upon the system in question. Using notion from mechanics, $v(t)$ are selected to be independent velocities.

A distribution denoted by Δ spanned by the vector fields s_i

$$(q) = span\{s_1(q),...,s_{n-m}(q)\} \tag{3.45}$$

is of dimension $\dim (q) = rankS(q)$ and any $\dot{q}$ that satisfies Equation (3.42) belongs to Δ.

Definition 3.12: For two vector fields f and g, the Lie bracket $[f,g](q)$ is a vector field defined by

$$[f,g](q) = \frac{\partial g}{\partial q} f(q) - \frac{\partial f}{\partial q} g(q).$$

By definition, $[f,g](q) = -[g,f](q)$ and $[f,g](q) = 0$ when f and g are constant vector fields. Also, the Jacobian identity holds: $[h,[f,g]]+[f,[g,h]]+[g,[h,f]]=0$.

A popular notation for iterated Lie brackets is as follows (Nijmeijer and van der Shaft 1990):

$$ad_f^0 g(q) = g(q), \quad ad_f g(q) = [f, g](q), \quad ad_f^k g(q) = \left[f, ad_f^{k-1} g \right](q), \quad k > 1.$$

Definition 3.13: A distribution Δ is said to be involutive if it is closed under Lie bracket operation, i.e., if $g_1 \in$ and $g_2 \in$, then $[g_1, g_2] \in$.

To verify the type of the constraints, we may analyze whether a distribution is involutive or not, as in Campion et al. (1991), or compute repeated Lie brackets on the vector fields s_i of the distribution Δ. To apply the latter method, let us recall the concept of filtration and degree of nonholonomy (Murray et al. 1994).

Definition 3.14: The filtration generated by the distribution (3.45) is defined as the sequence $\{ _i \}$ with $_i = _{i-1} + [_1, _{i-1}], i \geq 2, _1 =$ and $[_1, _{i-1}] = span\{[s_j, \alpha] \mid s_j \in _1, \alpha \in _{i-1}\}. j = 1, \ldots, n - m$.

Definition 3.15: A filtration is regular in a neighborhood ε of q_0 if $\dim _i(q) = \dim _i(q_0)$ for every $q \in \varepsilon$.

For a regular filtration, if $\dim _{i+1}(q) = \dim _i(q)$, then $_i(q)$ is involutive and $_{i+j}(q) = _i(q)$ for all $j \geq 0$. Because $\dim _1(q) = n - m$, $\dim _i(q) \leq n$, so the termination condition takes place after m steps. It agrees with the number of equations of constraints.

If the filtration generated by the distribution Δ is regular, it is possible to define the degree of nonholonomy.

Definition 3.16: The degree of nonholonomy of the distribution is the smallest integer κ that verifies the condition $_{\kappa+1}(q) = _{\kappa-1}(q)$.

The verification of this condition implies that $\kappa \leq m + 1$.

Based on the definitions above, the conditions for the integrability of the constraint equations, may be restated as follows:

1. All the constraints are holonomic for $\kappa = 1$: $_\kappa(q) = n - m$.
2. All the constraints are nonholonomic for $2 \leq \kappa \leq m$, and if $_\kappa(q) = n$.
3. The constraints are partially nonholonomic for $2 \leq \kappa \leq m$, and if $(n - m) + 1 \leq _\kappa(q) \leq n$.

Example 3.10

Take the two-wheeled mobile platform presented in Example 1.1 and verify that the three constraint equations are partially nonholonomic and really have the form of Equation (1.1). Let us use the formal tools as given above, not guessing because it is not always easy to guess which constraint can be integrated. The first of equations (1.1) represents the condition that the robot body does not slip sideways, and the two others represent the same condition for the right and left wheel, respectively. The three constraint equations have the forms

$$\dot{y}_C \cos - \dot{x}_C \sin - \dot{\,}d = 0,$$
$$\dot{x}_C \cos + \dot{y}_C \sin + \dot{\,}b = r\dot{\,}_r, \tag{3.46}$$
$$\dot{x}_C \cos + \dot{y}_C \sin - \dot{\,}b = r\dot{\,}_l,$$

where $\dot{x}_C, \dot{y}_C$ are components of the velocity of the mass center C. These constraints can be written in the form of Equation (3.42) with the matrix A:

$$A(q) = \begin{bmatrix} -\sin & \cos & -d & 0 & 0 \\ -\cos & -\sin & -b & r & 0 \\ -\cos & -\sin & b & 0 & r \end{bmatrix}. \tag{3.47}$$

Selecting angular velocities of the wheels to be control inputs, i.e., $v_1 = \dot{\,}_r, v_2 = \dot{\,}_l$, the nonholonomic constraints (3.46) can be presented in the form of Equation (3.44) as

$$\dot{q} = S(q)v(t) = \begin{bmatrix} c(b\cos - d\sin) & c(b\cos + d\sin) \\ c(b\sin + d\cos) & c(b\sin - d\cos) \\ c & -c \\ 1 & 0 \\ 0 & 1 \end{bmatrix} \begin{bmatrix} \dot{\,}_r \\ \dot{\,}_l \end{bmatrix}, \tag{3.48}$$

where $c = r/2b$.

Now, let us use the concept of filtration to verify the constraint type in (3.46). The computations are as follows: $(q) = {}_1 = span\{s_1(q), s_2(q)\}$ and $\dim {}_1 = rank[s_1(q), s_2(q)] = 2$. Then

$$s_3(q) = [s_1(q), s_2(q)] = [-rc\sin \quad rc\cos \quad 0 \quad 0 \quad 0]^T. \tag{3.49}$$

The vector $s_3(q)$ is linearly independent from $s_1(q)$ and $s_2(q)$, and then it is not in the distribution ${}_1$. This means that at least one of the constraints is nonholonomic. The distribution ${}_2$ is of the form

$${}_2 = span\{s_1(q), s_2(q), s_3(q)\} \quad \text{and} \quad \dim {}_2 = rank[s_1(q), s_2(q), s_3(q)] = 3.$$

Iterated Lie brackets have to be calculated to obtain

$$s_4(q) = [s_1(q), s_3(q)] = [-rc^2\cos \quad rc^2\sin \quad 0 \quad 0 \quad 0]^T$$

and

$$\dim {}_3 = rank[s_1(q), s_2(q), s_3(q), s_4(q)] = 4. \tag{3.50}$$

In the same way, it can be verified that

$$s_5(q) = [s_2(q), s_3(q)] = -s_4(q),$$

$$s_6(q) = [s_1(q), s_4(q)] = -c^2 s_3(q) \quad \text{and} \quad s_6(q) \in \ _3,$$

$$s_7(q) = [s_2(q), s_4(q)] = c^2 s_3(q) \quad \text{and} \quad s_7(q) \in \ _3,$$

$$s_8(q) = [s_3(q), s_4(q)] = 0 \quad \text{and} \quad s_8(q) \in \ _3.$$

Based on the calculations above, we can conclude that

$$\dim \ _3 = rank[s_1(q), s_2(q), s_3(q), s_4(q)] = 4,$$

and it is involutive. Based on definitions 3.15 through 3.16, the nonholonomy degree is 3, and the system is partially nonholonomic with one holonomic constraint. Subtracting the second from the third of equations (3.46), we obtain the holonomic constraint as in Equations (1.1).

3.3.2 Accessibility and Controllability

It is vital then to verify whether constraints on a system are holonomic or nonholonomic. It may not be verified "at first sight," and the geometric control theory tools may be needed. This is equivalent, from the control theory point of view, to the question of whether or not the dimension of the configuration space is reduced. In the case of k independent equations of constraints, linear in derivatives of n state variables, they specify a $(n-k)$ distribution Δ on the configuration manifold.

From the control perspective, a control is a function that allows the choice of a system state velocity at each instant by weighting of smooth vector fields. The control Lie algebra associated with Δ, usually denoted by $L(\Delta)$, is the smallest distribution that contains Δ and is closed under the Lie bracket operation.

Below we define basic notions of controllability and accessibility with some constructive results needed to investigate control properties of nonlinear systems. More details can be found in Bloch (2003).

Definition 3.17: The system (3.40) is said to be controllable if for any two points q_0 and q_1 in a smooth n-dimensional manifold M there exists an admissible control $u(t)$ defined on some interval $[0, T]$ such that the system (3.40) with the initial condition q_0 reaches the point q_1 in time T.

A related property, which is often easier to prove than controllability, is accessibility. It needs a definition of a reachable set.

Definition 3.18: Given $q_0 \in M$, we define a set $R(q_0, t)$ of all $q \in M$ for which there exists an admissible control u such that there is a trajectory of (3.40) with $q(0) = q_0$, $q(t) = q$. The reachable set from q_0 at time T is defined to be

$$R_T(q_0) = \bigcup_{0 < t < T} R(q_0, t)$$

$$(3.51)$$

The notion of a reachable set depends upon the choice of T. This definition specifies the set of points that may be reached by the system traveling on trajectories from the initial point in a time at most T.

Definition 3.19: The accessibility algebra **C** of (3.40) is the smallest Lie algebra of vector fields on M that contains the vector fields f and $g_1, \ldots, g_m$.

The accessibility algebra **C** of (3.40) is then a span of all possible Lie brackets generated by f and $g_1, \ldots, g_m$.

Definition 3.20: The accessibility distribution C of (3.40) is the involutive distribution generated by the vector fields in **C**:

$$C(q) = span\{X(q) \,|\, X \in \mathbf{C}\}. \tag{3.52}$$

Every element of C is a linear combination of repeated Lie brackets of the form

$$[X_k, [X_{k-1}, [\ldots, [X_1, X_0]\ldots]]],$$

where $X_i, i \in \{0, \ldots, k\}, k = 0, 1, \ldots$, belongs to the set $\{f, g_1, \ldots, g_m\}$.

Definition 3.21: The system of (3.40) on M is said to be locally accessible from $q_0 \in M$ if for any neighborhood ε of q_0 and all $T > 0$, the set $R_T(q_0)$ contains a nonempty open set.

If the system is locally accessible from any $q_0 \in M$, then it is locally accessible.

Let us note that generally, accessibility is far from proving controllability. In some cases, accessibility implies controllability.

Proposition 3.1

If the accessibility rank condition at $q_0 \in M$ is satisfied, i.e., if

$$\dim C(q_0) = n \tag{3.53}$$

then the system (3.40) is locally accessible from q_0. If the accessibility rank condition is satisfied at every $q \in M$, then the system (3.40) is locally accessible. If the system (3.40) is locally accessible, then $\dim C(q_0) = n$ for q in an open and dense subset of M.

The common name for the condition of Equation (3.53) is the Lie algebra rank condition (LARC) (Sussmann 1987).

Definition 3.22: The system (3.40) is locally strongly accessible from $q_0 \in M$ if for any neighborhood ε of q_0 and for any $T > 0$ sufficiently small, the set $R_T(q_0, T)$ contains a nonempty open set.

Definition 3.23: The strong accessibility algebra C_0 of (3.40) is the smallest subalgebra of the Lie algebra containing $g_1, \ldots, g_m$ and such that

$$[f, X] \in C_0 \quad \text{for all } X \in C_0.$$

Definition 3.24: The strong accessibility distribution C_0 of (3.40) is the involutive distribution generated by the vector fields in C_0:

$$C_0(q) = \text{span}\{X(q)|X \in C_0\}. \tag{3.54}$$

Every element of C_0 is a linear combination of repeated Lie brackets of the form

$$[X_k, [X_{k-1}, [\ldots, [X_1, g_j] \ldots]]],$$

for $j \in \{0, \ldots, m\}, X_i, i \in \{1, \ldots, k\}, k = 0, 1, \ldots$, belongs to the set $\{f, g_1, \ldots, g_m\}$. The drift vector field is not involved explicitly in this expression.

Proposition 3.2

If the strong accessibility rank condition at $q_0 \in M$ is satisfied, i.e., if

$$\dim C_0(q_0) = n, \tag{3.55}$$

then the system (3.40) is locally strongly accessible from q_0. If the strong accessibility rank condition is satisfied at every $q \in M$, then the system (3.40) is locally strongly accessible. If the system (3.40) is locally strongly accessible, then $\dim C_0(q_0) = n$ for q in an open and dense subset of M.

From the definitions above a couple of important conclusions can be drawn.

1. For a driftless system, i.e., when $f = 0$ in Equation (3.40), accessibility is equivalent to controllability, and the LARC condition can be used. It is an important implication, because many nonlinear control systems of a practical relevance are driftless.

2. Accessibility is not equivalent to controllability when the drift is present in Equation (3.40). Then, different notions of controllability have to be developed. One is the notion of small-time local controllability.

Definition 3.25: The system of Equation (3.40), i.e., $\dot{q} = f(q) + \sum_{i=1}^{m} g_i(q)u_i$, is small-time locally controllable (STLC) from $q_0 \in M$ if for any neighbor neighborhood ε of q_0 and for any $T > 0$, q_0 is an interior point of the set $R_T(q_0)$, i.e., a whole neighborhood of q_0 is reachable from q_0 at arbitrarily small time.

In Sussmann (1987), it is demonstrated that a sufficient condition for a system to be STLC is that it is accessible and the "bad" brackets are identically zero or can be "neutralized" by brackets of lesser degree, i.e., be a linear combination of such brackets. The "bad" bracket is the one for which the degree of the drift vector field, i.e., the number of times the drift vector field appears in the bracket, is odd, while the sum of the degrees of the control vector fields is even.

The controllability result for nonholonomic systems is based upon the Lie algebra rank condition (LARC). This condition is local. If the rank condition holds everywhere in the configuration space, then the system is globally controllable.

Example 3.11

Consider a kinematic control model for the Ishlinsky problem presented in Example 1.7, and examine its controllability properties (Jarzębowska and McClamroch 2000). This example can illustrate the way controllability may be proved using the LARC. It also demonstrates some challenges one may face when using the geometric control theory tools for some nontypical examples. Equations (1.9b) transformed into the nonlinear control form of Equations (3.44) are as follows:

$$\dot{x} = \frac{2az\sin\eta}{w\sin\eta - z\sin(\eta-\alpha)}\sin\alpha u_1 + R\sin\eta u_2,$$

$$\dot{y} = \frac{2a\sin\eta}{w\sin\eta - z\sin(\eta-\alpha)}[w - z\cos\alpha]u_1 - R\cos\eta u_2$$

$$\dot{\eta} = \frac{2a\sin\eta}{w\sin\eta - z\sin(\eta-\alpha)}\sin\alpha u_1, \tag{3.56}$$

$$\dot{}_1 = u_1,$$

$$\dot{}_2 = \frac{\sin\eta}{\sin(\eta-\alpha)}u_1,$$

$$\dot{} = u_2,$$

where w and z are defined as: $w = a_2 + x\sin\alpha - y\cos\alpha$, $z = a_1 - y$. The roll rates of the first lower cylinder and of the upper cylinder are control variables, i.e., $u_1 = \dot{}_1$, $u_2 = \dot{}$.

A control objective is to identify all possible maneuvers that can be accomplished by using the two control variables u_1 and u_2. To this end, we determine two control vector fields g_1, g_2 according to Equations (3.56) on open subsets of R^6 where (3.56) are well defined:

$$
g_1 = \begin{bmatrix} \dfrac{2az\sin\eta}{w\sin\eta - z\sin(\eta - \alpha)}\sin\alpha \\[2mm] \dfrac{2a\sin\eta}{w\sin\eta - z\sin(\eta - \alpha)}[w - z\cos\alpha] \\[2mm] \dfrac{2a\sin\eta}{w\sin\eta - z\sin(\eta - \alpha)}\sin\alpha \\[2mm] 1 \\[2mm] \dfrac{\sin\eta}{\sin(\eta - \alpha)} \\[2mm] 0 \end{bmatrix}, \quad g_2 = \begin{bmatrix} R\sin\eta \\ -R\cos\eta \\ 0 \\ 0 \\ 0 \\ 1 \end{bmatrix}. \quad (3.57)
$$

To examine controllability properties of our system, we use theoretical control tools as presented above. A distribution of interest is a span of the two control vector fields g_1, g_2 defined in Equation (3.57), that is, $= span\{g_1(q), g_2(q)\}$, where the span is taken over the set of smooth real-valued functions on R^6. Evaluated at any point $q \in R^6$, the distribution defines the linear subspace of the tangent space:

$$
_q = span\{g_1(q), g_2(q)\} \quad T_q R^6. \quad (3.58)
$$

This distribution is said to be locally regular at q if the dimension of the subspace $_q$ is constant in the neighborhood of q. It can be shown that the involutive closure of the distribution $^-$ is a Lie algebra denoted by $L(g_1, g_2)$. The rank of $L(g_1, g_2)$ at a point $q \in R^6$ is equal to the dimension of $_q$ as a linear space. The Lie algebra $_q = L(g_1, g_2)$ at q is referred to as the controllability Lie algebra at q. From this construction, it follows that motion in the direction of the Lie bracket $[g_1, g_2]$ is possible by an appropriate choice of the first and second control inputs. Similarly, it is possible to generate motion along any direction defined by an iterated Lie bracket associated with the vector fields g_1, g_2. It is possible to steer the system along any direction in the Lie algebra $L(g_1, g_2)$. If the rank of the controllability Lie algebra is six, then it is guaranteed that the Ishlinsky system is completely nonholonomic or, equivalently, it is completely controllable. This implies that the system can be maneuvered locally to all possible nearby states in the neighborhood of q. This requires evaluation of iterated Lie brackets of the control vector fields g_1, g_2. All possible Lie brackets associated with g_1, g_2 up to level three, are as follows:

$$
g_1, g_2, g_3 = [g_1, g_2], \quad g_4 = [g_2, [g_1, g_2]], \quad g_5 = [[g_1, g_2], g_1], \quad g_6 = [g_2, [g_2, [g_1, g_2]]],
$$

$$
g_7 = [g_2, [[g_1, g_2], g_1]], \quad g_8 = [g_1, [g_2, [g_1, g_2]]], \quad g_9 = [g_1, [[g_1, g_2], g_1]]. \quad (3.59)
$$

The vector fields g_1, g_2 are nonlinear for the Ishlinsky example, and computation of Lie brackets, involving iterated differentiation, leads to very complicated formulas, which are required in a symbolic form in order to obtain higher-level brackets. The software program Mathematica® is used to calculate the higher-level iterated Lie brackets symbolically. In particular, we use the singular value decomposition to determine if the distribution $\overline{}_{qi}$ evaluated at a selected cylinder configuration q is nonsingular. There are many possible choices of Lie brackets and configurations at which the distributions could be evaluated. Numerous calculations were completed, but a sample of results for the purpose of illustrating the general concepts is presented.

In all computations, the angle between the two lower cylinders $\alpha = 0.175$ rad is constant. The radii of all cylinders are identical, $a = R = 1$. To obtain the distribution of rank 6, it is necessary to include some level 3 Lie brackets. Three specific system configurations are selected, for which we identify the spanning distributions. These configurations are $q_1 = (0;0;1.05;0;0.175;0)$, $q_2 = (0;0;0.9;0;0.175;0.1)$, and $q_3 = (0;0;1.2;0;0.175;0.1)$. At these configurations, we identify distributions that have rank 6:

At the configuration q_1,

$$\overline{}_{14} = span\{g_1, g_2, g_3, g_8, g_5, g_6\}, \qquad \overline{}_{15} = span\{g_1, g_2, g_8, g_4, g_5, g_6\}$$

$$\overline{}_{16} = span\{g_1, g_2, g_3, g_7, g_5, g_6\}, \qquad \overline{}_{17} = span\{g_1, g_2, g_7, g_4, g_5, g_6\}.$$

At the configuration q_2,

$$\overline{}_{22} = span\{g_1, g_2, g_3, g_4, g_5, g_8\}, \qquad \overline{}_{25} = span\{g_1, g_2, g_8, g_4, g_5, g_6\}.$$

At the configuration q_3,

$$\overline{}_{34} = span\{g_1, g_2, g_3, g_7, g_5, g_8\}, \qquad \overline{}_{35} = span\{g_1, g_2, g_8, g_4, g_5, g_6\}.$$

It can be seen that these distributions are locally regular but not globally regular, that is, different brackets are required to obtain a spanning set at different configurations. The controllability properties of the system thus depend on the properties of the Lie algebra at a given configuration q. This is a significant feature of the Ishlinsky problem that has maneuver implications. The three cylinders can be maneuvered in an arbitrary way, locally in the state space, by controlling the rolling motions of the upper cylinder and of one of the lower cylinders. This is basically an existence result, demonstrating that such local arbitrary maneuvers are possible.

Methods are available for the construction of specific input functions that achieve a specified maneuver for the three cylinders. One construction involves the use of sinusoidal control input functions at multiple frequencies with the individual amplitudes and phases selected to guarantee incremental motion in the directions of the spanning Lie brackets. Such a construction is described in Murray et al. (1994) for models including the Ishlinsky example as a special case. We take advantage of it to design an open-loop control to steer the cylinders from an initial to a final position in such a way that a programmed constraint specifies the motion of the upper cylinder mass center G. Specifically, we want G to move along a straight line $y = \beta x$. Then, the kinematic control model consists of equations (3.56) and the constraint $y = \beta x$ with $\beta = 0.75$. The following control

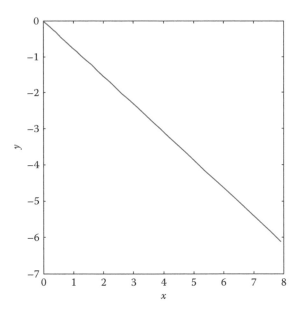

FIGURE 3.1
Motion of the upper cylinder center G.

inputs are designed as $u_1 = 0.01\sin(t/10), u_2 = 0.05t$. The constraint specification supports the construction of the inputs, that is, velocity components $\dot{x}$ and $\dot{y}$ have to be proportional. Based on this and on Equation (3.56), u_2 is designed to be a slow-growing time function and u_1 a sinusoidal input of a small amplitude and frequency. Figure 3.1 shows the motion of the upper cylinder center according to the prescribed trajectory.

The Ishlinsky model provides an excellent academic example of a non-Chaplygin nonholonomic system that can be used for various analytical studies. However, this example has potential engineering significance as a model for material handling problems involving multiple cylinders in a factory or warehouse.

Example 3.12

Consider now controllability of a planar diver kinematic control model in the flying phase of motion. The constraints on the model are of the same type as for the control model of a free-floating space manipulator, e.g., the one presented in Example 1.6. Three coordinates specify the diver motion: θ_1 is the orientation of the body and θ_2, θ_3 are the shape variables. (See Figure 3.2.)

If the initial angular momentum of the diver is denoted by K_0, then it is equal to

$$K_0 = [a_1 + 2b\cos\theta_2 + 2c\sin\theta_2 + 2d\cos\theta_3 + 2e\sin\theta_3 + 2h\cos(\theta_3 - \theta_2) + 2s\sin(\theta_3 - \theta_2)]\dot{\theta}_1 +$$

$$+ [a_2 + b\cos\theta_2 + c\sin\theta_2 + h\cos(\theta_3 - \theta_2) + s\sin(\theta_3 - \theta_2)]\dot{\theta}_2 + [a_3 + d\cos\theta_3 + e\sin\theta_3 +$$

$$+ h\cos(\theta_3 - \theta_2) + s\sin(\theta_3 - \theta_2)]\dot{\theta}_3, \qquad\qquad (3.60)$$

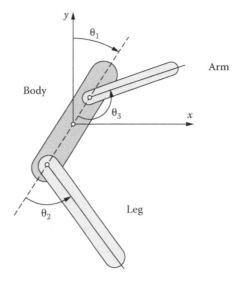

FIGURE 3.2
Planar model of a diver.

where a_i, b, c, d, e, h, s are constant coefficients. The angular momentum equation may be presented in a different form, i.e.,

$$K_0 = [K_1(\theta_2,\theta_3) \quad K_2(\theta_2,\theta_3) \quad K_3(\theta_2,\theta_3)] \begin{bmatrix} \dot{\theta}_1 \\ \dot{\theta}_2 \\ \dot{\theta}_3 \end{bmatrix}.$$

A kinematic control model can be designed selecting control inputs as angular velocities $u_1 = \dot{\theta}_2, u_2 = \dot{\theta}_3$. Then

$$\begin{bmatrix} \dot{\theta}_1 \\ \dot{\theta}_2 \\ \dot{\theta}_3 \end{bmatrix} = \begin{bmatrix} \dfrac{K_0}{K_1} \\ 0 \\ 0 \end{bmatrix} + \begin{bmatrix} \dfrac{-K_2}{K_1} \\ 1 \\ 0 \end{bmatrix} u_1 + \begin{bmatrix} \dfrac{-K_3}{K_1} \\ 0 \\ 1 \end{bmatrix} u_2 = f(\theta_2,\theta_3) + g_1(\theta_2,\theta_3)u_1 + g_2(\theta_2,\theta_3)u_2.$$

(3.61)

A drift vector $f(\theta_2,\theta_3)$ is a nonlinear function of shape variables. Let us verify whether the system model is STLC. The control model has a drift term, and in a general case one should use the Sussmann theorem to verify sufficient conditions for a system with drift to be STLC (Sussmann 1987). In this case, simple calculations show that

$$[g_1,g_2] = g_3 = \begin{bmatrix} \dfrac{1}{K_1^2}\left(K_3 \dfrac{\partial K_1}{\partial \theta_2} - K_1 \dfrac{\partial K_3}{\partial \theta_2} - K_2 \dfrac{\partial K_1}{\partial \theta_3} + K_1 \dfrac{\partial K_2}{\partial \theta_3} \right) \\ 0 \\ 0 \end{bmatrix}.$$

(3.62)

The distribution $= \{g_1, g_2, g_3\}$ has rank 3, so it spans the whole space without some isolated singular positions (θ_2, θ_3), for which the first term in g_3 is equal to zero. The control system is locally controllable then. For the simulation of human diving and controller design for the diver model, see Wooten and Hodgins (1995), Crawford and Sastry (1995).

Example 3.13

Consider a model of a rear wheel drive car. Its controllability properties demonstrate the essence of the control problems for nonholonomic systems. The kinematic control model of the four-wheel car with rear wheels driven is as follows:

$$
\begin{bmatrix} \dot{x} \\ \dot{y} \\ \dot{\theta} \\ \cdot \end{bmatrix} = \begin{bmatrix} \cos\theta \\ \sin\theta \\ \tan\ //l \\ 0 \end{bmatrix} v_1 + \begin{bmatrix} 0 \\ 0 \\ 0 \\ 1 \end{bmatrix} v_2, \tag{3.63}
$$

where the Cartesian coordinates (x, y) describe the rear wheel that substitutes the two wheels and it is located at the midpoint of the rear axle, θ measures the orientation of the car body with respect to the x axis, l is the distance between the wheels, and φ is the steering angle. Two control inputs are selected to be the driving and steering velocities.

The system model (3.63) is driftless, so any configuration q_0 is an equilibrium point when inputs are equal to zeros. In might seem that the easiest way to investigate controllability at q_0 is to consider the linear approximation that corresponds to Equation (3.63). It yields

$$
\dot{\tilde{q}} = g_1(q_0)v_1 + g_2(q_0)v_2 = S(q_0)v \tag{3.64}
$$

where $\tilde{q} = q - q_0$. The rank of the controllability matrix is 2. It means that the linearized system model is not controllable, and a linear controller would not work even locally. Also, this result implicates that exponential stability in the sense of Lyapunov cannot be achieved by a smooth feedback. Generally, it is not possible to stabilize the system at q_0 using a smooth time-invariant static-state feedback at all. This result is the Brockett theorem, which we cite in Section 3.3.3.

Now, let us use LARC to test the controllability of (3.63). It relies upon verifying whether

$$rank[g_1\ g_2\ [g_1, g_2]\ [g_1, [g_1, g_2]]\ [g_2, [g_1, g_2]]...] = 4.$$

For the system of Equation (3.63), the Lie brackets are as follows:

$$
[g_1, g_2] = \begin{bmatrix} 0 \\ 0 \\ -1/l\cos^2 \\ 0 \end{bmatrix}, \quad [g_1, [g_1, g_2]] = \begin{bmatrix} -\sin\theta/l\cos^2 \\ \cos\theta/l\cos^2 \\ 0 \\ 0 \end{bmatrix}.
$$

It is easy to verify that away from the model singularity for $= \pm\pi/2$, rank is 4. The angle can be modified by the input such that the system is controllable everywhere.

One can prove that the linearization of the system model (3.64) about a smooth trajectory results in a linear time-varying system that is controllable provided that some persistency conditions are satisfied by the reference trajectory (for details, see De Luca et al. 1998).

3.3.3 Stabilizability

A stabilization problem relies upon obtaining feedback laws that can guarantee that equilibrium of a closed-loop system is asymptotically stable. For a linear time-invariant system, if all unstable eigenvalues are controllable, then the origin can be asymptotically stabilized by a linear time-invariant static-state feedback. The linearization of a nonholonomic system about any equilibrium is not asymptotically stabilizable by any static smooth (even continuous) time-invariant state feedback. It means that linear stabilization control tools cannot be used even locally. Formally, this property is formulated by the Brockett necessary condition (Brockett 1983). Moreover, there is no dynamic continuous time-invariant feedback which makes closed-loop nonholonomic dynamics asymptotically stable.

It may be stated that nonholonomic systems exhibit control properties that have no counterpart in holonomic systems. Other approaches for stabilization of nonholonomic systems have been developed. In this section, we recall basic stabilization properties of this class of control systems based on the classical dynamics for nonholonomic systems as presented in Chapter 2, Section 2.3.3. The Lagrange equations with multipliers (2.90) written in a vector notation have the form of Equation (2.155). For control applications, a vector Q of generalized forces has to be replaced by a control vector $B(q)\tau$:

$$M(q)\ddot{q} + C(q,\dot{q}) = J^T(q)\lambda + B(q)\tau \qquad (3.65)$$

$$J(q)\dot{q} = 0. \qquad (3.66)$$

We assume that $B(q)$ is a full rank $(n \times r)$ matrix function.

The constraints of Equation (3.66) define a $(2n-m)$-dimensional smooth submanifold $M = \{(q,\dot{q}) \,|\, J(q)\dot{q} = 0\}$ of the phase space. It plays a significant role in the formulation of control and stabilization problems for the nonholonomic dynamics presented by Equations (3.65) and (3.66). This dynamics represents a well-posed model that is associated with a unique solution, at least locally (Bloch et al. 1992). We are interested in a class of equilibrium solutions of Equations (3.65) and (3.66). It is (q_e, λ_e), and q_e is referred to as an equilibrium configuration. The set of equilibrium configurations is given by $\{q \,|\, C(q,0) - J^T(q)\lambda = 0 \text{ for some } \lambda \in R^m\}$.

For stabilization applications, assume that the control $\tau(q,\dot{q})$ satisfies $\tau : M \to R^r$, and it is a smooth function. If the initial conditions satisfy $(q_0,\dot{q}_0) \in M$, then there exists a unique solution, at least local, of the initial value problem of Equations (3.65) and (3.66) that satisfies $(q(t),\dot{q}(t)) \in M$ for each t. Then, the set of equilibrium configurations is presented by

$$\{q | C(q,0) - J^T(q)\lambda = B(q)\tau(q,0) \text{ for some } \lambda \in R^m\}.$$

As remarked in Bloch et al. (1992), the equilibrium manifold has dimension at least m. However, for certain cases of nonholonomic systems like a ball rolling on an inclined surface, there may not be even a single equilibrium configuration. Specific assumptions about the matrix J let us introduce an equilibrium manifold of dimension m by an appropriate choice of input.

Following Bloch, Reyhanoglu, and McClamroch (1992), a stability definition of the closed-loop system (3.65) and (3.66) with the control input $\tau(q,\dot{q})$ may be formulated.

Definition 3.26: For the input $\tau(q,\dot{q})$ and $M_s = \{(q,\dot{q}) | \dot{q} = 0\}$ embedded in M, M_s is locally stable if for any neighborhood $T \supset M_s$ there is a neighborhood of V of M_s with $T \supset V \supset M_s$ such that if $(q_0,\dot{q}_0) \in V \cap M$ then the solution S of the closed-loop dynamics satisfies $(S(t,q_0,\dot{q}_0),\dot{S}(t,q_0,\dot{q}_0)) \in T \cap M$ for all $t \geq 0$.

Also, if $(S(t,q_0,\dot{q}_0),\dot{S}(t,q_0,\dot{q}_0)) \to (q_s,0)$ as $t \to \infty$ for some $(q_s,0)$ M_s, then M_s is said to be a locally asymptotically stable equilibrium manifold of the closed-loop dynamics.

The definition of local asymptotic stability formulated in Section 3.2 corresponds to M_s being a single equilibrium solution.

Definition 3.27: The dynamics of (3.65) and (3.66) is locally asymptotically stabilizable to a smooth equilibrium manifold M_s in M, if there exists a feedback function $\tau : M \to R^r$ such that for the associated closed-loop dynamics M_s is locally asymptotically stable.

Despite the Brockett condition, a number of feedback strategies that locally asymptotically stabilize equilibrium of the nonholonomic dynamics are developed. They can be classified as time-varying stabilization strategies, discontinuous time-invariant strategies, and hybrid stabilization strategies (see Kolmanovsky and McClamroch 1995).

Example 3.14

Consider the dynamics (3.65) and (3.66) and design a smooth feedback that asymptotically stabilizes it to an equilibrium submanifold of M specified by

$$M_s = \{(q,\dot{q}) \ | \dot{q} = 0, g(q) = 0\},$$

where $g(q)$ is some smooth $(n\text{-}m)$ vector function. In Bloch, Reyhanoglu, and McClamroch (1992), it is shown that with the appropriate assumptions, there exists a smooth feedback $\tau : M \to R^r$ that the closed loop is locally asymptotically stable to M_s. Such feedback can also be designed.

First, note that the nonholonomic first order constraints (3.66) can be presented as

$$\dot{q}_2(t) = -J_2^{-1}(q_1,q_2)J_1(q_1,q_2)\dot{q}_1(t), \qquad (3.67)$$

where $J_2(q_1,q_2)$ is an $(m \times m)$ locally nonsingular matrix function.

The state-space control formulation (3.2) of the nonholonomic dynamics, see Sections 3.4 and 3.5 for details, may be presented as

$$\dot{x}_1 = x_3,$$

$$\dot{x}_2 = -J_2^{-1}(x_1,x_2)J_1(x_1,x_2)x_3, \qquad (3.68)$$

$$\dot{x}_3 = \upsilon,$$

where the new input $\upsilon \in R^{n-m}$ that is defined for $\tau \in R^r$, which satisfies $\ddot{q}_1 = \upsilon$ and new state variables $x_1 = q_1, x_2 = q_2, x_3 = \dot{q}_1$ were defined.

The state-space equations possess the property that if $q_1(t)$ and $\dot{q}_1(t)$ are exponentially decaying, then the solution $\dot{q}_2(t) = -J_2^{-1}(q_1,q_2)J_1(q_1,q_2)\dot{q}_1(t)$ is bounded. Also, the first time derivative of $g(t)$ is given by

$$\dot{g} = \frac{\partial g(q)}{\partial q}\left[I - J_2^{-1}J_1\right]^T \dot{q}_1 = \frac{\partial g(q)}{\partial q}G(q)\dot{q}_1.$$

Then, the constraints can be presented as $\dot{q} = G(q)\dot{q}_1$. The second time derivative of $g(q)$ is

$$\ddot{g} = \frac{\partial}{\partial q}\left(\frac{\partial g(q)}{\partial q}G(q)\dot{q}_1\right)G(q)\dot{q}_1 + \frac{\partial g(q)}{\partial q}G(q)\upsilon. \qquad (3.69)$$

With the assumptions as above for the nonholonomic dynamics and its solutions, the following may be formulated: The nonholonomic system (3.65) and (3.66) is locally asymptotically stabilizable to the smooth equilibrium submanifold $M_s = \{(q,\dot{q}) \mid \dot{q} = 0, g(q) = 0\}$ using smooth feedback, if the transversality condition

$$\det\left(\frac{\partial g(q)}{\partial q_1}\right)\det\left(\frac{\partial g(q)}{\partial q}G(q)\right) \neq 0$$

holds. This smooth feedback can be designed as

$$\upsilon = -\left(\frac{\partial g(q)}{\partial q}G(q)\right)^{-1}\left[\frac{\partial}{\partial q}\left(\frac{\partial g(q)}{\partial q}G(q)\dot{q}_1\right)G(q)\dot{q}_1 + K_1\frac{\partial g(q)}{\partial q}G(q)\dot{q}_1 + K_2 g(q)\right] \qquad (3.70)$$

and k_1 and k_2 are constant positive definite $(n\text{-}m \times n\text{-}m)$ matrices.

The proof of the above consists in using the transversality condition, based on which the change of variables from (q_1,q_2,q_3) to $(g,q_2,\dot{g})$ is a diffeomorphism. This change of variables results in $\upsilon = \dot{x}_3 = \ddot{g}$. The control law selected as above leads to a kind of error dynamics of the form

$$\ddot{g} + K_1\dot{g} + K_2 g = 0. \qquad (3.71)$$

It indicates that $(g, \dot{g}) \to 0$ as $t \to \infty$, i.e., g is asymptotically stable. The other variables satisfy the second of equations (3.68) so they remain bounded. It means that the system is asymptotically stable with $(q(t), \dot{q}(t)) \to M_s$ as $t \to \infty$.

Concluding, the nonholonomic dynamics system can be smoothly asymptotically stabilized to the m dimensional equilibrium manifold $M_s = \{(q, \dot{q}) \,|\, \dot{q} = 0, g(q) = 0\}$. The transversality condition depends upon the partitioning of the configuration variables in the constraint equation (3.66).

Example 3.15

Take the reduced dynamics of a unicycle derived in Chapter 2, Example 2.14, i.e.,

$$\dot{z}_1 = z_5,$$
$$\dot{z}_2 = z_6,$$
$$\dot{z}_3 = z_5 r \cos z_2,$$
$$\dot{z}_4 = z_5 r \sin z_2, \tag{3.72}$$
$$\dot{z}_5 = \tau_1 / (mr^2 + I_\theta),$$
$$\dot{z}_6 = \tau_2 / I,$$

and analyze its properties. The following proposition may be formulated (Jarzębowska 2007).

Proposition 3.3

The constraint manifold for the unicycle is a six-dimensional constraint manifold $M = \{(\theta, \, , x, y, \dot{\theta}, \, \dot{}, \dot{x}, \dot{y}) \,|\, \dot{x} = r\dot{\theta}\cos \, , \dot{y} = r\dot{\theta}\sin \, \}$.

The configuration is an equilibrium $(q_e, 0)$ if the controls are zero, that is, $t = 0$ then $z_e = (z_{1e}, z_{2e}, z_{3e}, z_{4e}, 0, 0)$.

The unicycle dynamic control model has the following properties:

1. The system (3.72) is strongly accessible at z_e because the space spanned by the vectors $g_1, g_2, \; g_3 = [g_1, f], \; g_4 = [g_2, f], \; g_5 = [g_2, [f, g_3]], \; g_6 = [g_2, [f, [g_1, [f, g_4]]]]$ is of dimension 6 at z_e.
2. The system (3.72) is strongly accessible because the accessibility rank condition holds for any z from the unicycle state-space.
3. The system (3.72) is small-time locally controllable at z_e because the brackets satisfy sufficient conditions for small-time local controllability.
4. There is no time-invariant continuous feedback law that asymptotically stabilizes the closed-loop system to z_e.

The proof can be demonstrated by direct calculations of the indicated Lie brackets.

3.3.4 Differential Flatness

The most typical control approach to nonlinear systems is to invert the system dynamics to compute the inputs required to perform a predefined task. This inverse dynamics approach assumes that the system dynamics is known. In practice, uncertainty, noise, and various kinds of disturbances are always present, and they must be accounted for in order to achieve an acceptable system performance. The use of feedback control formulations allows the system to respond to errors and can help in its stabilization around the desired operation conditions.

The basic assumption used in most control techniques is to exploit the mathematical structure of the system to obtain solutions to the inverse dynamics and feedback control problems. The most common structure is linear where the approximation is performed by the system linearization. Then, properties of linear control problems may be used. By using different linearizations about different operation points, good control results may be obtained.

However, modern systems are more and more complex, and the use of linear structure alone may not be sufficient. This is specifically true for the inverse dynamics problems, where the solution is not unique, and the desired task may span multiple operating regions.

Looking for different than linear types of structures to exploit, a class of systems, called differentially flat systems, has been distinguished. Flat systems are the generalization of linear systems, but techniques used to control them are different than the ones for linear systems.

The concept of flatness was first defined in Fliess et al. (1992). It is defined in the differential algebra setting, where a system is viewed as a differential field generated by a set of variables, i.e., states and inputs.

The system is said to be differentially flat if a set of variables, called the flat outputs, can be found such that the system is algebraic over the differential field generated by the set of the flat outputs. In other words, a system is flat if we can find a set of outputs (equal in number to the number of inputs) such that all states and inputs can be determined from these outputs without integration. Then, if the system states are $q \in R^n$ and inputs are $u \in R^m$, this system is flat if we can find outputs $y \in R^m$ of the form

$$y = h(q, u, \dot{u}, \ldots, u^{(r)}) \tag{3.73}$$

such that

$$q = (y, \dot{y}, \ldots, y^{(s)}) \tag{3.74a}$$

$$u = \phi(y, \dot{y}, \ldots, y^{(s)}). \tag{3.74b}$$

An active research goes on around flat systems and applications of their properties to control. In van Nieuwstadt et al. (1998), flatness has been defined in a more geometric setting to enable tools from nonlinear control. In Rouchon et al. (2003) and Fliess et al. (1999), the flat system description is presented using a Lie–Bäcklund framework. This point of view offers a compact theoretic framework, and it may be related to feedback linearization. A more rigorous description of flat systems together with their relations to feedback is presented in Section 3.8. A monograph that examines flatness properties in a variety of controlled dynamical systems is available (Sira-Ramirez and Agrawal 2004), among many other publications on this topic. There, the flatness property is examined for linear time-invariant single-input single-output (SISO) and multi-input multi-output (MIMO) systems, time-varying linear systems, and nonlinear systems, among other models of a practical importance. Also, non-differentially flat systems are presented there.

In summary, differential flatness is a property of some controlled dynamic systems that allows simplifying the feedback controller design problem. Flatness is equivalent to controllability. The flatness property allows for a complete parameterization of all system variables in terms of a finite set of independent variables, called flat outputs, and a finite number of their time derivatives. Finding the flat outputs is, generally speaking, hard, because no systematic method exists for their determination, except in a linear systems case and an affine nonlinear single input case. In practice, they can be found by inspection, guessing, or intuition.

An interest in studying flatness grows because many classes of systems, also nonlinear, important in applications are flat. The truth is that any flat system can be feedback linearized using dynamic feedback. Flatness is a system property, and it indicates that the nonlinear structure of the system is well characterized and one can exploit this structure in designing control algorithms for motion planning, trajectory generation, or stabilization.

Another advantage of flatness comparing to dynamic feedback linearization is that it is a geometric property of a system, independent of a coordinate selection. Specifically, in a traditional control approach to systems, when one classifies a system as linear in a state-space, this is meaningless from the geometric point of view because the system is linear in the specific coordinate representation. In this way, flatness can be considered the proper geometric notion of linearity even though the system may be nonlinear in any selected coordinates.

3.4 Kinematic Control Models

Control for holonomic systems at a kinematic level is trivial, but at a dynamic level it is quite challenging. The kinematic control problem for holonomic systems usually relies upon the inverse kinematics, for which well-established control methods are developed. The kinematic control problem for nonholonomic systems is not trivial.

Consider first constraint kinds and their sources from the control perspective. In the control setting, nonholonomic constraints are viewed in a different way than in analytical mechanics. There are two situations when nonholonomic constraints arise:

1. Explicit kinematic constraints are put upon systems.
2. Dynamic constraints are preserved by Lagrange's or Hamilton's equations.

Underactuated systems, which are usually treated separately, are classified as systems with constraints on controls or second order nonholonomic constraints (Oriolo and Nakamura 1991; Reyhanoglu et al. 1996).

The first conclusion based on this classification is that, excluding underactuated systems, nonholonomic constraints that control theory incorporates into its dynamics are of first orders and of the material type or they specify conservation laws. The control theory approach to constraints can be illustrated by an example of a car-parking maneuver between two cars. If to ignore constraints, a path from a given initial position, say in a street, to a final position, say between two cars, would consist of two line segments, which is of course not feasible for a car. It is convenient then, from the control theory perspective, to convert the problem with nonholonomic constraints into a steering problem. We are not interested in directions a system cannot move, but in directions it can. Thus, instead of the constraint formulation (2.23), that is, $B_1(q)\dot{q} = 0$, where we put $b_1(t,q) = 0$ and B_1 does not depend explicitly on time, which is true for most material kinematic constraint equations, we choose a basis for the right null space of the constraints, denoted by $g_i(q) \in R^n, i = 1,\ldots,n-k$, where k is the number of constraint equations. The control problem for a nonholonomic system can be restated as finding input functions $u_i(t) \in R^{n-k}$, such that constraints on a system are presented as

$$\dot{q} = g_1(q)u_1 + \ldots + g_{n-k}(q)u_{n-k} \tag{3.75}$$

and the system can be driven from a given q_0 to q_1. It can be shown that if in Equation (2.23) the matrix B_1 elements are smooth and linearly independent, so are the g_i (Bloch 2003; Murray et al. 1994). The constraints from two groups that nonlinear control theory considers can be presented in the form of Equation (3.75), which is a driftless version of Equation (3.40).

Herein, we introduce an extension of the list of constraint sources. It is based on the programmed constraint concept. The motivation is as follows. Robotic systems, which are a representative class of mechanical systems in our development, are designed to perform work. Usually, this is the end-effector at which work is performed. It can be writing, painting, scribing, grinding, and carrying objects and other tasks. It is natural then to focus on the end-effector of a robot and specify tasks in terms of its desired motions. Also, during the robot motion, its other parts may perform specified motions or be constrained. Specifically, the end-effector or the wheeled robot may move along a specified trajectory or have some desired velocity (Chang and Chen 1996; Fukao et al. 2000; Koh and Cho 1999; Scheuer and Laugier 1998; Shekl and Lumelsky 1998; Ye-Hwa 1988; Zotov and Tomofeyev 1992; Zotov 2003). Wheeled robots are also subjected to material nonholonomic constraints. Other constraints, like the one that specifies an allowable value of the acceleration that prevents a robot wheel from slippage and mechanical shock during motion (Koh and Cho 1999), have to be taken into account when a control strategy is designed.

We distinguish then a new group of constraints that are tasks that can be specified by equations. The source of these constraints is not in other bodies. They are non-material constraints put upon systems in order to specify tasks they have to perform. The programmed constraint concept presented in Chapter 2 is a proper one to describe tasks. In our approach then, robots and manipulators are viewed as constrained systems, and constraints may originate from different sources. We introduce the following classification.

Constraints on mechanical systems can be divided into four primary groups (Jarzębowska 2007):

1. Material constraints, which are position or kinematic. They can be holonomic or nonholonomic specified by Equations (2.19) or (2.22).
2. Programmed constraints, which are non-material and they are put on systems in order to specify their desired motions. They can be presented by Equation (2.160).
3. Constraints that may come from dynamic, design, control, or operation specifications including underactuated systems; they also can be presented by Equation (2.160).
4. Constraints that may specify obstacles in a robot or manipulator workspace.

Material constraints, especially nonholonomic, are well known in dynamics and control, and appropriate tools are developed to study systems with such constraints (Bloch 2003; Kwatny and Blankenship 2000). Dynamic models for systems with material constraints are developed based on classical mechanics methods, mostly on Lagrange's equations with multipliers or Kane's approach (Kane and Levinson 1996; Tanner and Kyriakopulos 2001). It means that only first order constraints are merged into these models. Programmed constraints and constraints from groups 3 and 4 are taken into account when control strategies are to be designed to obtain motions that satisfy these constraints. The above modeling procedure and control design are typical ways in which tasks are executed within a nonlinear control theory framework.

Constraints that we regard as programmed are formulated mostly as position constraints that concern trajectories or put by the driving coordinates, which also have the form of Equations (2.19) (de Jalon and Bayo 1994; Nikravesh 1988). Sometimes, they are referred to as task constraints or performance goals (Nikravesh 1988). Some tasks are specified by velocity constraints for space vehicles (Vafa 1991). Systems subjected to high order constraints are discussed in Chapter 7 following Jarzębowska (2007).

A significant example of constraints from group 3 is a constraint for an underactuated mechanical system. It comes from the requirement that one or more degrees of freedom are not actuated. The equation that specifies dynamics of the degree of freedom that is not actuated is shown to be a second order nonholonomic. Specialized nonlinear control tools are developed to control these systems because control techniques developed for first order nonholonomic systems cannot be directly applied (Reyhanoglu et al. 1996).

There is a need for a theoretical control framework that could explicitly incorporate constraints on mechanical systems that originate from all four groups. We stress the typical procedures of tracking control designs in nonlinear control because we propose a different approach to motion tracking for constrained systems. A development of this approach starts from a definition of a unified constraint formulation.

Since now on, constraints can be holonomic or nonholonomic, material or programmed.

Holonomic constraints specify positions or trajectories and have the form

$$A(t,q) = 0, \tag{3.76}$$

where $A(t,q)$ is an a-dimensional vector. We assume that $a < n$, i.e., the number of equations of constraints is less than the dimension of q.

Nonholonomic first order constraints are mostly linear in velocity, i.e.,

$$B_1(t,q)\dot{q} = 0, \tag{3.77}$$

where $B_1(t,q)$ is a $(k \times n)$-dimensional matrix with $n > k$. It is assumed that constraint equations are independent, i.e., that the rank of $B_1(t,q)$ is k. Most often, they are material constraints that can be converted into the steering formulation (3.40). In a traditional controller design process, only the material constraints (3.77) are merged into dynamic control models.

The high order programmed constraints can be extended to encompass a unified formulation of a constraint, which is Equation (2.160), i.e.,

$$B(t,q,\dot{q}, \ldots ,q^{(p-1)})q^{(p)} + s(t,q,\dot{q}, \ldots ,q^{(p-1)}) = 0. \tag{3.78}$$

Constraints from all four groups can be presented as (3.78). Combining the above with the constraint classification from Section 2.2, we formulate the following proposition.

Proposition 3.4

The equations of constraints in the form of Equation (3.78) can specify both material and programmed constraints.

Equation (3.78) is referred to as a unified formulation of equations of constraints.

Proof: The proof is based upon reasoning that the type of constraint equations does not influence generation of equations of motion of a system subjected to these constraints. The only concern is the constraint order and if the constraints are ideal. By "type of constraint equations," we mean whether constraints are material or non-material. When $p = 0$, we get a constraint in the form of Equation (3.76), that is, a position constraint that may be a material constraint that describes, for example, a constant distance between link ends, or be a programmed constraint that specifies a desired trajectory for the end-effector. When $p = 1$, a constraint equation is in the form of Equation (3.77), and it can be a material constraint that describes the condition of rolling without slipping. However, it can be a programmed constraint that specifies, for example, the end-effector motion with some desired velocity. For both examples of constraint types for $p = 1$, equations of motion are generated in the same way provided that constraints are ideal. Material constraints are of orders equal to zero or one. Constraint equations for $p > 1$ are of the non-material type. When needed, two or more such constraint equations, each of a different type, may be listed in Equation (3.78). The constraint formulation (3.78) can be used then to specify constraint equations of any order and type. q.e.d.

The high order constraint equations (3.78) can be transformed into the state-space representation. To this end, we introduce a new p-vector $x = (x_1,...,x_p)$ such that $x_1 = q, \dot{x}_1 = x_2,..., \dot{x}_{p-1} = x_p$. We assume that t is not present explicitly in Equation (3.78). If it is, we reorder coordinates, assigning $x_0 = t$. With the new vector x, Equation (3.78) can be written as $(p-1 + k)$ first order equations:

$$\dot{x}_1 = x_2,$$
$$\dot{x}_2 = x_3,$$
$$\vdots \qquad \vdots \tag{3.79}$$
$$\dot{x}_{p-1} = x_p,$$
$$B(x_1,...,x_p)\dot{x}_p = -s(x_1,...,x_p)$$

or in a matrix form

$$C(x)\dot{x} = b(x), \tag{3.80}$$

where C is a $(p-1+k) \times p$ matrix, and b is a $(p-1+k)$-dimensional vector. Let $f(x)$ be a particular solution of Equation (3.80) so $C(x)f(x) = b(x)$. Let $g(x)$ be a $p \times (n-k)$ full rank matrix whose column space is in the null space of $C(x)$, that is, $C(x)g(x) = 0$. Then, the solution of Equation (3.80) is given by

$$\dot{x} = f(x) + g(x)u(t) \tag{3.81}$$

for any smooth vector $u(t)$. The problem with the constraint equation (3.80) has been converted into a control problem for a system with high order constraints. In general, a drift term is present in Equation (3.81) or the constraints may be non-Chaplygin. Because Equation (3.81) is a state-space representation of the unified constraint formulation (3.78), we refer to it as a unified state-space control formulation. The kinematic control model (3.81) is only formally equivalent to (3.40), because in (3.81), u_i, $i = 1, \dots, m$, may not have a physical interpretation of velocities. They may be accelerations or their time derivatives. The following examples demonstrate what $u_i's$ are and that programmed and material constraints combined together within Equation (3.78) may have neither a special form nor be Chaplygin.

Example 3.16

Consider one of the well-known tracking control algorithms based upon a kinematic model, that is, the Samson algorithm (Samson 1993; Samson and Ait-Abderrahim 1991b), and apply it to the unicycle model. The Samson algorithm requires the kinematic control model to be transformed to the control dynamics:

$$\begin{bmatrix} \dot{x}_e \\ \dot{y}_e \\ \dot{\ }_e \end{bmatrix} = \begin{bmatrix} 0 & \omega_d & 0 \\ -\omega_d & 0 & 0 \\ 0 & 0 & 0 \end{bmatrix} \begin{bmatrix} x_e \\ y_e \\ \ _e \end{bmatrix} + \begin{bmatrix} 0 \\ \sin\ _e \\ 0 \end{bmatrix} V_d + \begin{bmatrix} 1 & 0 \\ 0 & 0 \\ 0 & 1 \end{bmatrix} \begin{bmatrix} u_1 \\ u_2 \end{bmatrix}, \tag{3.82}$$

where $\dot{x}_e, \dot{y}_e, \dot{\ }_e$ are the tracking errors, and the original control inputs in the unicycle kinematic control model (compare to Example 3.2) are transformed as

$$V = V_d \cos\ _e - u_1,$$
$$\omega = \omega_d - u_2. \tag{3.83}$$

Linearizing Equations (3.82) about a desired trajectory, the control dynamics yields

$$\begin{bmatrix} \dot{x}_e \\ \dot{y}_e \\ \dot{\ }_e \end{bmatrix} = \begin{bmatrix} 0 & \omega_d & 0 \\ -\omega_d & 0 & V_d \\ 0 & 0 & 0 \end{bmatrix} \begin{bmatrix} x_e \\ y_e \\ \ _e \end{bmatrix} + \begin{bmatrix} 1 & 0 \\ 0 & 0 \\ 0 & 1 \end{bmatrix} \begin{bmatrix} u_1 \\ u_2 \end{bmatrix}. \tag{3.84}$$

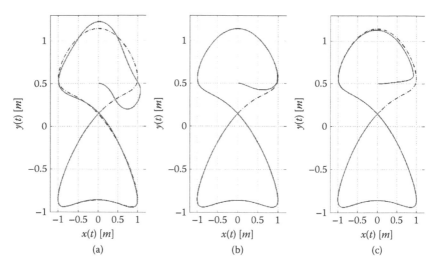

FIGURE. 3.3 (SEE COLOR INSERT.)
Tracking a desired trajectory by the Samson algorithm for $k_2 = 1$ and different k_1: (a) $k_1 = 0.1$, (b) $k_1 = 0.5$, and (c) $k_1 = 1$.

The control inputs are designed to be

$$u_1 = -k_1 x_e ,$$

$$u_2 = -k_2 sign[V_d(t)]y_e - k_3 \theta_e ,$$

(3.85)

with k_1, k_2, and k_3 being the control gains. For the simulation study (Szymański 2011), the gains are selected to be $k_1 = k_3$. A desired trajectory is specified by

$$x_d = \sin\left(\frac{t}{10}\right), \quad y_d = \sin\left(\frac{t}{20}\right) + \frac{1}{7}\cos\left(\frac{t}{5}\right)$$

(3.86)

Tracking the desired trajectory specified by Equations (3.86) for a kinematic model of a mobile robot Pioneer 3-DX is presented in Figure 3.3.

Example 3.17

Let us consider a unicycle model from Example 2.12. Impose on the unicycle a programmed constraint that specifies a circular trajectory, for which its radius $R(t)$ is assumed to be constant, that is, $x^2 + y^2 - const = \beta_1 = 0$. The unicycle is also subjected to the nonholonomic constraints (2.163), so three constraint equations are imposed, and one control input is available. We show that a kinematic control model for the unicycle has the form of Equation (3.81). To this end, select the forward velocity to be a control input (i.e., $u_1 = r\dot\theta$), and Equations (2.163) can be written in a control form

$$\dot{x} = u_1 \cos\theta , \quad \dot{y} = u_1 \sin\theta .$$

(3.87)

Next, we transform the programmed constraint and present all the constraints in the form of Equation (3.77). Specifically, we differentiate the programmed constraint equation twice, augment this differentiated constraint by stabilizing terms $\alpha\dot\beta_1 + \gamma\beta_1$, and reuse the material constraint equation to obtain $\ddot{}$, i.e.,

$$x\dot{x} + y\dot{y} + \gamma\beta_1 = u_1(x\cos + y\sin) + \gamma\beta_1 = 0, \tag{3.88}$$

and the second differentiation yields

$$u_1(\dot x\cos + \dot y\sin) + \dot u_1(x\cos + y\sin) + u_1(y\cos - x\sin)\dot{} + \alpha\dot\beta_1 + \gamma\beta_1 = 0. \tag{3.89}$$

The constraint Equations (3.87) and Equation (3.88) are reused again to transform the first and second terms on the left-hand side of Equation (3.89) to obtain

$$u_1^2 - \dot u_1 \frac{\gamma\beta_1}{u_1} + u_1(y\cos - x\sin)\dot{} + \alpha\dot\beta_1 + \gamma\beta_1 = 0.$$

The kinematic control model for the programmed motion of the unicycle is then

$$\dot x = u_1\cos,$$

$$\dot y = u_1\sin, \tag{3.90}$$

$$\dot{} = \frac{-\alpha\dot\beta_1 - \gamma\beta_1 - u_1^2 + \gamma\beta_1\dot u_1 / u_1}{u_1(y\cos - x\sin)}.$$

Define new variables $x_1 = u_1, x_2 = x, x_3 = y, x_4 = $ for (3.90), where u_1 is an extra state and $u_1 \neq 0$. In the new variables, (3.90) has the form

$$\dot x_1 = v,$$

$$\dot x_2 = x_1\cos x_4,$$

$$\dot x_3 = x_1\sin x_4, \tag{3.91}$$

$$\dot x_4 = \frac{-\alpha\dot\beta_1 - \gamma\beta_1 - x_1^2}{x_1(x_3\cos x_4 - x_2\sin x_4)} + \frac{\gamma\beta_1}{x_1^2(x_3\cos x_4 - x_2\sin x_4)} v,$$

with a redefined input $\dot u_1 = v$.
Equation (3.91) is of the form (3.81) with the drift and control vectors given by

$$f(x) = \left(0, x_1\cos x_4, x_1\sin x_4, \frac{-\alpha\dot\beta_1 - \gamma\beta_1 - x_1^2}{x_1(x_3\cos x_4 - x_2\sin x_4)}\right), g(x) = \left(1, 0, 0, \frac{\gamma\beta_1}{x_1^2(x_3\cos x_4 - x_2\sin x_4)}\right).$$

The system input must now be obtained by integration; thus, the control law contains an integrator yielding a dynamic controller. The programmed constraint restricts the unicycle configuration space, and singular positions can be avoided for $y\cos \neq x\sin$, i.e., the unicycle restricted position is $(0,0)$ from any direction and positions along x or y axis.

Example 3.18

Consider a two-link manipulator from Example 2.16. The programmed constraint for the desired change of the trajectory curvature for the end-effector is specified by Equation (2.211): $\ddot{\Theta}_2 = F_1 - F_2\dot{\Theta}_1$. We define new variables $x_1 = \Theta_1$, $x_2 = \Theta_2$, $x_3 = \dot{\Theta}_1$, $x_4 = \dot{\Theta}_2$, $x_5 = \ddot{\Theta}_1$, $x_6 = \ddot{\Theta}_2$, and select a lower link angular velocity to be a control input $u = \dot{\Theta}_1$. A kinematic control model based on (2.211) takes the form

$$\dot{x}_1 = x_3,$$

$$\dot{x}_2 = x_4,$$

$$\dot{x}_3 = x_5,$$

$$\dot{x}_4 = x_6, \tag{3.92}$$

$$\dot{x}_5 = v,$$

$$\dot{x}_6 = F_1 - F_2 v.$$

This is the control model (3.81) with $f(x) = (x_3, x_4, x_5, x_6, 0, F_1)$ and $g(x) = (0,0,0,0,1,-F_2)$, and a redefined input is $v = \ddot{u}$. The control input must be obtained by integration. Note that $F_1 = F_1(x_1, x_2, x_3, x_4, x_5, x_6)$ and $F_2 = F_2(x_1, x_2)$, so we have a more general case, in which the vector field g might depend on u.

3.5 Dynamic Control Models

Dynamic control models actually used, which we refer to as classical dynamic control models, are based on Lagrange's equations with multipliers. They have the form

$$M(q)\ddot{q} + C(q,\dot{q}) + D(q) = J^T(q)\lambda + E(q)\tau,$$

$$J(q)\dot{q} = 0, \tag{3.93}$$

where $M(q)$ is a $(n \times n)$ positive definite symmetric inertia matrix, $J(q)$ is a full rank $(k \times n)$ matrix, $2 \le n - k < n$, λ is a k-dimensional vector of Lagrange's multipliers, $E(q)\tau$ is a n-dimensional vector of generalized forces applied to a system, and τ is a r-dimensional vector of control inputs. For model-based control applications, Equations (3.93) have to be transformed to the reduced-state form. It can be accomplished in several ways. We start from Lagrange's equations with multipliers (2.155), which we write as

$$\frac{d}{dt}\left(\frac{\partial T}{\partial \dot{q}}\right) - \frac{\partial T}{\partial q} = J^T(q)\lambda + Q(q,\dot{q}),$$

$$J(q)\dot{q} = 0, \tag{3.94}$$

where we assume that $Q(q,\dot{q})$ stands for all external forces applied to a system.

To eliminate constraint forces from Equations (3.94), we project them onto the linear subspace generated by the null space of $J(q)$. Because $(J^T(q)\lambda) \cdot \delta q = 0$, Lagrange's equations become

$$\left[\frac{d}{dt}\left(\frac{\partial T}{\partial \dot{q}}\right) - \frac{\partial T}{\partial q} - Q\right] \cdot \delta q = 0, \tag{3.95}$$

where $\delta q \in R^n$ and satisfies $J(q)\delta q = 0$. We partition the coordinate vector q and the $J(q)$ matrix such that $q = (q_1, q_2) \in R^{n-k} \times R^k$, and $J = [J_1(q)\ J_2(q)]$, $J_2(q) \in R^{k \times k}$ is invertible. Then the relation $\delta q_2 = -J_2^{-1}(q)J_1(q)\delta q_1$ holds. Inserting it into Equation (3.95), we obtain

$$\left[\frac{d}{dt}\left(\frac{\partial T}{\partial \dot{q}_1}\right) - \frac{\partial T}{\partial q_1} - Q_1\right] - J_1^T J_2^{-T}\left[\frac{d}{dt}\left(\frac{\partial T}{\partial \dot{q}_2}\right) - \frac{\partial T}{\partial q_2} - Q_2\right] = 0. \tag{3.96}$$

Equations (3.96) are second order differential equations in terms of q. They can be simplified by reusing the constraint equation $\dot{q}_2 = -J_2^{-1}(q)J_1(q)\dot{q}_1$ to eliminate $\dot{q}_2$ and $\ddot{q}_2$. The evolution of q_2 can be retrieved by reapplication of the constraint equations.

We can demonstrate that Equations (3.96) with the constraint $J(q)\dot{q} = 0$ are equivalent to Nielsen's equations in Maggi's form (2.119). It is enough to show that

$$\frac{d}{dt}\left(\frac{\partial T}{\partial \dot{q}_\sigma}\right) = \frac{\partial^2 T}{\partial \dot{q}_\sigma \partial t} + \sum_{\rho=1}^{n}\frac{\partial^2 T}{\partial \dot{q}_\sigma \partial q_\rho}\dot{q}_\rho + \sum_{\rho=1}^{n}\frac{\partial^2 T}{\partial \dot{q}_\sigma \partial \dot{q}_\rho}\ddot{q}_\rho, \tag{3.97a}$$

$$\dot{T} = \frac{\partial T}{\partial t} + \sum_{\rho=1}^{n}\frac{\partial T}{\partial q_\rho}\dot{q}_\rho + \sum_{\rho=1}^{n}\frac{\partial T}{\partial \dot{q}_\rho}\ddot{q}_\rho. \tag{3.97b}$$

Based on relation (3.97b), we have

$$\frac{\partial \dot{T}}{\partial \dot{q}_\sigma} = \frac{\partial^2 T}{\partial t \partial \dot{q}_\sigma} + \sum_{\rho=1}^{n}\frac{\partial^2 T}{\partial q_\rho \partial \dot{q}_\sigma}\dot{q}_\rho + \sum_{\rho=1}^{n}\frac{\partial^2 T}{\partial \dot{q}_\rho \partial \dot{q}_\sigma}\ddot{q}_\rho + \frac{\partial T}{\partial q_\sigma} \tag{3.98}$$

and comparing (3.97a) and (3.98), we obtain that

$$\frac{d}{dt}\left(\frac{\partial T}{\partial \dot{q}_\sigma}\right) = \frac{\partial \dot{T}}{\partial \dot{q}_\sigma} - \frac{\partial T}{\partial q_\sigma}. \tag{3.99}$$

Relations (3.99) inserted into Equation (3.96) and developed for each q_σ, $\sigma = 1, \ldots, n$, yield that terms in brackets in (3.96) are equal to

$$\left(\frac{\partial \dot{T}}{\partial \dot{q}_\sigma} - 2\frac{\partial T}{\partial q_\sigma}\right).$$

Thus, Equations (3.96) are equivalent to Equations (2.119), which are equivalent to the generalized programmed motion equations (GPME) for $p = 1$.

The dynamic control model (3.93) can be transformed to the state-space representation obtained by an extension of the kinematic control model (3.75) as

$$\dot{q} = g(q)v = g_1(q)v_1 + \cdots + g_{n-k}(q)v_{n-k}, \quad i = 1,\ldots,n-k, \quad 2 \le n-k < n \quad (3.100a)$$

$$v_i^{r_i} = u_i, \quad (3.100b)$$

where $r_i,\ldots,r_m$ denote an order of time differentiation, and v is the output of a linear system consisting of chains of integrators. This reduction procedure resulted in Equations (3.100a,b) is proposed in Campion et al. (1991). This model is referred to as a dynamic control model because in applications from mechanics $r_i = 1$, $i = 1,\ldots,n-k$, controls are typically generalized forces, and the model consists of the constraint equation (3.100a) and the dynamic equations of motion (3.100b), which reduce to $\dot{v} = u$. To demonstrate this, consider the dynamics (3.93). Equation $J(q)\dot{q} = 0$ constrains the velocity $\dot{q}$ at each q to the null space of $J(q)$. Let the vector fields $g_1,\ldots,g_m$, $m = n-k$, form the basis for the null space of $J(q)$ at each q, and let $g(q) = (g_1(q),\ldots,g_m(q))$. Then $J(q)g(q) = 0$ for each q and the second of equations (3.93) can be presented as (3.100a) for some appropriately defined m-dimensional vector $v = (v_1,\ldots,v_m)$. Components of v may or may not have physical interpretations as velocities. By differentiating Equation (3.100a), we obtain $\ddot{q} = g(q)\dot{v} + \dot{g}(q)v$. Substituting the above into the first of equations (3.93), and premultiplying by $g^T(q)$, we obtain

$$g^T(q)M(q)g(q)\dot{v} + F(q,\dot{q}) = g^T(q)E(q)\tau,$$

in which $F(q,\dot{q}) = g^T(q)[M(q)\dot{g}(q)v + C(q,\dot{q}) + D(q)]$. We assume that the map $g^T(q)E(q)$ is onto what means that we require that independent degrees of freedom of the system are actuated. Then we can apply feedback linearization $U(\dot{q},q,u): R^n \times R^n \times R^m \to R^m$ such that $\dot{v} = u$, where $u = (u_1,\ldots,u_m)$ is a m-dimensional vector control. In this way, we get Equation (3.100b) with $r_i = 1$.

A class of dynamic nonholonomic systems for which the vector fields have a special form is studied in Bloch et al. (1992). The unicycle belongs to this class, for example. For this class of systems, for the reduction procedure that leads to Equations (3.100a,b), it is possible to select a basis vector field for the null space of $J(q)$ such that $g(q)$ can be

$$g(q) = \begin{bmatrix} \tilde{g}(q) \\ I_k \end{bmatrix}, \quad (3.101)$$

where $\tilde{g}(q)$ is a $k \times (n-k)$ matrix, and I_k is a $(n-k) \times (n-k)$ identity matrix. Partition of the vector q as $q = (z, y)$, where $z = (z_1, ..., z_k), y = (y_1, ..., y_{n-k})$ results in the dynamic extension of Equation (3.75), which is

$$\ddot{y}_i = u_i,$$

$$\dot{z} = \sum_{i=1}^{n-k} \tilde{g}_i(z, y)\dot{y}_i. \qquad i = 1, ..., n-k \qquad (3.102)$$

Equations (3.102) are said to be in the dynamic Chaplygin form if $\tilde{g}_i(q)$, $i = 1, ..., n-k$, depend only on the base vector y but not on the fiber vector z.

3.6 Feedback Linearization of Nonlinear Systems

Feedback linearization is an important technique for analyzing nonlinear systems. The basic question is whether a nonlinear control system of the form of Equation (3.40), i.e.,

$$\dot{q} = f(q) + \sum_{i=1}^{m} g_i(q)u_i \qquad (3.103)$$

can be transformed by some feedback of the form $u_j = a(q) + \sum_{i=1}^{m} b_{ji}(q)v_i$ to a linear system. The answer to this question is interesting, because when a system can be transformed into a linear one, one may apply standard linear control techniques.

For nonlinear systems with outputs, two types of exact linearization can be considered. One is the possibility of transforming the whole set of differential equations into a linear system, and that is referred to as full-state linearization. The second is the input–output linearization. Necessary and sufficient conditions are established for the solution of both problems using static feedback. Only sufficient conditions exist for the solution using dynamic feedback (Isidori 1989). Feedback linearization methods are now standard in many textbooks and papers on control (e.g., Isidori 1989; Nijmeijer and van der Schaft 1990; Pomet 1992, 1995).

Let us rewrite the system of (3.103) into a form suitable for further discussion, i.e.,

$$\dot{q} = f(q) + G(q)u, \qquad (3.104)$$

where $G(q)$ is the matrix whose i-th column is $g_i(q)$.

Definition 3.28: The system of Equation (3.104) is static feedback linearizable around q_0, which is an equilibrium of f, that is, $f(q_0) = 0$, if there exists a state transformation $z = \quad (q)$ and a regular static state feedback $u = a(q) + b(q)v$, with $\quad (q)$, $a(q)$, and $b(q)$ defined in a neighborhood of q_0, $a(q)$, and $b(q)$ being smooth mappings, and with $\quad (q_0) = 0$, $a(q_0) = 0$, and $b(q)$ a $(m \times m)$ nonsingular matrix, such that the feedback transformed system

$$\dot{q} = \left[f(q) + G(q)a(q) \right] + \left[G(q)b(q) \right] v$$

transforms under $z = \quad (q)$ into a linear controllable system

$$\dot{z} = A(z)z + B(z)v. \tag{3.105}$$

Theorem 3.6

Consider the system of Equation (3.104) with $f(q_0) = 0$, and assume that the strong accessibility rank condition (3.55) holds at q_0. This system is static feedback linearizable if and only if the distributions $\quad_{1}, \ldots, \quad_{n}$ defined by

$$_{k} = span \left\{ ad_f^r g_1(q), \ldots, ad_f^r g_m(q) \ \middle| \ r = 0, 1, \ldots, k-1 \right\} \qquad k = 1, 2, \ldots \tag{3.106}$$

are all involutive and of constant dimensions in a neighborhood of q_0.

Corollary 3.1

A single input system (3.103) with $f(q_0) = 0$ is static feedback linearizable around q_0 if and only if dim $_{n}(q_0) = n$ and $_{n-1}(q_0)$ is involutive.

Definition 3.29: For a control system of the form $\dot{q} = f(q, u)$, if there exists a dynamic feedback with the state z, input (q, v) and output u of the form $\dot{z} = g(z, q, v)$, $u = \gamma(z, q, v)$, which when applied to the original system, gives an extended system with the state (q, z) and input v, of the form $\dot{q} = f(q, \gamma(z, q, v))$, $\dot{z} = g(z, q, v)$ and which, by the transformation of coordinates $\Phi = \phi(q, z)$ of the extended state of the system, transforms into a linear controllable system, then the original system is referred to as dynamic feedback linearizable.

Proposition 3.5

A single input nonlinear system is dynamic feedback linearizable if and only if it is static feedback linearizable.

Dynamic feedback linearization as defined above is equivalent to the concept of differential flatness discussed in Section 3.3.4.

The exact dynamic feedback linearization consists in finding a dynamic state feedback compensator of the form

$$\dot{\xi} = a(q,\xi) + b(q,\xi)u,$$
$$v = c(q,\xi) + d(q,\xi)u, \tag{3.107}$$

with r-dimensional state ξ and m-dimensional external input u, such that the closed-loop system (3.44) and (3.107) is equivalent to a linear controllable system, under the state transformation $\Phi = \phi(q,\xi)$. Only necessary or sufficient conditions for the solution of the dynamic feedback linearization problem exist. Control algorithms that are developed based on input–output decoupling can be found in Isidori (1989).

The starting point to perform the input–output linearization is based on the definition of an appropriate m-dimensional output $\eta = h(q)$, which complies with a control goal, e.g., tracking a desired trajectory. Then, one has to differentiate the output until the input appears in a nonsingular way. During this differentiation, an addition of integrators on a subset of inputs may be needed to avoid the differentiation of the original inputs. This dynamic extension algorithm builds up the state ξ of the dynamic compensator (3.107). The algorithm terminates when the system model is invertible from the chosen output. In addition, if the number of the output differentiations equals $(n + r)$ of the extended state-space, full input–output linearization is obtained. We say then that the closed-loop system is equivalent to a set of decoupled input–output chains of integrators from u_i to η_i, $i = 1, \dots, m$.

Example 3.19

Consider a kinematic control model of a unicycle presented in Example 3.2:

$$\begin{bmatrix} \dot{x} \\ \dot{y} \\ \dot{\theta} \end{bmatrix} = \begin{bmatrix} \cos \\ \sin g \\ 0 \end{bmatrix} u_1 + \begin{bmatrix} 0 \\ 0 \\ 1 \end{bmatrix} u_2, \tag{3.108}$$

where the two control inputs are selected to be $u_1 = \dot{r\theta}, u_2 = \ \dot{\ }$.
It can be verified easily that the LARC condition is satisfied globally:

$$rank[g_1\ g_2\ [g_1, g_2]] = 3 = n. \tag{3.109}$$

Based on the discussion from this section and Section 3.3.2, we may conclude that the unicycle model cannot be transformed into a linear controllable system using static-state feedback. For the model (3.108), the controllability condition

(3.109) implies that the necessary conditions for full state linearization are not satisfied because the distribution generated by the vector fields g_1, g_2 is not involutive. However, we still may transform the system model by the input–output linearization and decoupling. The choice of linearizing outputs is not unique. Let us define the linearizing output vector by $\eta = (x, y)$. Differentiating η with respect to time, we obtain that

$$\dot{\eta} = \begin{bmatrix} \cos & 0 \\ \sin & 0 \end{bmatrix} \begin{bmatrix} u_1 \\ u_2 \end{bmatrix}. \tag{3.110}$$

From Equation (3.110), only the relation between $\dot{\eta}$ and u_1 may be obtained. Still, the velocity u_2 cannot be recovered from Equation (3.110), so we add an integrator on the forward velocity vector, whose state is denoted by ξ, as follows:

$$\ddot{\eta} = \dot{\xi} \begin{bmatrix} \cos \\ \sin \end{bmatrix} + \xi \cdot \begin{bmatrix} -\sin \\ \cos \end{bmatrix} = \begin{bmatrix} \cos & -\xi\sin \\ \sin & \xi\cos \end{bmatrix} \begin{bmatrix} a \\ u_2 \end{bmatrix}, \tag{3.111}$$

where a is the linear acceleration of the unicycle, $\xi = u_1$ and $\dot{\xi} = a$. Assuming that $\xi \neq 0$, we obtain

$$\begin{bmatrix} a \\ u_2 \end{bmatrix} = \begin{bmatrix} \cos & -\xi\sin \\ \sin & \xi\cos \end{bmatrix}^{-1} \begin{bmatrix} v_1 \\ v_2 \end{bmatrix} = \begin{bmatrix} \cos & -\xi\sin \\ \sin & \xi\cos \end{bmatrix}^{-1} \begin{bmatrix} \ddot{x} \\ \ddot{y} \end{bmatrix}, \tag{3.112}$$

where $\ddot{\eta} = [v_1 \quad v_2]^T$. Based on Equation (3.112), we get

$$\ddot{x} = v_1, \qquad\qquad$$
$$\ddot{y} = v_2. \tag{3.113}$$

Since the dynamic compensator is one dimensional, select new coordinates as

$$z_1 = x,$$
$$z_2 = y,$$
$$z_3 = \dot{x} = \xi\cos , \tag{3.114}$$
$$z_4 = \dot{y} = \xi\sin .$$

The extended system (3.114) is fully linearized and described by two chains of second order input–output integrators:

$$\ddot{z}_1 = v_1,$$
$$\ddot{z}_2 = v_2. \tag{3.115}$$

We may formulate a control goal as tracking a desired trajectory specified by (x_d, y_d). To the linearized system (3.115), we may apply a feedback controller of the form

$$v_1 = \ddot{x}_d + k_{p1}(x_d - x) + k_{d1}(\dot{x}_d - \dot{x}),$$
$$v_2 = \ddot{y}_d + k_{p2}(y_d - y) + k_{d2}(\dot{y}_d - \dot{y}),$$

(3.116)

which guarantees global exponential convergence to the desired output trajectory. The proportional-derivative (PD) gains in Equations (3.116) are chosen as $k_{pi} > 0, k_{di} > 0$.

Getting back to "old variables" can be done as

$$\dot{\xi} = v_1 \cos + v_2 \sin,$$
$$u_1 = \xi,$$
$$u_2 = \frac{-v_1 \sin + v_2 \cos}{\xi}.$$

(3.117)

It can be seen that this algorithm of dynamic feedback linearization has a singularity at $\xi = 0$. It is equivalent to the situation when the unicycle is not rolling. This singularity is structural for nonholonomic systems and has to be avoided during control.

Tracking a desired trajectory given by Equations (3.86) in Example 3.16 using the feedback linearization based controller (3.116) for the model of the mobile robot Pioneer 3-DX is presented in Figure 3.4. More details about the selection of initial conditions and control gains can be found in Szymański (2011).

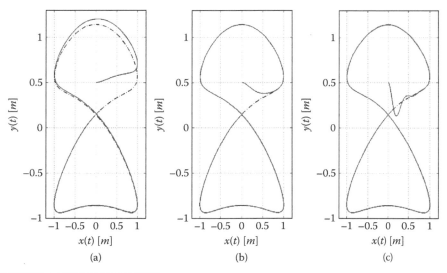

FIGURE 3.4 (SEE COLOR INSERT.)
Tracking a desired trajectory by a feedback linearization–based algorithm for $k_d = 1$ and different k_p: (a) $k_p = 0.05$, (b) $k_p = 0.3$, and (c) $k_p = 2$.

3.7 Model-Based Control Design Methods

Controllers based on dynamic models are referred to as model based. We discuss them briefly without any distinction between their specializations to holonomic or nonholonomic systems. Such exposition of the overview relates to the tracking control strategy we present in Chapter 7 because it is based on model-based controllers, it is designed for nonholonomic systems, and it can adapt controllers dedicated to holonomic systems.

Model-based controllers can be roughly classified as computed torque controllers and passivity-based controllers (Lewis et al. 2004; Ortega and Spong 1989).

Computed torque controllers execute a control goal by feedback linearization of nonlinear system dynamics. Feedback linearization can be demanding in terms of computation time and input magnitudes; hence, it can be difficult to use feedback to globally convert a nonlinear system into a linear one. For some controller schemes, it is necessary to secure that the inverse of the inertia matrix exists. However, for robot manipulators, unboundedness of the input is rarely a problem because the inertia matrix is bounded; hence, control torques always remain bounded. Experimental results show that computed torque controllers have very good performance characteristics (see, e.g., in Murray et al. 1994; Oriolo et al. 2002). Exponential stability can be proved for computed torque controllers, and exponentially stable systems automatically enjoy various robustness and optimality properties (Khalil 2001).

One of the main disadvantages of computed torque controllers is the need for online computation of system dynamics during tracking. To remedy this problem, versions of computed torque controllers are designed, for which information about the system dynamics along the motion we track is utilized. The Wen–Bayard version of the computed torque controller can be an example of such a design (Wen and Bayard 1988). For other versions of computed torque controllers, see Lewis et al. (2004) and references therein. To the best of the author's knowledge, a computed torque controller in its version proposed by Wen and Bayard was not applied to track motion of nonholonomic systems.

Passivity-based controllers reshape a system's mechanical energy in order to achieve a tracking goal (Berghuis 1993; Lewis et al. 2004). Many versions of passivity-based controllers are designed, for example, controllers by Slotine and Li, Sadegh and Horowitz, Paden and Panja, and others (for details, see Berghuis 1993, Lewis et al. 2004, and references therein). A nice property of passivity-based tracking controllers is that they can be easily modified to adaptive versions (Berghuis 1993). These are, among many others, a controller by Slotine and Li, a composite parameter adaptation law by Slotine and Li, Sadegh and Horowitz adaptive version of a passivity-based controller, or an adaptive version of a controller by Bayard and Wen (Bayard and Wen 1988).

A significant drawback of feedback linearization–based controllers is that they require an exact mathematical model of the system dynamics. This dynamics contains many parameters, and most of them like inertia and geometrical parameters can be measured with a limited accuracy. Also, parameters of a payload may vary from operation to operation. Unknown parameters in a dynamic model are one source of uncertainties—parametric uncertainties. Other uncertainties referred to as structural uncertainties are also present. Good examples are friction effects in robot joints. They vary over the robot lifetime and depend on lubrication, temperature of operation, level of robot wear, and frequency of its operation time and operation velocities, and many other factors (Armstrong-Helouvry 1991; Jarzębowska and Greenberg 2000). Two basic approaches, i.e., robust control and adaptive control can compensate for system uncertainties.

Robust controllers are applicable to a smaller class of systems comparing with adaptive controllers, but they are simpler in implementation and no time is required to adjust them to a specific system. Robust controllers enable keeping online computations to a minimum. They can cope well with additive bounded disturbances in a system. Significant disadvantages of robust controllers are that they require a priori known bounds on uncertainties, and that even in the absence of additive bounded disturbances, they cannot guarantee asymptotic stability of the tracking error (Berghuis 1993; Lewis et al. 2004; Spong and Vidyasagar 1989).

In the adaptive approach, a controller attempts to "learn" and "adjust" an uncertain parameter in a system. Adaptive controllers are classified with respect to the underlying methodology, i.e., feedback linearization or passivity based. A different classification is based upon an adaptation mechanism. This is direct adaptation referred to as tracking error driven, indirect adaptation, which is predictor error driven, or composite adaptation, which is the combination of the two (Ortega and Spong 1989; Slotine and Li 1996). Improvements, modifications, and some experimental evaluation of adaptive controllers in both robustifying and simplifying can be found in Berghuis (1993) and Whitcomb et al. (1991). Adaptive feedback-linearizing controllers have one serious drawback—effective controller gains that are gains from tracking errors to the control torque rely on parameter estimates. This indicates that there is an interaction between the error compensating loop and the adaptation loop. From the controller trimming point of view, feedback linearization–based controllers are more attractive than passivity-based schemes.

A significant feature of adaptive controllers is that their implementation does not require a priori known uncertain constant parameters such as inertia parameters or friction coefficients. Disadvantages of adaptive controllers are the large amount of online computation and the lack of robustness to additive bounded disturbances in a system. Specifically, a regression matrix must be calculated online because it depends on measurements of joint positions and velocities, and in some controller schemes on accelerations. It turns out that the best features of robust and adaptive controllers can be matched and some disadvantages eliminated. New controllers developed in this way can significantly reduce online computation and can compensate for actuator dynamics or joint flexibility. A review of advanced control design techniques is presented in Lewis et al. (2004). They all, to the best of the author's knowledge, are dedicated to holonomic systems only.

Among adaptive control strategies, a model reference adaptive control (MRAC) (Slotine and Li 1996) is the one that relies on the reference dynamics. It is composed of four parts: the system to be controlled, a reference model for the specification of a desired output, a feedback controller, and an adaptation mechanism. The reference model is used to specify an ideal response of an adaptive control system to an external command. In the MRAC approach, a choice of the reference model is part of the adaptive control design. This choice should satisfy two criteria—it should reflect the performance specification in a control task, and a desirable behavior should be available for the adaptive control system. In MRAC systems, the goal is to design a controller to make a closed-loop system like a predefined reference model in the face of system's uncertainties. Other tracking strategies that use reference models are reported in Pandian and Hanmandlu (1993) and Chen et al. (1996).

Model-based controllers in both non-adaptive and adaptive versions suffer from several drawbacks that can be summarized as rather a large amount of online computation, the need for measurement of velocities and accelerations, and guarantee that the inverse of the mass matrix exists. When we control a system with many degrees of freedom, simulation and implementation may become hard and time consuming because of these reasons. These practical concerns made researchers rethink theoretical developments for previous robot controllers. One of the main ideas was to employ information about a desired motion of a system, that is, use position, velocity, and acceleration signals along this motion. Both nonadaptive and adaptive controllers were redesigned such that they could use these signals. An extensive review of control algorithms, which depend on desired motion information, is available in Lewis et al. (2004). They are referred to as advanced control laws.

In contrast to the model-based approach to adaptive control of nonholonomic systems, a performance-based adaptive control methodology is proposed (Colbaugh et al. 1999). In this approach, the adaptive control law adjusts controller gains based on system performance rather than on the knowledge of its dynamics.

Most of the proposed dynamic control strategies for nonholonomic systems are developed by assuming that a full dynamic model is precisely known and the entire system state is measurable. A few works report solutions to control problems for nonholonomic systems with uncertainties (Oriolo et al. 2002).

A few papers address robustness in control of nonholonomic systems. In Rouchon et al. (2003), robust stabilization of wheeled mobile robots in the chained forms is obtained. In Lucibello and Oriolo (2001), exponential convergence to a desired equilibrium is obtained for a case where small perturbations in a kinematic model were added. In Oriolo et al. (1998), a kind of learning controller for a nonholonomic system is proposed to cope with nonidealities and uncertainties in the system. These works use kinematic control models of nonholonomic systems.

Specific research on control of nonholonomic systems, which is worth noting, is developed by Zotov, and it concerns "programmed motion" tracking which is trajectory tracking developed for mobile robots (Zotov and Tomofeyev 1992; Zotov 2003). Work by Gutowski and Radziszewski (1969) that concerns programmed motion is cited on the occasion of programmed constraints in Chapter 2. To the best of the author's knowledge, there are no other works that consider "programmed constraints" and design control strategies to get them satisfied.

Research concentrated on systems subjected to high order constraints, which do not come from nature but are put by a designer, are the main focus of works by Jarzębowska (2002, 2005, 2006, 2007, 2008).

In spite of a popular opinion that tracking is simpler than stabilization, in the literature there is no unified tracking control method for constrained systems. Tracking means trajectory tracking in most cases. Other constraints when put on a system motion are satisfied using control strategies developed for specific applications. In Chapter 7, we present one unified model-based strategy for tracking any motion specified by equations of constraints. This is the model-based tracking strategy, because tracking at the kinematics level for systems with high order constraints may not be possible simply because of the constraint order. We do not design new tracking or force controllers. We employ existing controllers to new architecture of the model-based tracking control strategy.

3.8 Flatness-Based Control Design Methods

3.8.1 Basic Notions of Equivalence and Flatness

In Section 3.3.4, the property of differential flatness as well as its meaning in the context of control was discussed. In this section, we briefly present relations between flatness and feedback design. Readers interested in more details may consult Fliess et al. (1992, 1995, 1999), Rouchon et al. (2003), Martin et al. (1997, 2003), Sira-Ramirez and Agrawal (2004), and Levine (2009).

Definition 3.30: A system is a pair $(\mathcal{M}, F)$ where $\mathcal{M}$ is a smooth manifold, that may be of infinite dimension, and F is a smooth vector field on $\mathcal{M}$.

To see how a traditionally presented control system fits this definition, consider a system

$$\dot{q} = f(q), \qquad q \in X \quad R^n \tag{3.118}$$

By definition, this is a pair (X, f), where X is an open set of R^n and f is a smooth vector field on X. A solution of Equation (3.118) is a mapping $t \mapsto q(t)$ such that $\dot{q}(t) = f(q(t))$ for all t.

Also, if $q \mapsto h(q)$ is a smooth function on X, and $t \mapsto q(t)$ is a trajectory of Equation (3.118), then the total derivative of $h(q)$ is

$$\frac{d(h(q(t)))}{dt} = \frac{\partial h}{\partial q}(q(t)) \cdot \dot{q}(t).$$

Now, let us add a control input, and consider a control system of the form of Equation (3.2):

$$\dot{q} = f(q, u), \tag{3.119}$$

where f is smooth on an open subset $X \times U \in R^n \times R^m$. It can be seen that f is no longer a vector field on X. It is a collection of vector fields on X parameterized by u: for all $u \in U$ the mapping $q \mapsto f_u(q) = f(q, u)$ is a vector field on X.

It is possible to associate to Equation (3.119) a vector field with "the same" solutions as follows. Given a smooth solution of Equation (3.119), i.e., $t \mapsto (q(t), u(t))$ with values in $X \times U$ such that for all t $\dot{q}(t) = f(q(t), u(t))$, we can consider the infinite mapping

$$t \mapsto \xi(t) = (q(t), u(t), \dot{u}(t), \ddot{u}(t), \ldots) \tag{3.120}$$

taking values in $X \times U \times R^\alpha_m$ with $R^\alpha_m = R^m \times R^m \times \ldots$ being a product of an infinite, countable number of copies of R^m. A point of R^α_m is of the form $(u^1, u^2, \ldots)$ with $u^i \in R^m$. This mapping satisfies, for each t, the relation $\dot{\xi}(t) = (f(q(t)), u(t), \dot{u}(t), \ddot{u}(t), \ldots)$, so it may be thought of as a trajectory of an infinite vector field $(q, u, u^1, \ldots) \mapsto F(q, u, u^1, \ldots) = f((q, u), u^1, u^2, \ldots)$ on $X \times U \times R^\infty_m$.

Conversely, any mapping $t \mapsto \xi(t) = (q(t), u(t), u^1(t)), \ldots)$ which is a trajectory in this infinite vector field, takes the form $(q(t), u(t), \dot{u}(t), \ddot{u}(t), \ldots)$ with $\dot{q}(t) = f(q(t), u(t))$, so it corresponds to the solution of Equation (3.119). F is then a vector field and no longer a parameterized family of vector fields. The construction as above let us think about the control system (3.119) in "space" $X \times U \times R^\alpha_m$ together with a "smooth" vector field F on this space.

In the same way, as in the case of an uncontrolled system (3.118), a "time derivative" of a "smooth" function $(q, u, u^1, ...) \mapsto h(q, u, u^1, ..., u^k)$ may be determined.

Locally, a control system (3.119) looks like an open set of R^α (α not necessarily finite) with coordinates $(\xi_1, ..., \xi_\alpha)$ together with the vector field $\xi \mapsto F(\xi) = (F_1(\xi), ..., F_\alpha(\xi))$, where all components F_i depend only on a finite number of coordinates such that $\dot{\xi} = F(\xi(t))$.

However, there is an important difference with this description adjustment. We lose the notion of the state dimension. Take two systems:

$$\dot{q} = f(q, u), \quad (q, u) \in X \times U \quad R^n \times R^m \tag{*}$$

and

$$\dot{q} = f(q, u),$$
$$\dot{u} = v. \tag{**}$$

They have the same description $(X \times U \times R_m^\infty, F)$ with

$$F(q, u, u^1, ...) = f((q, u), u^1, u^2, ...)$$

in the new formalism: $t \mapsto (q(t), u(t))$ is a trajectory of (*) if and only if $t \mapsto (q(t), u(t), \dot{u}(t))$ is a trajectory of (**).

Definition 3.31: The system $\left(R_m^\infty, F_m \right)$ with coordinates $(y, y^1, y^2, ...)$ and the vector field $F_m(y, y^1, y^2, ...) = (y^1, y^2, y^3, ...)$ is the trivial system.

The trivial system describes any system made of m chains of integrators of arbitrary length, and, in particular, the direct transformation $y = u$. Most often, the system $F(q, u) := (f(q, u), u^1, u^2, ...)$ is identified with $\dot{q} = f(q, u)$.

The main reason for introducing the formalism to describe control systems is to proceed to notions of equivalence and flatness. Roughly speaking, two systems are "equivalent" if there exists an invertible transformation exchanging their trajectories. This statement will have an interpretation in terms of dynamic feedback. To present formal definitions of equivalence and flatness, we need some preliminary concepts and definitions first.

Consider two systems $(\mathcal{M}, F)$ and $(\mathcal{N}, G)$, and a smooth mapping: $\Psi: \mathcal{M} \to \mathcal{N}$. If $t \mapsto \xi(t)$ is a trajectory of $(\mathcal{M}, F)$, i.e., for each ξ $\dot{\xi} = F(\xi(t))$, the composed mapping $t \mapsto \zeta(t) = \Psi(\xi(t))$ satisfies the chain rule

$$\dot{\zeta} = \frac{\partial \Psi}{\partial \xi}(\xi(t)) \cdot \dot{\xi}(t) = \frac{\partial \Psi}{\partial \xi}(\xi(t)) \cdot F(\xi(t)).$$

If the vector fields F and G are Ψ-related, then $\dot{\zeta}(t) = G(\Psi(\xi)) = G(\zeta(t))$, which means that $t \mapsto \zeta(t) = \Psi(\xi(t))$ is a trajectory of $(\mathcal{N}, G)$. If the mapping Ψ has a smooth inverse Φ, then F and G are also Ψ-related, and the correspondence between the trajectories of the two systems is one-to-one. Such an invertible Ψ relating F and G is referred to as an endogenous transformation.

Definition 3.32: Two systems $(\mathcal{M}, F)$ and $(\mathcal{N}, G)$ are equivalent at $(p, r) \in \mathcal{M} \times \mathcal{N}$ if there exists an endogenous transformation from a neighborhood of p to a neighborhood of r.

The systems $(\mathcal{M}, F)$ and $(\mathcal{N}, G)$ are equivalent if they are equivalent at every pair of points (p, r) of a dense open subset of $\mathcal{M} \times \mathcal{N}$.

Now, consider two systems $(X \times U \times R_m^\infty, F)$ and $(Y \times V \times R_s^\infty, G)$ that describe the dynamics

$$\dot{x} = f(x, u), \qquad (x, u) \in X \times U \quad R^n \times R^m \tag{3.121}$$

$$\dot{y} = g(y, v), \qquad (y, v) \in Y \times V \quad R^r \times R^s \tag{3.122}$$

The vector fields F and G are defined by

$$\begin{aligned} F(x, u, u^1, \ldots) &:= (f(x, u), u^1, u^2, \ldots), \\ G(y, v, v^1, \ldots) &:= (g(y, v), v^1, v^2, \ldots). \end{aligned} \tag{3.123}$$

If the systems are equivalent, the endogenous transformation Ψ takes the form

$$\Psi(x, u, u^1, \ldots) := (\psi(x, \bar{u}), \beta(x, \bar{u}), \dot{\beta}(x, \bar{u}), \ldots),$$

where $\bar{u} = (u, u^1, \ldots, u^k)$, and k is some finite integer. The transformation Ψ is completely specified by ψ and β, which specify y and v in terms of x and $\bar{u}$. The inverse Φ of Ψ takes the form

$$\Phi(y, v, v^1, \ldots) := (\ (y, \bar{v}), \alpha(y, \bar{v}), \dot{\alpha}(y, \bar{v}), \ldots)$$

Since the mappings ψ and φ are inverse, we may write that

$$y = \psi(\ (y, \bar{v}), \bar{\alpha}(y, \bar{v})),$$

$$v = \beta(\ (y, \bar{v}), \bar{\alpha}(y, \bar{v})),$$

$$x = (\psi(x, \bar{u}), \bar{\beta}(x, \bar{u})),$$

$$u = \alpha(\psi(x, \bar{u}), \bar{\beta}(x, \bar{u})).$$

Then, if $t \mapsto (x(t), u(t))$ is a trajectory of Equation (3.121), then $t \mapsto (y(t), v(t)) = (\ (x(t), \bar{u}(t)), \alpha(x(t), \bar{u}(t)))$ is a trajectory of Equation (3.122), and vice versa.

An important property of endogenous transformations is that they preserve the input dimension.

Theorem 3.7

If two systems $(X \times U \times R_m^\infty, F)$ and $(Y \times V \times R_s^\infty, G)$ are equivalent, then they have the same number of inputs, i.e., $m = n$.

The proof can be found in Rouchon et al. (2003).

Definition 3.33: The system $(\mathcal{M}, F)$ is flat at $p \in \mathcal{M}$ if it is equivalent at p to a trivial system.

A flat system $(X \times U \times R_m^\infty, F)$ that describes the dynamics (3.121) is equivalent to the trivial system (R_s^∞, F_s) with the endogenous transformation Ψ of the form

$$\Psi(x, u, u^1, \ldots) := (h(x, \bar{u}), \dot{h}(x, \bar{u}), \ddot{h}(x, \bar{u}), \ldots). \tag{3.124}$$

The inverse transformation Φ is $\Phi(\bar{y}) := (\ (\bar{y}), \beta(y), \dot{\beta}(y), \ldots)$.

Definition 3.34: Let $(\mathcal{M}, F)$ be a flat system and Ψ the endogenous transformation putting it into a trivial system. The mapping h in Equation (3.124) is called a flat output.

3.8.2 Flatness in Control Applications

Equipped with the ideas, the definitions, and the theorems as above, we may relate flatness to control problems. Let us focus on feedback linearization as used for tracking and on constructing flat outputs.

Consider again the dynamics of Equations (3.121) and (3.122). Using the formalism presented earlier, they are described by the systems $(X \times U \times R_m^\infty, F)$ and $(Y \times V \times R_s^\infty, G)$ with F and G defined by (3.123). Assume that the systems are equivalent, that is, they have the same trajectories. An important question is whether we may transfer from $\dot{x} = f(x, u)$ to $\dot{y} = g(y, v)$ and back by a dynamic feedback (compare to the one described in Section 3.6):

$$\dot{z} = a(x, z, v),$$

$$u = \kappa(x, z, v),$$

with $z \in Z \quad R^r$.

Theorem 3.8

For two equivalent systems $\dot{x} = f(x, u)$ and $\dot{y} = g(y, v)$, the first can be transformed by dynamic feedback and a coordinate change into

$$\dot{y} = g(y, v),$$

$$\dot{v} = v^1, \ldots, v^\gamma = w$$

for some γ, and the second can be transformed by dynamic feedback and a coordinate change into

$$\dot{x} = f(x, u)$$

$$\dot{u} = u^1, \dots, u^\lambda = w$$

for some λ.

The proof can be found in Rouchon et al. (2003). The consequence of Theorem 3.8 is that a flat dynamics can be linearized by dynamic feedback and a coordinate change. Theorem 3.8 provides the existence of a feedback such that

$$\dot{x} = f(x, \kappa(x, z, w)),$$

$$\dot{z} = a(x, z, w) \tag{3.125}$$

transform, up to a coordinate change, to

$$\dot{y} = g(y, v)$$

$$\dot{v} = v^1, \dots, v^\gamma = w. \tag{3.126}$$

The system of Equations (3.126) is equivalent to $\dot{y} = g(y, v)$, which is equivalent to $\dot{x} = f(x, u)$, that is, to Equation (3.119). Then, we may formulate a definition.

Definition 3.35: We say that the feedback

$$u = \kappa(x, z, w),$$

$$\dot{z} = a(x, z, w), \tag{3.127}$$

is endogenous if the open-loop dynamics $\dot{x} = f(x, u)$ is equivalent to the closed-loop dynamics

$$\dot{x} = f(x, \kappa(x, z, w)),$$

$$\dot{z} = a(x, z, w).$$

Using Definition 3.35, we may formulate the following theorem.

Theorem 3.9

Two dynamics $\dot{x} = f(x, u)$ and $\dot{y} = g(y, v)$ are equivalent if and only if $\dot{x} = f(x, u)$ can be transformed by endogenous feedback and a coordinate change into

$$\dot{y} = g(y, v),$$

$$\dot{v} = v^1, \dots, v^\gamma = w$$

for some large γ and vice versa.

Two consequences of Theorem 3.9 may be formulated:

1. A dynamics is flat if and only if it is linearizable by endogenous feedback and a coordinate change.
2. An endogenous feedback can be generated by another endogenous feedback. Consider the dynamics (3.125) where u and z are specified by Equations (3.127), and it is an endogenous feedback. Then, it can be transformed by endogenous feedback and a coordinate change into

$$\dot{x} = f(x, u),$$
$$\dot{u} = u^1, \ldots, u^\lambda = w. \tag{3.128}$$

for some large enough integer λ.

The properties as listed above show that the controllability property is preserved by equivalence. However, feedback defined as in control theory (compare to Section 3.6 and to Nijmeijer and van der Schaft 1990) is not necessarily endogenous.

3.8.3 Flatness-Based Control Design—Examples

The flatness property is helpful in trajectory tracking problems for nonlinear systems. Let us stress that flatness is not another way of performing feedback linearization, but it is a structural property of a system, and it helps to reveal the system properties necessary for an application of a feedback controller, e.g., feedback linearization, backstepping, or passivity-based control.

In this section, we present a couple of examples of controller designs for nonlinear systems that are flat. There is no general methodology for determining flatness; however, years of extensive research and experience gathered allow us to distinguish many flat systems of practical importance.

Consider first multivariable nonlinear systems that are exactly linearizable by static state feedback and state coordinate transformation. They are mostly holonomic systems and constitute a large class of systems.

Example 3.20

Consider a dynamic model of a two-link manipulator, similar to the one presented in Example 1.5. The difference is that the manipulator works in the vertical plane so gravity forces have to be taken into account. The manipulator dynamics is of the form

$$M(q)\ddot{q} + C(q, \dot{q})\dot{q} + D(q) + f(\dot{q}) = \tau, \tag{3.129}$$

where the matrix $M(q)$ is symmetric, positive definite, and square; $C(q, \dot{q})$ is the matrix of centripetal and Coriolis torques; the vector $D(q)$ is the two-dimensional vector of gravity forces; and $f(\dot{q})$ represents friction forces of the Coulomb and viscous type. Selecting q_1 and q_2 to be joint coordinates, the matrices in Equation (3.129) are

$$
M(q) = \begin{bmatrix} m_{11} & m_{12} \\ m_{21} & m_{22} \end{bmatrix} = \begin{bmatrix} a_1 + 2a_2 \cos q_2 & a_3 + a_2 \cos q_2 \\ a_3 + a_2 \cos q_2 & a_3 \end{bmatrix},
$$

$$
C(q,\dot{q}) = \begin{bmatrix} c_{11} & c_{12} \\ c_{21} & c_{22} \end{bmatrix} = \begin{bmatrix} -2a_2 \dot{q}_2 \sin q_2 & -a_2 \dot{q}_2 \sin q_2 \\ a_2 \dot{q}_1 \sin q_2 & 0 \end{bmatrix},
$$

$$
D(q) = \begin{bmatrix} d_1 \\ d_2 \end{bmatrix} = \begin{bmatrix} a_4 \sin q_1 + a_5 \sin(q_1 + q_2) \\ a_5 \sin(q_1 + q_2) \end{bmatrix}, \quad f(\dot{q}) = \begin{bmatrix} f_1 \\ f_2 \end{bmatrix} = \begin{bmatrix} a_6 \dot{q}_1 + a_8 sign(\dot{q}_1) \\ a_7 \dot{q}_2 + a_9 sign(\dot{q}_2) \end{bmatrix},
$$

with $a_1 = m_1 r_1^2 + m_2 \left(l_1^2 + r_2^2 \right) + I_1 + I_2$, $a_2 = m_2 l_1 r_2$, $a_3 = m_2 r_2^2 + I_2$, $a_4 = (m_1 r_1 + m_2 l_1)g$, $a_5 = m_2 r_2 g$, $a_6 = b_1$, $a_7 = b_2$, $a_8 = f_{C1}$, $a_9 = f_{C2}$. The geometric parameters are as follows: $a_4 = (m_1 r_1 + m_2 l_1)g$ are the link masses; l_1, l_2 are the link lengths; r_1, r_2 are the distances to mass centers; I_1, I_2 are the moments of inertia; b_1, b_2 are the viscous friction coefficients; and f_{C1}, f_{C2} are the Coulomb friction coefficients.

Introducing new state variables as $x_1 = q_1, x_2 = q_2, x_3 = \dot{q}_1, x_4 = \dot{q}_2$, we may transform Equation (3.129) into

$$
\dot{x}_1 = x_3,
$$

$$
\dot{x}_2 = x_4,
$$

$$
\dot{x}_3 = \frac{m_{22}[\tau_1 - c_{11}x_3 - c_{12}x_4 - (d_1 + f_1)] + m_{12}[-\tau_2 + c_{21}x_3 + c_{22}x_4 + (d_2 + f_2)]}{m_{11}m_{22} - m_{12}m_{21}}, \quad (3.130)
$$

$$
\dot{x}_4 = \frac{m_{11}[\tau_2 - (d_2 + f_2) - c_{21}x_3 - c_{22}x_4] + m_{21}[-\tau_1 + (d_1 + f_1) + c_{11}x_3 + c_{12}x_4]}{m_{11}m_{22} - m_{12}m_{21}}.
$$

The system of Equations (3.130) is differentially flat, with the flat outputs being the link angles x_1, x_2. Therefore, it is equivalent, under the change of input coordinates, to the set of second order integrator systems:

$$
\ddot{x}_1 = v_1, \quad \ddot{x}_2 = v_2. \quad (3.131)
$$

For the linearized system (3.131), a controller can be designed, for example, a kind of a PD controller of the following form (compare to Example 3.19):

$$
v_1 = \ddot{x}_{1d} + k_{p1}(x_{1d} - x_1) + k_{d1}(\dot{x}_{1d} - \dot{x}_1),
$$
$$
v_2 = \ddot{x}_{2d} + k_{p2}(x_{2d} - x_2) + k_{d2}(\dot{x}_{2d} - \dot{x}_2), \quad (3.132)
$$

which provides an exponential convergence of the tracking error. Quantities x_{1d}, x_{2d} stand for the desired trajectory coordinates, and k_{d1}, k_{d2} and k_{p1}, k_{p2} stand for the control gains.

A multivariable system that is not static state feedback linearizable may be linearized by dynamic state feedback. When we know that a system is flat, the differential parameterization of control inputs indicates which of the inputs is to be dynamically extended and to what order.

Example 3.21

Consider the unicycle model presented in Section 2.4, in Figure 2.8. Its kinematical control model is then

$$\dot{x} = u_1 \cos \quad ,$$
$$\dot{y} = u_1 \sin \quad , \tag{3.133}$$
$$\dot{} = u_2.$$

The system is flat with the flat outputs given by the coordinates of the wheel contact point with the ground x and y. Let $x = X, y = Y$, and then we obtain from Equations (3.133)

$$= \arctan\left(\frac{\dot{Y}}{\dot{X}}\right),$$
$$u_1 = \sqrt{\dot{X}^2 + \dot{Y}^2}, \tag{3.134}$$
$$u_2 = \frac{\ddot{Y}\dot{X} - \dot{Y}\ddot{X}}{\dot{X}^2 + \dot{Y}^2}.$$

We can find from Equations (3.134) that u_1 needs a first order extension. Putting $\dot{u}_1 = v$, we obtain that

$$\begin{bmatrix} v \\ u_2 \end{bmatrix} = \begin{bmatrix} \dfrac{\dot{X}}{\sqrt{\dot{X}^2 + \dot{Y}^2}} & \dfrac{\dot{Y}}{\sqrt{\dot{X}^2 + \dot{Y}^2}} \\ \dfrac{-\dot{Y}}{\dot{X}^2 + \dot{Y}^2} & \dfrac{\dot{X}}{\dot{X}^2 + \dot{Y}^2} \end{bmatrix} \begin{bmatrix} \ddot{X} \\ \ddot{Y} \end{bmatrix}. \tag{3.135}$$

Relation (3.135) is invertible anywhere except at $\dot{X} = \dot{Y} = 0$, i.e., a singularity occurs for $u_1 = 0$. Such a singularity is common in many nonholonomic systems and is called structural. A remedy for the singularity relies upon the "time-reparameterization." For details, see Sira-Ramirez and Agrawal (2004) and Fliess et al. (1999).

Example 3.22

Consider the two-wheeled mobile platform model presented in Example 1.1. The platform wheels are powered independently, and their angular velocities are equal to ω_l and ω_r. The constraint equations (1.1) can be presented using ω_l and ω_r as

$$\dot{x} = \frac{r}{2}(\omega_l + \omega_r)\cos\ ,$$

$$\dot{y} = \frac{r}{2}(\omega_l + \omega_r)\sin\ , \qquad (3.136)$$

$$\dot{\ } = \frac{r}{2b}(\omega_r - \omega_l).$$

To derive Equations (3.136), we used relations between the forward velocity of the platform $v = \frac{r}{2}(\omega_l + \omega_r)$ and its angular velocity $\dot{\ } = \frac{r}{2b}(\omega_r - \omega_l)$.
The following input transformations lead to a simpler form of Equations (3.136):

$$u_1 = \frac{r}{2}(\omega_l + \omega_r), \quad u_2 = \frac{r}{2b}(\omega_r - \omega_l),$$

which results in

$$\dot{x} = u_1\cos\ ,$$

$$\dot{y} = u_1\sin\ , \qquad (3.137)$$

$$\dot{\ } = u_2.$$

The kinematic control model (3.137) is the same as for the unicycle. From Example 3.21, we know that the system model is flat with flat outputs being the coordinates x,y describing motion of the midpoint of the platform axis. All system variables can be parameterized in terms of x and y, and the relations are given by (3.134). However, if we would like to design a tracking controller for the platform to perform any maneuver, e.g., parking, a singularity in a control model comes out in the same way as for the unicycle model from the previous example. To remedy the situation, we may proceed as follows (Sira-Ramirez and Agrawal 2004). Multiply the expression for u_1 by dt and denote by ds the differential $ds = \sqrt{(dx)^2 + (dy)^2}$. Then,

$$u_1 dt = \sqrt{\dot{x}^2 + \dot{y}^2}\, dt = ds$$

$$u_1 = \frac{ds}{dt}. \qquad (3.138)$$

Also, we have that $ds = (\rho + b)d\ $ so

$$\frac{d\ }{ds} = 1/(\rho + b), \qquad (3.139)$$

where ρ is a distance from the center of curvature to the slowest wheel, measured in the direction of the platform axis, and $2b$ is the distance between the platform wheels as shown in Figure 1.2. From Equations (3.137) and (3.139), it follows that

$$u_2 = \frac{d}{dt} = \frac{d}{ds}\frac{ds}{dt} = \frac{1}{(\rho+b)}\frac{ds}{dt}. \tag{3.140}$$

It can be seen from (3.140) that the velocity along the path ds/dt and the inverse of the radius of curvature may play the role of new inputs free of singularities. If to define

$$u_1 = {}_1\frac{ds}{dt}, \qquad u_2 = {}_2\frac{ds}{dt}$$

and insert these new inputs into Equations (3.137), we obtain

$$\frac{dx}{dt} = {}_1\frac{ds}{dt}\cos ,$$

$$\frac{dy}{dt} = {}_1\frac{ds}{dt}\sin , \tag{3.141}$$

$$\frac{d}{dt} = {}_2\frac{ds}{dt}.$$

Eliminating the time differential dt from Equations (3.141) and dividing their both sides by ds, yields

$$\frac{dx}{ds} = {}_1\cos ,$$

$$\frac{dy}{ds} = {}_1\sin , \tag{3.142}$$

$$\frac{d}{ds} = {}_2.$$

The transformed kinematic control model (3.142) for the platform is of the same form as the "old" one given by Equations (3.137). However, in (3.142) the new "time coordinate" is s.

The matrix that relates the flat outputs and inputs ${}_1, {}_2$ is not globally invertible, i.e.,

$$\begin{bmatrix} \dfrac{dx}{ds} \\ \dfrac{dy}{ds} \end{bmatrix} = \begin{bmatrix} \cos & 0 \\ \sin & 0 \end{bmatrix}\begin{bmatrix} {}_1 \\ {}_2 \end{bmatrix}. \tag{3.143}$$

Differentiating, both sides of Equations (3.143) with respect to time, introducing a new input $v_1 = \frac{d\lambda_1}{dt}$, and a new state variable $\xi = \lambda_1$, we obtain

$$
\begin{bmatrix} \dfrac{d^2x}{ds^2} \\[2mm] \dfrac{d^2y}{ds^2} \end{bmatrix} = \begin{bmatrix} \cos\theta & -\xi\sin\theta \\ \sin\theta & \xi\cos\theta \end{bmatrix} \begin{bmatrix} v_1 \\ \lambda_2 \end{bmatrix}. \tag{3.144}
$$

Equations (3.144) relate the inputs v_1, λ_2 with the states x, y, ξ. The matrix determinant depends upon ξ, which is equal to 1 when the platform follows the prespecified path. From (3.144), it can be obtained that

$$
\begin{bmatrix} v_1 \\ \lambda_2 \end{bmatrix} = \begin{bmatrix} \cos\theta & -\xi\sin\theta \\ \sin\theta & \xi\cos\theta \end{bmatrix}^{-1} \begin{bmatrix} u_1 \\ u_2 \end{bmatrix}. \tag{3.145}
$$

Now, select $x_d(s)$ and $y_d(s)$, which are the coordinates of a reference trajectory, parameterized by the arc length s. For a control model (3.145), a controller may be designed as

$$
\begin{aligned}
u_1 &= \ddot{x}_d(s) + k_{p1}(x_d(s) - x(s)) + k_{d1}(\dot{x}_d(s) - \dot{x}(s)), \\
u_2 &= \ddot{y}_d(s) + k_{p2}(y_d(s) - y(s)) + k_{d2}(\dot{y}_d(s) - \dot{y}(s)).
\end{aligned} \tag{3.146}
$$

Example 3.23

Consider now a dynamic control model of the unicycle from Example 3.21. Its motion equations derived using the Lagrange equations with multipliers (2.91) have the form

$$
\begin{aligned}
\ddot{x} &= \lambda\sin\theta + u_1\cos\theta, \\
\ddot{y} &= -\lambda\cos\theta + u_1\sin\theta, \\
\dot{\theta} &= u_2, \\
\dot{x}\sin\theta &- \dot{y}\cos\theta = 0.
\end{aligned} \tag{3.147}
$$

Flat outputs are x and y. Changing t by the arc length s of the path in the same way as in Example 3.22, we obtain

$$
\begin{aligned}
\frac{dx}{ds} &= \cos\theta, \\
\frac{dy}{ds} &= \sin\theta, \\
\dot{s} &= u_1, \\
\rho\ddot{s} &+ \frac{d\rho}{ds}\dot{s}^2 = u_2.
\end{aligned} \tag{3.148}
$$

Notice that relations (3.148) remain valid even if both $u_1 = u_2 = 0$.

This example of a system subjected to a single nonholonomic constraint may be generalized to a mechanical system subjected to m nonholonomic constraints provided that it is fully actuated, i.e., there may be applied $(n\text{-}m)$ independent control inputs, where n is a number of configuration variables.

3.8.4 Concluding Remarks—Verifying Flatness

In Sections 3.8.1 through 3.8.3, we presented some basic theory necessary to take advantage of the flatness property for control applications. Also, we presented four examples of flat systems, and we demonstrated how to use flatness for control design. We have not presented examples of systems that are known to be non-flat. These may be found in Sira-Ramirez and Agrawal (2004).

However, the main problem for today, and the author is not aware of any solution to it, is to design a general test for verifying flatness. This problem remains open. Because it is not known whether flatness can be verified with a finite test, it is difficult to prove that a system is not flat.

In Chapter 7, we propose a new tracking control strategy that does not rely on flatness and enables tracking a variety of motions, including trajectory tracking as a peculiar case.

3.9 Other Control Design Techniques for Nonlinear Systems

The most common approach to control nonlinear systems is feedback linearization discussed in Section 3.6. The goal of feedback linearization is to develop a control input u such that it renders either the input–output linear map or results in a linearization of the full state of a system. However, it is not free from several drawbacks that motivated researchers to look for alternative or complementary control techniques. One disadvantage of feedback linearization arises when there are uncertain parameters or unknown functions in the coupling matrix. A variety of decoupling schemes are developed for various MIMO nonlinear systems using the feedback linearization approach. However, two main drawbacks are the singularity in the decoupling matrix, and the decoupling algorithm is not valid when a MIMO system is numerically ill posed or exhibits nearly singular behavior (for details, see, e.g., Godbole and Sastry 1995). Another instance where feedback linearization may be used and may cause problems is flight control. Exact input–output linearization is often used to control specific output variable sets in nonlinear flight control problems. The first fact is that the direct application of feedback linearization requires the second and third derivatives of uncertain aerodynamic coefficients and does not guarantee internal stability for non-minimum phase systems. Another difficulty related to the application of feedback linearization to flight control is that an accurate and complete aircraft dynamics is needed. It is difficult to identify accurately aerodynamic coefficients because they are nonlinear functions of several physical variables. To overcome these problems, gain scheduling with a linear H_∞ design is used; however, it can guarantee the desired performance only in the presence of small perturbations and slow variances. As a remedy, the backstepping-based controller may be applied to systems like aircraft dynamics (Harkegard and Glad 2000).

To place feedback linearization, backstepping, and sliding mode control (SMC), which we discuss in this book, in a proper context among other control techniques, the following classification can be useful:

1. Control techniques that attempt to treat a system as linear in a limited range of operation and use linear design techniques for each region: gain scheduling, adaptive control.

2. Control techniques that attempt to introduce auxiliary nonlinear feedback in such a way that a system can be treated as linear for purposes of control design: feedback linearization, Lyapunov-based methods, Lyapunov redesign, backstepping, sliding mode control.

Gain scheduling is an approach that uses a family of linear controllers, each of which provides satisfactory control for a different operating point of the system. One or more observable gains (scheduling variables) are used to determine what operating region the system is currently in and to enable an appropriate linear controller. A good example of the gain scheduling–based control is a flight control system for aircraft. The attitude and Mach number might be the scheduling variables with different linear controller parameters available for various combinations of these two variables. More details about the gain scheduling–based control can be found in Slotine and Li (1996).

Adaptive control is an excellent technique for modifying the controller parameters to compensate slow time-varying behavior or uncertainties. It does not need a priori information about the bounds on the uncertainties or time-varying parameters. Robust control can guarantee that if the changes are within the specified bounds, the control law need not be changed. Adaptive control is precisely concerned with control law changes. If to consider the aircraft control example again, as the aircraft flies, its mass decreases due to fuel consumption. One needs a control law that adapts to this change. More details about adaptive control can be found in Slotine and Li (1996) and Lewis et al. (2004).

Lyapunov-based methods let us verify stability through the Lyapunov function. Methods depend upon structures of the control system equations. An example of the use of Lyapunov-based design may be a swing-up control of a pendulum. The pendulum control through the energy shaping relies upon finding the expression for the total energy of the pendulum. Assuming some energy in the pendulum upright position, we select the Lyapunov function which is $V = \frac{1}{2}(E - E_{up})^2 \geq 0$, and then we have to find the swing strategy such that $\frac{dV}{dt} \leq 0$.

Let us discuss briefly the backstepping and sliding mode control approaches. The selection is motivated by their power in control of both holonomic and nonholonomic control systems.

3.9.1 Backstepping

Backstepping based on the Lyapunov function design governed by the LaSalle–Yoshizawa theorem is a recursive design for systems with nonlinearities not constrained by linear bounds. For nonlinear systems, backstepping was a real milestone as it appeared in the 1990s. Its use as a powerful design tool was initiated by Tsinias (1989, 1991), by Byrnes and Isidori (1989), and by Sontag and Sussmann (1988). The true potential of backstepping was discovered when it was extended to nonlinear systems with structured uncertainty. An adaptive backstepping approach that activated global stabilization in the presence of unknown parameters was developed by Kanellakopoulos (Krstic et al. 1995). For more details about the backstepping technique, see Krstic et al. (1995).

The exact controller in backstepping is calculated in the last step by using virtual control laws calculated in past steps. It may be interpreted as an addition of an integrator after each step. Comparing backstepping with feedback linearization, it offers more flexibility in dealing with nonlinearities. Specifically, using Lyapunov functions, their influence upon a system performance may be analyzed, and the "stabilizing nonlinearities" may be kept while the others can be canceled or dominated by a control signal. Let us recall that feedback linearization cancels nonlinearities so that the closed-loop system is rendered linear. The consequence of not canceling all nonlinearities means that the control law may be of a simpler form compared to one developed within feedback linearization.

To apply backstepping, a system dynamics needs to have a strict feedback structure, i.e.,

$$\dot{x}_1 = f_1(x_1, x_2),$$
$$\vdots$$
$$\dot{x}_{n-1} = f_{n-1}(x_1, ..., x_{n-1}, x_n),$$
$$\dot{x}_n = f_n(x_1, ..., x_n, u). \tag{3.149}$$

This structure is needed due to the way in which a backstepping control law $u(x)$ is recursively constructed with the goal of steering x to the origin.

The backstepping procedure design is as follows. First, x_2 is considered virtual control of the scalar subsystem $\dot{x}_1$. A stabilizing function $\phi_1(x_1)$ can be determined such that $\dot{x}_1 = f_1(x_1, \phi_1(x_1))$ has the desired properties. However, x_2 is not available for control. Then, backstepping offers a constructive way of forwarding the unattainable control demand $x_2 = \phi_1(x_1)$ to a new virtual control $x_3 = \phi_2(x_1, x_2)$. If it can be satisfied, x_1 and x_2 would be brought to the origin. This recursive procedure is repeated until the actual control variable u is reached after n steps, which means that the control law $u = \phi_{n-1}(x)$ is constructed. A Lyapunov function is designed as well as the control law to prove the stability of the closed-loop system.

Example 3.24

Consider a system

$$\dot{x}_1 = f_1(x_1) + g(x_1)x_2,$$

$$\dot{x}_2 = u. \tag{3.150}$$

Our goal is to design a state feedback $u(x)$ that stabilizes (3.150) at $x = 0$ with $f(0) = 0$. Suppose that the partial system

$$\dot{x}_1 = f_1(x_1) + g(x_1)v_1$$

can be stabilized by $v_1 = \phi(x_1)$, and we can select a Lyapunov function candidate $V_1(x_1)$ such that

$$\dot{V}_1(x_1) = \frac{dV_1}{dx_1}(f(x_1) + g(x_1)\phi(x_1)) \leq -W(x_1)$$

for some positive definite function W.
Equations (3.150) can be written as

$$\dot{x}_1 = f_1(x_1) + g(x_1)\phi(x_1) + g(x_1)[x_2 - \phi(x_1)],$$

$$\dot{x}_2 = u. \tag{3.151}$$

Introduce a new state variable $\xi = x_2 - \phi(x_1)$ and a control variable $v = u - \dot{\phi}$. Then, Equations (3.151) take the form

$$\dot{x}_1 = f_1(x_1) + g(x_1)\phi(x_1) + g(x_1)\xi,$$

$$\dot{\xi} = v, \tag{3.152}$$

where

$$\dot{\phi}(x_1) = \frac{d\phi}{dx_1}\dot{x}_1 = \frac{d\phi}{dx_1}[f_1(x_1) + g(x_1)x_2].$$

Select now a new Lyapunov function candidate $V_2(x_1, x_2)$ such that

$$V_2(x_1, x_2) = V_1(x_1) + \xi^2/2.$$

Its time derivative yields

$$\dot{V}_2(x_1, x_2) = \frac{dV_1}{dx_1}(f(x_1) + g(x_1)\phi(x_1)) + \frac{dV_1}{dx_1}g(x_1)\xi + \xi v \leq -W(x_1) + \frac{dV_1}{dx_1}g(x_1)\xi + \xi v. \quad (3.153)$$

Selecting

$$v = -\frac{dV_1}{dx_1}g(x_1) - k\xi, k > 0,$$

yields that Equation (3.153) satisfies

$$\dot{V}_2(x_1, x_2) \leq -W(x_1) - k\xi^2. \tag{3.154}$$

Based on (3.154), we may conclude that $x = 0$ is asymptotically stable for Equations (3.150) with the control law $u(x) = \dot{\phi}(x) + v(x)$. Additionally, if $V_1(x_1)$ is radially unbounded, then the system of Equations (3.150) is globally stable.

The backstepping approach may be summarized in the "backstepping lemma."

Lemma 3.4

Let $z = (x_1, ..., x_{k-1})^T$ and

$$\dot{z} = f(z) + g(z)x_k,$$
$$\dot{x}_k = u. \tag{3.155}$$

Assume that $\phi(0) = 0$, $f(0) = 0$, and

$$\dot{z} = f(z) + g(z)\phi(z)$$

is stable, and $V(z)$ is a Lyapunov function candidate with $\dot{V}(z) \leq -W(z)$. Then, the control u can be selected as

$$u = \frac{d\phi}{dz}[f(z) + g(z)x_k] - \frac{dV}{dz}g(z) - (x_k - \phi(z)). \tag{3.156}$$

It stabilizes $x = 0$ with the Lyapunov function selected as

$$V_1(z) = V(z) + (x_k - \phi(z))^2/2.$$

The backstepping lemma can be applied to stabilize systems in the form of Equations (3.149). It can be applied recursively to a system of the state-space control form $\dot{x} = f(x) + g(x)u$ as discussed in Chapter 3, which is in a strict feedback form. Backstepping generates stabilizing feedbacks $\phi_k(x_1,...,x_k)$ (equivalent to u in Lemma 3.4) and Lyapunov functions of the forms $V_k(x_1,...,x_k) = V_{k-1}(x_1,...,x_{k-1}) + (x_k - \phi_{k-1})^2/2$ by stepping back from x_1 to u. The backstepping approach results then in the final state feedback controller $u = \phi_k(x_1,...,x_k)$.

Main advantages of backstepping may be summarized as follows:

1. Global stability can be achieved.
2. The transient performance can be guaranteed and explicitly analyzed.
3. Cancellations of "useful" nonlinearities can be avoided, compared to feedback linearization.

Example 3.25

Consider a control system

$$\dot{x}_1 = x_1^2 + x_2,$$
$$\dot{x}_2 = x_3, \tag{3.157}$$
$$\dot{x}_3 = u$$

and design a backstepping controller u. The design steps are as follows. In the first step, verify the strict feedback form of the system (3.157). In the second step, propose first a subsystem:

$$\dot{x}_1 = x_1^2 + \phi(x_1),$$
$$\dot{x}_2 = u_1 \tag{3.158}$$

where $\phi(x_1) = -x_1^2 - x_1$ stabilizes the first equation. Selecting $V_1(x_1) = x_1^2/2$ and based on Lemma 3.4, we may find that

$$u_1 = \phi_2(x_1, x_2) = (-2x_1 - 1)\left(x_1^2 + x_2\right) - x_1 - \left(x_2 + x_1^2 + x_1\right),$$

$$V_2(x_1, x_2) = x_1^2/2 + \left(x_2 + x_1^2 + x_1\right)^2/2.$$

In the third step, apply Lemma 3.4 to Equations (3.157) to obtain

$$u = u_2 = \frac{d\phi_2}{dz}[f(z) + g(z)x_k] - \frac{dV_2}{dz}g(z) - (x_k - \phi_2(z))$$

$$= \frac{d\phi_2}{dx_1}\left(x_1^2 + x_2\right) + \frac{d\phi_2}{dx_2}x_3 - \frac{dV_2}{dx_2} - (x_3 - \phi_2(x_1, x_2)). \tag{3.159}$$

The controller (3.159) globally stabilizes the system of Equations (3.157).

3.9.2 Sliding Mode Control

Sliding mode control (SMC) is a type of variable structure control where the dynamics of a nonlinear system is altered via an application of a high-frequency switching control. The time-varying state feedback control scheme involves two steps:

1. Selecting a hypersurface or a manifold such that the system trajectory exhibits desirable behavior when confined to this manifold.
2. Finding feedback gains so that the trajectory intersects and stays on the manifold.

Main advantages of the SMC may be summarized as follows:

1. Controllers with switching elements are easy to realize.
2. Outputs of the controller can be restricted to a certain region of operation.
3. The controllers can also be used inside the region of operation to ensure the closed-loop dynamics.

Historically, the variable structure control (VSC) was first proposed and elaborated in the early 1950s by Emelyanov and his co-researchers. The most distinguished feature of VSC is its ability to result in very robust control systems; the system is completely insensitive to parametric uncertainty and external disturbances or "invariants." The SMC is the major mode of operation in variable structure systems.

The SMC technique became popular because most of the real-life processes in mechanical, electrical, and aerospace engineering, and other areas when characterized by differential equations have discontinuity. The discontinuity is due to certain peculiarities in system behavior. The simplest case is the Coulomb friction in mechanical systems, which is not defined in points where velocity equals zero. If such discontinuities are deliberately introduced on certain surfaces in a system state-space, then motions in a sliding mode may occur in the system. The discontinuous nature of the control action in the feedback channels results in switching between two distinctively different system structures such that a new type of system motion, called the sliding mode, exists in a manifold. This results in superb system performance, which includes insensitivity to parameter variations and a complete rejection of disturbances. Because the variations of dynamic characteristics in control systems cause a significant problem in control, discontinuous control systems provide an effective tool for solving such control problems.

The first step in the SMC design consists in finding a switching control law such that it can drive a system state trajectory to some surface called a sliding surface or a sliding manifold. The Lyapunov approach is used to design the switching controller.

A Lyapunov control function, which is a generalized Lyapunov function, is defined in terms of the surface. It characterizes motion of the state trajectory to the sliding manifold. For each selected switched control surface, one can select the "gains" such that the derivative of the Lyapunov function is negative definite, i.e., it generates motion of the state trajectory to the sliding manifold. When the manifold is defined, a switched controller is designed such that the state is driven to the sliding manifold and maintained on it. The resulting closed-loop system is discontinuous.

Consider a single input nonlinear system

$$q^{(n)} = f(q,t) + b(q,t)u(t), \tag{3.160}$$

where $q(t)$ is a state vector, $u(t)$ is a control input, and $f(q,t)$ and $b(q,t)$ may be nonlinear functions of state and time, and they may not be exactly known but bounded by a known continuous function of q.

The control problem for Equation (3.160) is to obtain the state $q(t)$ to track a desired $q_d(t)$ in the presence of model uncertainties. We may define a time-varying surface $S(t)$ in R^n by a scalar equation

$$s(q,t) = \left(\frac{d}{dt} + \delta\right)^{n-1} \tilde{q}(t), \tag{3.161}$$

where $\tilde{q}(t) = q(t) - q_d(t)$ and δ is a strict positive constant that plays the role of the system bandwidth. With Equation (3.161), the tracking problem of n-dimensional q is replaced by the first order stabilization problem for s. The surface $S(t)$ is a sliding surface. The system's behavior on the surface is referred to as a sliding regime. It can be seen from Equation (3.161) that one differentiation of s is enough to get u explicitly. Bounds on s can be translated on $\tilde{q}(t)$ so the scalar s can be a tracking performance measure. When $\tilde{q}(0) = 0$, initial conditions are zero, the transformations of performance measures are

$$\forall t \geq 0, \quad |s(t)| \leq \quad \text{then } \forall t \geq 0, \quad |\tilde{q}^{(i)}(t)| \leq (2\delta)^{(i)}\varepsilon, \tag{3.162}$$

where $\varepsilon = /\delta^{(n-1)}$, $i = 1, \dots, n-1$. In the case when $\tilde{q}(0) \neq 0$ bounds (3.162) are obtained asymptotically.

Example 3.26

Consider the second order system $\ddot{x} = f(x) + g(x)u$ that can be presented as

$$\dot{x}_1 = x_2,$$
$$\dot{x}_2 = f(x) + g(x)u, \tag{3.163}$$

where $f(x)$ and $g(x)$ are nonlinear functions and are not necessarily continuous, $g(x) > 0$. Note that x_1 is stable if $\dot{x}_1 = -ax_1$, $a > 0$. Define the coordinate with respect to the stable manifold

$$s = x_2 + ax_1 \tag{3.164}$$

so that

$$\dot{x}_1 = x_2 = -ax_1 + s. \tag{3.165}$$

Equation (3.165) is stable if $s = 0$. The time derivative of Equation (3.164) is

$$\dot{s} = \dot{x}_2 + a\dot{x}_1 = f(x) + g(x)u + ax_2. \tag{3.166}$$

To verify stability, select the Lyapunov function candidate to be $V = \frac{1}{2}s^2$. Then,

$$\dot{V} = s\dot{s} = s[f(x) + g(x)u + ax_2]$$

and $\dot{V}$ is negative definite when

$$f(x) + g(x)u + ax_2 \begin{cases} < 0 & \text{for} \quad s > 0 \\ = 0 & \text{for} \quad s = 0 \\ > 0 & \text{for} \quad s < 0 \end{cases} \tag{3.167}$$

The control input u that ensures stability may be selected as

$$u \begin{cases} < u_1(x) & \text{for} \quad s > 0 \\ = u_1(x) & \text{for} \quad s = 0 \\ > u_1(x) & \text{for} \quad s < 0 \end{cases} \tag{3.168}$$

and $u_1 = \frac{-f(x) - ax_2}{g(x)}$. Based on relations (3.168), the control law u may be selected as

$$u = u_1 - k\,\text{sign}(s), \qquad k > 0. \tag{3.169}$$

Problems

1. The controller (Equation 3.146) is designed in the "changed" time variable. Present it in the original time domain t.
2. Derive the equations of motion for the unicycle in Example 3.23 and design a feedback controller for the parameterized equations (3.148).

3. Design a feedback controller for the unicycle in Example 3.23 and present it in the original time domain *t*.
4. Design a sliding mode controller, based on the theoretical Example 3.26, for the pendulum dynamics from Example 3.8. Add friction to the pendulum motion.

References

Arimoto, S. 1990. Design of robot control systems. *Adv. Robotics* 4:79–91.

Armstrong-Helouvry, B. 1991. *Control of machines with friction.* New York: Kluwer Academic.

Astolfi, A. 1996. Discontinuous control of nonholonomic systems. *Syst. Contr. Lett.* 27:37–45.

Bayard, D., and J. Wen. 1988. New class of control laws for robotic manipulators: Adaptive case. *Int. J. Contr.* 47:1387–1406.

Berghuis, H. 1993. Model-based robot control: From theory to practice. PhD diss., University of Twente, Netherlands.

Bloch, A.M., and N.H. McClamroch. 1989. Control of mechanical systems with classical nonholonomic constraints. In *Proc. IEEE Conf. Decision Contr.*, 201–205.

Bloch, A.M., and N.H. McClamroch. 1990. Controllability and stabilizability properties of a nonholonomic control system. In *Proc. IEEE Conf. Decision Contr.*, 268–273.

Bloch, A.M., M. Reyhanoglu, and N.H. McClamroch. 1992. Controllability and stabilization of nonholonomic dynamic systems. *IEEE Trans. Automat. Contr.* 37(11): 1746–1757.

Bloch, A.M., P.S. Krishnaprasad, J.E. Marsden, and R.M. Murray. 1994. Nonholonomic mechanical systems with symmetry. *Technical Report CIT/CDS 94-013*, California Institute of Technology, Pasadena.

Bloch, A.M., and P.E. Crough. 1995. Nonholonomic control systems on Riemannian manifolds. *SIAM J. Contr.* 33(1):126–148.

Bloch, A.M. 2003. *Nonholonomic mechanics and control.* New York: Springer-Verlag.

Brockett, R.W. 1983. Asymptotic stability and feedback stabilization. In *Differential geometric control theory*, ed. R.W. Brockett, R.S. Millman, and H.J. Sussmann. Boston, MA: Birkhauser, 181–194.

Bushnell, L., D. Tilbury, and S. Sastry. 1995. Steering three-input nonholonomic systems: The fire truck example. *Int. J. Robot. Res.* 14:366–381.

Byrnes, C.I., and A. Isidori. 1989. New results and examples in nonlinear feedback stabilization. *Syst. Contr. Lett.* 12:437–442.

Campion, G., B. d'Andrea-Novel, and G. Bastin. 1991. Controllability and state feedback stabilizability of nonholonomic mechanical systems. In *Advanced robot control*, ed. C.Canudas-de-Wit, 106–124. LNCIS 162. New York: Springer.

Canudas-de-Wit, C., and O.J. Sordalen. 1992. Exponential stabilization of mobile robots with nonholonomic constraints. *IEEE Trans. Automat. Contr.* 37(1):1791–1797.

Canudas-de-Wit, C., H. Khennouf, C. Samson, and O. Sordalen. 1993. Nonlinear control design for mobile robots. In *Recent trends in mobile robotics*, ed. Y. F. Zheng,

Vol. 11, 121–156. World scientific series in robotics and automated systems. River Edge, NJ: World Scientific.

Chang, Y.-Ch., and B.-S. Chen. 1996. Adaptive tracking control design of nonholonomic mechanical systems. In *Proc. IEEE Conf. Decision Contr.* 4739–4744. Kobe, Japan.

Chelouah, A. 1997. Extensions of differential flat fields and Liouvillian systems. In *Proc. IEEE Conf. Decision Contr.* 4268–4273. San Diego, CA.

Chen, B.-S., T.-S. Lee, and W.-S. Chang. 1996. A robust Hμ model reference tracking design for nohnolonomic mechanical control systems. *Int. J. Contr.* 63(2):283–306.

Cheng, M.-P., and Ch.-Ch. Tsai. 2003. Dynamic modeling and tracking control of a nonholonomic wheeled mobile manipulator with two robotic arms. In *Proc. IEEE Conf. Decision Contr.* 2932–2936. Maui, Hawaii.

Colbaugh, J.R., E. Barany, and M. Trabatti. 1999. Control of nonholonomic mechanical systems using reduction and adaptation. *Robotica* 17:249–260.

Coron, J.-M., and J.-B. Pomet. 1992. A remark on the design of time-varying stabilizing feedback laws for controllable systems without drift. In *Proc. NOLCOS Symp.* 413–417. Bordeaux, France.

Crawford, L.S., and S.S. Sastry. 1995. Biological motor control approaches for a planar diver. In *Proc. Conf. Decision Contr.* 3881–3886.

de Jalon, J.G. and E. Bayo. 1994. *Kinematic and dynamic simulation of multibody systems*. Mech. Eng. Series, Berlin: Springer–Verlag.

De Luca, A., G. Oriolo, L. Paone, and P.R. Giordano. 2002. Experiments in visual feedback control of a wheeled mobile robot. In *Proc. IEEE Conf. Robot. Automat.* 2073–2078. Washington DC.

De Luca, A., G. Oriolo, and C. Samson. 1998. Feedback control of a nonholonomic car-like robot. In *Robot Motion Planning and Control*, ed. J.-P. Laumond, 171–253. London: Springer.

del Rio F., G. Jimenez, J.L. Sevillano, C. Amaya, and A.C. Ballcells, 2002. A new method for tracking memorized path: applications to unicycle robots. In *Proc. MED 2002*, Lisbon, Portugal.

Encarnacao, P., and A. Pascoal. 2000. 3D path following for autonomous underwater vehicles. In *Proc. 39th IEEE Conf. Decision Contr.* Sydney, Australia.

Fliess, M., J. Levine, P. Martin, and P. Rouchon. 1995. Flatness and defect of nonlinear systems: Introductory theory and examples. *Int. J. Contr.* 61:1327–1361.

Fliess, M., J. Levine, Ph. Martin, and P. Rouchon. 1992. Sur les systems lineairies differentiellement plats. C.R. Acad. Sci. Paris I:315–619.

Fliess, M., J. Levine, Ph. Martin and P. Rouchon. 1999. A Lie-Backlund approach to equivalence and flatness. *IEEE Trans. Automat. Contr.* 44(5):922–937.

Fossen, T.I. 1994. *Guidance and control of ocean vehicles*. New York: John Wiley & Sons.

Fukao, T., H. Nakagawa, and N. Adachi. 2000. Adaptive tracking control of a nonholonomic mobile robot. *IEEE Trans. Robot. Automat.* 16(5):609–615.

Getz, N. 1994. Control of balance for a nonlinear nonholonomic non-minimum phase model of a bicycle. In *Proc. Am. Control Conf.* 148–151. Baltimore, MD, USA.

Godbole, D.N., and S.S. Sastry. 1995. Approximate decoupling and asymptotic tracking for MIMO systems. *IEEE Trans. Automat. Contr.* 40(3).

Gutowski, R., and B. Radziszewski. 1969. The behavior of the solutions of equations of motion of the mechanical system with program constraints. *Bull. Acad. Polon. Sci. Ser. Sci. Tech.* 17(2):17–25.

Harkegard, O., and Torkel Glad, S. 2000. A backstepping design for flight path angle control. In *Proc. IEEE Conf. Decision Contr.* 3570–3575. Sydney, Australia.

Hespanha, J.P., and A.S. Morse. 1999. Stabilization of nonholonomic integrators via logic-based switching. *Automatica* 35(3):385–393.

Isidori, A. 1989. *Nonlinear control systems*. 2nd ed. Berlin: Springer.

Jarzębowska, E. 2002. On derivation of motion equations for systems with nonholonomic high-order program constraints. *Multibody Syst. Dyn.* 7(3):307–329.

Jarzębowska Jarzębowska, E. 2005. Dynamics modeling of nonholonomic mechanical systems: Theory and applications. *Nonlin. Anal.* 63(5–7):185–197.

Jarzębowska, E. 2006. Control oriented dynamic formulation of robotic systems with program constraints. *Robotica* 24(1):61–73.

Jarzębowska, E. 2007. *Model-based tracking control strategies for constrained mechanical systems*. Palermo: International Society for Advanced Research.

Jarzębowska, E. 2008. Advanced programmed motion tracking control of nonholonomic mechanical systems. *IEEE Trans. Robot.* 24(6):1315–1328.

Jarzębowska, E., and J. Greenberg. 2000. On an improved vehicle steering system friction force model. Techn. Report No. AJ 441, *Ford Technical Journal*.

Jarzębowska, E., and N.H. McClamroch. 2000. On nonlinear control of the Ishlinsky problem as an example of a nonholonomic non-Chaplygin system. In *Proc. Am. Control Conf.* 3249–3253. Chicago, IL.

Jiang, Z., and H. Nijmeijer. 1999. A recursive technique for tracking control of nonholonomic systems in chained form. *IEEE Trans. Robot. Automat.* 44(2):265–279.

Kane, T.R., and D.L. Levinson. 1996. The use of Kane's dynamical equations in robotics. *Int. J. Robot. Res.* 2(3):3–21.

Khalil, H. *Nonlinear systems*. 2001. Englewood Cliffs, NJ: Prentice Hall.

Kim, M.S., J.H. Shin, and J.J. Lee. 2000. Design of a robust adaptive controller for a mobile robot. In *Proc. IEEE/RSJ Int. Conf. Intell. Robot. Syst.* 1816–1821.

Koh, K.C., and H.S. Cho. 1999. A Smooth path tracking algorithm for wheeled mobile robots with dynamic constraints. *J. Intell. Robot. Syst.* 24:367–385.

Kokotovic, P.V., and H. Sussmann. 1989. A positive real condition for global stabilization of nonlinear systems. *Syst. Contr. Lett.* 19:177–185.

Kolmanovsky, I., M. Reyhanoglu, and N.H. McClamroch. 1994. Discontinuous feedback stabilization of nonholonomic systems in extended power form. In *Proc. IEEE Conf. Decision Contr.* 3469–3474.

Kolmanovsky, I., and N.H. McClamroch. 1995. Developments in nonholonomic control problems. *IEEE Contr. Syst. Mag.* 15:20–36.

Krstic, M., I. Kanellakopoulos, and P. Kokotovic. 1995. *Nonlinear and adaptive control design*. New York: John Wiley & Sons.

Kwatny, H.G., and G.L. Blankenship. 2000. *Nonlinear control and analytical mechanics, a computational approach*. Boston: Birkhauser.

Lafferiere, G., and H. Sussmann. 1991. Motion planning for controllable systems without drift. In *Proc. IEEE Int. Conf. Robot. Automat.* 1148–1153.

Laiou, M.C., and A. Astolfi. 1999. Quasi-smooth control of chained systems. In *Proc. Am. Control Conf.* 3940–3944. San Diego, CA.

LaSalle, P., and S. Liefschetz. 1961. *Stability by Lyapunov's direct method*. New York: Academic Press.

Laumond, J.-P., and J.-J. Risler. 1996. Nonholonomic systems: Controllability and complexity. *Theoretical Comp. Sci.* 157:101–114.

Levine, J. 2009. *Analysis and control of nonlinear systems: A flatness-based approach*. Dordrecht: Springer.

Lewis, F., D.M. Dowson, and Ch.T. Abdallach. 2004. *Robot manipulator control. Theory and practice*. 2nd ed. New York: Marcel Dekker.

Lucibello, P., and G. Oriolo. 2001. Robust stabilization via iterative state steering with an application to chained-form systems. *Automatica* 37(1):71–79.

M'Closkey, T.R., and R.M. Murray. 1994. Extending exponential stabilizers for nonholonomic systems from kinematic controllers to dynamic controllers. In *Proc. IFAC Symp. Robot Control*, 249–256, Italy.

M'Closkey, T.R., and R.M. Murray. 1997. Exponential stabilization of driftless nonlinear control systems using homogeneous feedback. *IEEE Tran. Automat. Contr.* 42(8):614–628.

Martin, Ph., R.M. Murray, and P. Rouchon. 1997. Flat systems. In *Plenary Lectures and Minicourses of the 1997 European Control Conference*, ch. 8, ed. G. Bastin, and M. Gevers, 211–264. Brussels: Belgium.

Martin, Ph., R.M. Murray, and P. Rouchon. 2003. Flat systems, equivalence and trajectory generation. *CDS Technical Rep., CDS 2003-2008*. California Institute of Technology, Pasadena.

Micaelli, C., and C. Samson. 1993. Trajectory tracking for unicycle type and two-steering-wheels mobile robots. *Technical Report No. 2097*, INRIA, Sophia-Antipolis, France.

Morin, P., J.-B. Pomet, and C. Samson. 1998. Developments in time-varying feedback stabilization of nonlinear systems. In *Proc. IFAC Symp. Nonl. Contr. Syst. Design* 587–592.

Morin, P., and C. Samson 2003. Practical stabilization of driftless systems on Lie groups: The transverse function approach. *IEEE Trans. Automat. Contr.* 48(9):1496–1508.

Murray, R.M., Z.X. Li, and S.S. Sastry. 1994. *A mathematical introduction to robotic manipulation*. Boca Raton, FL: CRC Press.

Nijmeijer, H., and A. van der Schaft. 1990. *Nonlinear dynamical control systems*. New York: Springer-Verlag.

Nikravesh, P. 1988. *Computer aided analysis of mechanical systems*. Englewood Cliffs, NJ: Prentice Hall.

Oriolo, G., and Y. Nakamura. 1991. Control of mechanical systems with second order nonholonomic constraints: Underactuated manipulators. In *Proc. Conf. Decision Contr.* 2398–2403. Brighton, England.

Oriolo, G., S. Panzieri, and G. Ulivi. 1998. An iterative learning controller for nonholonomic robots. *Int. J. Robot. Res.* 17(9):954–970.

Oriolo, G., A. De Luca, and M. Vendittelli. 2002. WMR control via dynamic feedback linearization: Design, implementation, and experimental validation. *IEEE Trans. Contr. Systems Techn.* 10(6):835–852.

Ortega, R., and M.W. Spong. 1989. Adaptive motion control of rigid robots: A tutorial. *Automatica* 25:877–888.

Pandian, S.R., and M. Hanmandlu. 1993. Adaptive generalized model-based control of robot manipulators. *Int. J. Contr.* 58(4):835–852.

Papadopoulos, E.G., and G.C. Chasparis. 2002. Analysis and model-based control of servomechanisms with friction. In *Proc. IEEE/RSJ Int. Conf. Intell. Robot. Syst.* 68–73. Lausanne: Switzerland.

Pomet, J.-B. 1992. Explicit design of time-varying stabilizing control laws for a class of controllable systems without drift. *Syst. Contr. Lett.* 18:147–158.

Pomet, J.-B. 1995. A differential geometric setting for dynamic equivalence and dynamic linearization. In *Geometry in nonlinear control and differential inclusions*, ed. B. Jakubczyk, W. Respondek, and T. Rzeżuchowski, no. 32. Institute of Mathematics, Polish Academy of Science. Warsaw: Banach Center Publications.

Reyhanoglu, M., A. van der Schaft, N.H. McClamroch, and I. Kolmanovsky. 1996. Nonlinear control of a class of underactuated systems. In *Proc. 35th Conf. Decision Contr.* 1682–1687. Kobe: Japan.

Rouchon, P., P. Martin, and R.M. Murray. 2003. Flat systems, equivalence and trajectory generation. *CDS Technical Report*, California Institute of Technology, Pasadena.

Samson, C. 1990. Velocity and torque feedback control of a nonholonomic cart. In *Int. Workshop in Adaptive and Nonlinear Control.* Issues in Robotics.

Samson, C., and K. Ait-Abderrahim. 1991a. Feedback control of a nonholonomic wheeled cart in Cartesian space. In *Proc. IEEE Conf. Robot. Automat.* 1136–1141. Sacramento: CA.

Samson, C., and K. Ait-Abderrahim. 1991b. Mobile robot control part 1: Feedback control of a nonholonomic mobile robot. *Technical Report no. 1281, INRIA*, Sophia-Antipolis: France.

Samson, C. 1993. Time-varying feedback stabilization of car-like wheeled mobile robots. *Int. J. Robot. Res.* 12(1):55–64.

Samson, C. 1995. Control of chained systems: Application to path following and time-varying point stabilization of mobile robots. *IEEE Trans. Automat. Contr.* 40:64–77.

Sarkar, N., X. Yun, and V. Kumar. 1993. Dynamic path following: A new control algorithm for mobile robots. In *Proc. 32nd IEEE Conf. Decision Contr.* 2670–2675.

Sarkar, N., X. Yun, and V. Kumar. 1994. Control of mechanical systems with rolling constraints: Application to dynamic control of mobile robots. *Int. J. Robot. Res.* 13(1):55–69.

Scheuer, A., and Ch. Laugier. 1998. Planning sub-optimal and continuous-curvature paths for car-like robots. In *Proc. IEEE/RSJ Int. Conf. Intell. Robot. Syst.*, 25–31.

Shekl, A.M., and V.L. Lumelsky. 1998. Motion planning for nonholonomic robots in a limited workspace. In *Proc. IEEE/RSJ Int. Conf. Intell. Robot. Syst.* 1473–1478.

Sira-Ramirez, H., R. Castro-Linares, and E. Liceago-Castro. 2000. A liouvillian systems approach for the trajectory planning-based control of helicopter models. *Int. J. Robust Nonl. Contr.* 10:301–320.

Sira-Ramirez, H., and S. Agrawal. 2004. *Differentially flat systems.* New York: Marcel Dekker.

Slotine, J.J., and W. Li. 1989. Composite adaptive control of robot manipulators. *Automatica* 25:509–519.

Slotine, J.J., and W. Li. 1996. *Applied nonlinear control.* Englewood Cliffs, NJ: Prentice Hall.

Sontag, E.D. 1990. *Mathematical control theory.* New York: Springer.

Sontag, E.D. 2000. The ISS philosophy as a unifying framework for stability-like behavior. In *Nonlinear Control in 2000, Conference Proc.* 1:443–465. Paris.

Sontag, E.D., and H. Sussmann. 1988. Further comments on the stabilizability of the angular velocity of a rigid body. *Syst. Contr. Lett.* 42:437–442.

Sordalen, O.J., and O. Egeland. 1995. Exponential stabilization of nonholonomic chained systems. *IEEE Trans. Automat. Contr.* 40(1):35–49.

Spong, M.W., and M. Vidyasagar. 1989. *Robot control and dynamics.* New York: Wiley.

Sussmann, H.J. 1987. A general theorem on local controllability. *SIAM J. Contr. Optim.* 25:158–194.

Szymański, M. 2011. *Control of a mobile robot Pioneer 3-DX*. B.S. diss., Warsaw University of Technology, Warsaw, Poland.

Tanner, H.G., and K.G. Kyriakopulos. 2001. Mobile manipulator modeling with Kane's approach. *Robotica* 19:675–690.

Tanner, H.G. 2003. ISS properties of nonholonomic mobile robots. In *Proc. IEEE/RSJ Int. Conf. Intell. Robot. Syst.* 3799–3804. Las Vegas, NV.

Tian, Y.-P., and S. Li. 1999. Smooth time-varying exponential stabilization of nonholonomic systems. In *Proc. IEEE Conf. Decision Contr.* 1912–1917. Sydney: Australia.

Tian, Y.-P., and S. Li. 2002. Exponential stabilization of nonholonomic dynamic systems by smooth time-varying control. *Automatica* 38:1139–1146.

Tsinias, J. 1989. Sufficient Lyapunov-like conditions for stabilization. *Math. Contr. Sign. Syst.* 2:343–357.

Tsinias, J. 1991. Existence of control Lyapunov functions and applications to state feedback stabilizability of nonlinear systems. *SIAM J. Contr. Optim.* 29:457–473.

Vafa, Z. 1991. Space manipulator motion with no satellite attitude disturbances. In *Proc. IEEE Int. Conf. Robot. Automat.* 1770–1775.

van Nieuwstadt, M.J. 1997. Trajectory generation for nonlinear control systems. *Techn. Report CDS-96-011*, Department of Mechanical Engineering, California Institute of Technology, Pasadena.

van Nieuwstadt, M.J., M. Rathinam, and R.M. Murray. 1998. Differential flatness and absolute equivalence. *SIAM J. Contr. Optimization* 36(4):1225–1239.

Vidyasagar, M. 1992. *Nonlinear systems analysis*. Englewood Cliffs, NJ: Prentice Hall.

Walsh, G., D. Tilbury, S. Sastry, R.M. Murray, and J.P. Laumond. 1994. Stabilization of trajectories for systems with nonholonomic constraints. *IEEE Trans. Automat. Contr.* 39(1):216–222.

Wen, J., and D.S. Bayard. 1988. A new class of control laws for robotic manipulators. Part 1: Non-adaptive case. *Int. J. Contr.* 47(8):1361–1385.

Wen, J. and K. Kreutz-Delgado. 1991. The attitude control problem. *IEEE Trans. Automat.Contr.*, 36:1148–1162.

Whitcomb, L.L., A.A. Rizzi, and D.E. Koditchek. 1991. Comparative experiments with a new adaptive controller for robot arms. In *Proc. IEEE Conf. Robot. Automat.* 2–7. Sacramento, CA.

Wooten, W.L., and J.K. Hodgins. 1995. Simulation of human diving. In *Proc. Graphics Interface* 1–9. Doex. Quebec City, Quebec, Canada.

Yang, J.-M., and J.-H. Kim. 1999. Sliding mode control for trajectory tracking of nonholonomic wheeled mobile robots. *IEEE Trans. Robot. Automat.* 15(3):578–587.

Yazdapanah, M.J., and G.H. Khosrowshahi. 2003. Robust control of mobile robots using the computed torque plus Hμ compensation method. In *Proc. Conf. Decision Contr.* 2920–2925. Maui: HI.

Ye-Hwa. 1988. Second order constraints for equations of motion of constrained systems. *IEEE/ASME Trans. of Mechatr.* 3(3):240–248.

Yun, X., and N. Sarkar. 1998. Unified formulation of robotic systems with holonomic and nonholonomic constraints. *IEEE Trans. Robot. Automat.* 14(4):640–650.

Zotov, Yu.K., and A.V. Tomofeyev. 1992. Controllability and stabilization of programmed motions of reversible mechanical and electromechanical systems. *J. Appl. Math. Mech.* (trans. from Russian), 56(6):873–880.

Zotov, Yu.K. 2003. Controllability and stabilization of programmed motions of an automobile-type transport robot. *J. Appl. Maths. Mech.* (trans. from Russian), 67(3):303–327.

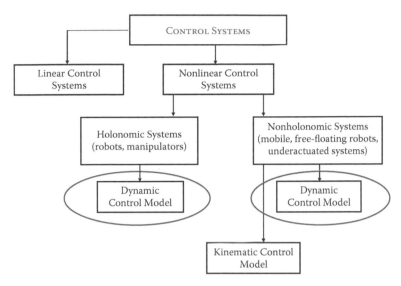

COLOR FIGURE 1.1
The scope of the book—dynamic control models for nonlinear systems.

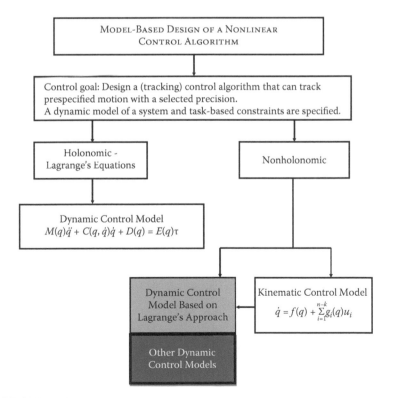

COLOR FIGURE 1.13
Model-based control design for nonlinear systems.

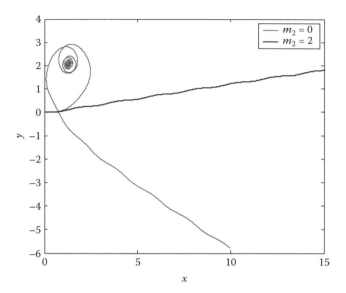

COLOR FIGURE 2.6
Motion patterns of the uncontrolled motion of the roller-racer.

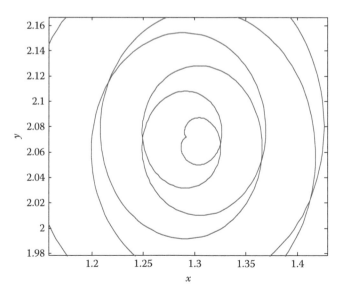

COLOR FIGURE 2.7
Magnification of the roller-racer motion with $m_2 = 0$.

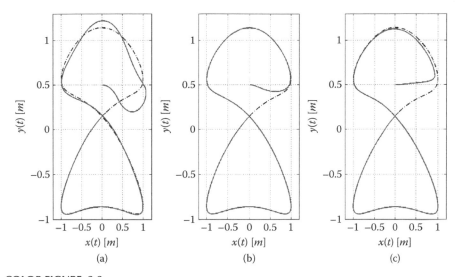

COLOR FIGURE. 3.3
Tracking a desired trajectory by the Samson algorithm for $k_2 = 1$ and different k_1: (a) $k_1 = 0.1$, (b) $k_1 = 0.5$, and (c) $k_1 = 1$.

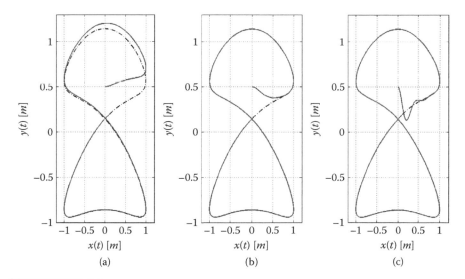

COLOR FIGURE 3.4
Tracking a desired trajectory by a feedback linearization–based algorithm for $k_d = 1$ and different k_p: (a) $k_p = 0.05$, (b) $k_p = 0.3$, and (c) $k_p = 2$.

4

Stabilization Strategies for Nonlinear Systems

In nonlinear control, one of the basic theorems is the Brockett theorem (Brockett 1983). It provides necessary conditions for feedback stabilization.

Theorem 4.1

Consider a nonlinear system $\dot{q} = f(q, u)$ with $f(q_0, 0) = 0$ and f continuously differentiable in a neighborhood of $(q_0, 0)$. Necessary conditions for the existence of a continuously differentiable control law for asymptotically stabilizing $(q_0, 0)$ are as follows:

1. The linearized system has no uncontrollable modes associated with eigenvalues with positive real part.
2. There exists a neighborhood E of $(q_0, 0)$ such that for each $\xi \in E$ there exists a control $u_\xi(t)$ defined for all $t > 0$ that drives the solution of $\dot{q} = f(q, u_\xi)$ from the point $q = \xi$ at $t = 0$ to $q = q_0$ at $t = \infty$.
3. The mapping φ: $E \times R^m \rightarrow R^n$, with E being a neighborhood of the origin, defined by $:(q, u) \rightarrow f(q, u)$ should be onto an open set of the origin.

The proof can be found in Brockett (1983) and Bloch (2003).

Stabilization strategies are not within the scope of this book; however, they are related to formulations of control problems, specifically tracking. A short exposition of this topic is presented in this section. For more details, see Coron (1992); Samson (1992); Kolmanovsky and McClamroch (1995); De Luca, Oriolo, and Samson (1998); M'Closkey and Murray (1994); and references therein.

Despite the Brockett theorem, the completely nonholonomic system may be asymptotically stabilized using feedback strategies. These strategies can be classified as discontinuous time-invariant stabilization, time-varying stabilization, and hybrid stabilization. Let us illustrate them by examples; interested readers are referred to the references at the end of this chapter.

Example 4.1

Consider a kinematic control model of a unicycle from Section 3.1.1, Example 3.2:

$$\dot{x} = u_1 \cos ,$$
$$\dot{y} = u_1 \sin , \qquad (4.1)$$
$$\cdot = u_2,$$

where control inputs are the forward velocity u_1 and angular velocity u_2. Equations (4.1) can be transformed to power form by the following state and control transformation:

$$x_1 = x\cos\ + y\sin\ ,$$

$$x_2 = \ ,$$

$$x_3 = x\sin\ - y\cos\ , \qquad (4.2)$$

$$w_1 = u_1 - u_2 x_3,$$

$$w_2 = u_2,$$

which yields

$$\dot{x}_1 = w_1,$$

$$\dot{x}_2 = w_2, \qquad (4.3)$$

$$\dot{x}_3 = x_1\dot{x}_2.$$

Let us apply a time-invariant feedback law developed using the sliding mode approach (Bloch and Drakunov 1994). The idea of this discontinuous feedback is to force the trajectory to slide along the manifold of codimension one toward the equilibrium. To stabilize the system (4.3) to the origin, the following control laws are used:

$$w_1 = -x_1 + 2x_2 \mathrm{sign}\left(x_3 - \frac{x_1 x_2}{2}\right), \qquad (4.4a)$$

$$w_2 = -x_2 - 2x_1 \mathrm{sign}\left(x_3 - \frac{x_1 x_2}{2}\right) \qquad (4.4b)$$

Select a Lyapunov function candidate as

$$V = \frac{1}{2}\left(\frac{x_1^2}{4} + x_2^2\right).$$

Its derivative along the closed-loop trajectories of Equations (4.3) is equal to $\dot{V} = \frac{1}{4}x_1 w_1 + x_2 w_2 = -2V$. It can be verified that $V(0) = V(0)e^{-2t} \to 0$ as $t \to \infty$ and $x_1 \to 0$, $x_2 \to 0$ as $t \to \infty$. Also, let

$$= x_3 - \frac{x_1 x_2}{2}$$

and then $\dot{\ } = -2V\mathrm{sign}(\ )$. The angle $|\ (t)|$ is nondecreasing and can reach zero in finite time if $V(x_1(0), x_2(0)) > |\ (0)|$. If initial conditions do not meet this condition, then an extra control may be used to force the trajectory to the region where it is satisfied. Then, the controls of Equations (4.4a) and (4.4b) may be applied. Notice that the control laws (4.4a) and (4.4b) are transformed to conform to the control model (4.3) so they differ from the original formulation in Bloch and Drakunov (1994). The sliding mode approach is available for some classes of higher-dimensional kinematic control system models (for details, see, e.g., Bloch and Drakunov 1994).

Example 4.2

Present now a time-varying stabilization algorithm proposed in Pomet (1992), which is known as Pomet's algorithm. It is based upon Lyapunov's direct method.

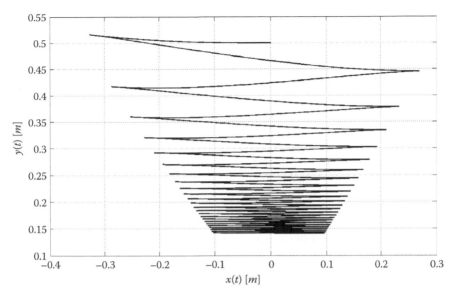

FIGURE 4.1
Trajectory of the axis midpoint of a Pioneer 3-DX robot.

To illustrate Pomet's algorithm in action, take the control model (4.3) for a unicycle from Example 4.1. The smooth feedback control laws are as follows:

$$w_1(x,t) = -x_1 + x_3 \sin t - x_2 \cos t, \tag{4.5a}$$

$$w_2(x,t) = x_1 x_3 - x_1(x_1 + x_3 \cos t)\cos t. \tag{4.5b}$$

Select a Lyapunov function candidate as

$$V(x,t) = \frac{1}{2}(x_1 + x_3 \cos t)^2 + \frac{1}{2}x_2^2 + \frac{1}{2}x_3^2.$$

It can be verified that the origin is a globally asymptotically stable equilibrium of the closed-loop system (4.3). The simulation result for the stabilization to the origin of the mobile robot Pioneer 3-DX by Pomet's algorithm is presented in Figure 4.1. For more details, see Szymański (2011).

Example 4.3

The smooth time-periodic feedbacks as in Example 4.2 result in non-exponential convergence to equilibrium states. A constructive procedure that provides a non-smooth, in the sense non-differentiable, feedback law with exponential convergence to equilibrium can be found in M'Closkey and Murray (1994).

Consider again the kinematic control model of a unicycle in the power form:

$$\begin{aligned} \dot{x}_1 &= w_1, \\ \dot{x}_2 &= w_2, \\ \dot{x}_3 &= x_1 \dot{x}_2. \end{aligned} \tag{4.6}$$

A non-smooth time-periodic feedback law that results in exponential convergence to the equilibrium has the form

$$w_1 = -x_1 + \frac{x_3}{\rho(x)}\cos t, \qquad x \neq 0 \tag{4.7a}$$

$$w_2 = -x_2 - \frac{x_3^2}{\rho^3(x)}\sin t, \qquad x \neq 0 \tag{4.7b}$$

and $w_1(0,t) = w_2(0,t) = 0$.

In Equations (4.7a) and (4.7b) $\rho(x) = \left(x_1^4 + x_2^4 + x_3^4\right)^{1/4}$. The closed-loop system is globally exponentially stable with respect to the homogenous norm $\rho(x)$. It can be verified that there exist constants $\lambda_1 > 0$ and $\lambda_2 > 0$ such that $\rho(x(t)) \leq \lambda_1 \rho(x(0)) \exp(-\lambda_2 t)$.

Another group of stabilizing controllers are hybrid. They combine continuous-time with discrete-event or discrete-time features (see, e.g., Bloch 2003; Kolmanovsky, Reyhanoglu, and McClamroch 1996; Sordalen and Egeland 1995; or a good introduction in de Wit et al. 1993). Time-periodic feedback laws for stabilization of dynamic models for nonholonomic systems can be derived from kinematic models (see, e.g., Walsh and Bushnell 1995).

Let us consider another example of stabilizing a nonholonomic system to the equilibrium manifold. This example illustrates an important property of nonholonomic systems, i.e., there exists an equilibrium manifold rather than an isolated equilibrium point for them (Bloch 2003; Hui and Goldenberg 1991). Example 4.4 uses the theory presented in Section 3.1.

Example 4.4

Consider the dynamics model of a unicycle from Example 2.14, and write it in the form

$$m\ddot{x} = \lambda_1,$$

$$m\ddot{y} = \lambda_2,$$

$$I_\theta \ddot{\theta} = \tau_1 - \lambda_1 r\cos \quad - \lambda_2 r\sin \quad,$$

$$I \ \ddot{} = \tau_2, \tag{4.8}$$

$$\dot{x} = r\dot{\theta}\cos \quad,$$

$$\dot{y} = r\dot{\theta}\sin \quad.$$

Selecting the desired orientation angles as φ_d and θ_d, and the control laws as

$$\tau_1 = k_{\theta p}(\theta_d - \theta) - k_\theta \dot{\theta},$$

$$\tau_2 = k_{\ p}(\ _d - \ ) - k_{\ }\dot{}, \tag{4.9}$$

the closed-loop dynamics yields

$$m\ddot{x} = \lambda_1,$$

$$m\ddot{y} = \lambda_2,$$

$$I_\theta \ddot{\theta} = k_{\theta p}(\theta_d - \theta) - k_\theta \dot{\theta} - \lambda_1 r \cos \quad -\lambda_2 r \sin \quad , \tag{4.10}$$

$$I \ddot{} = k_p(\quad_d - \quad) - k \dot{}_1.$$

Selecting the Lyapunov function candidate as

$$V = \frac{1}{2}m\dot{x}^2 + \frac{1}{2}m\dot{y}^2 + \frac{1}{2}I_\theta \dot{\theta}^2 + \frac{1}{2}I \dot{}^2 + \frac{1}{2}k_{\theta p}e_\theta^2 + \frac{1}{2}k_p e^2, \tag{4.11}$$

where $e_\theta = \theta_d - \theta, e = \quad_d - \quad$, its time derivative can be reduced to

$$\dot{V} = -k_{\theta p}\dot{\theta}^2 - k_p \dot{}^2. \tag{4.12}$$

We may conclude that since the derivative of V is negative semi-definite, the system is stable. Based on Equation (4.12), the derivative of V becomes zero when $\dot{\theta}$ and $\dot{}$ tend to zero. By letting these velocities go to zero in Equations (4.10), we can see that $\dot{V}$ vanishes only on the manifold defined by

$$M = \{(x, y, \theta, \quad) \mid \theta = \theta_d, \quad = \quad_d, \dot{x} = \dot{y} = \dot{\theta} = \dot{} = 0\}. \tag{4.13}$$

This is the equilibrium manifold of the system, and it is asymptotically stable.

One significant conclusion may be stated based on Equation (4.13), i.e., on the equilibrium M, the positions x and y of the unicycle are arbitrary. It means that when the unicycle starts from different positions and different control gains are selected, the convergence to a different final position will be achieved even if the angles φ_d and θ_d remain the same.

Example 4.5

Let us relate this example to the formulation of a control objective. Nonholonomic systems have a reputation of being "hard to control"; however, the formulation of a control objective influences the stabilization problem. According to Definition 3.10, the control model (3.40) is not locally asymptotically stabilizable (LAS) due to Brockett's theorem. However, for this control model, we may formulate a control objective, for which we may design a continuous static state feedback that makes it LAS (Jarzębowska 2007). To show this, consider the free-floating space robot model presented in Example 1.6. The angular momentum conservation yields the constraint equation

$$[J + (m_1 + m_2)l_1^2 + m_2 l_2^2]\dot{\phi} + [(m_1 + m_2)l_1^2 + m_2 l_2^2]\dot{\theta}_1 + m_2 l_2^2 \dot{\theta}_2 + m_2 l_1 l_2 \cos\theta_2 (2\dot{\phi} + 2\dot{\theta}_1 + \dot{\theta}_2) = K_0$$

which can be written as

$$B_1 \; ^{\cdot} + B_{1\theta1}\dot{\theta}_1 + B_{1\theta2}\dot{\theta}_2 = K_0, \tag{4.14}$$

where K_o is the initial angular momentum that may or may not be zero, and

$$B_1 = J + (m_1 + m_2)l_1^2 + m_2 l_2^2 + 2m_2 l_1 l_2 \cos\theta_2, \; B_{1\theta1} = (m_1 + m_2)l_1^2 + m_2 l_2^2 + 2m_2 l_1 l_2 \cos\theta_2,$$

$$B_{1\theta2} = m_2 l_2^2 + m_2 l_1 l_2 \cos\theta_2.$$

In Equation (4.14), J is the inertia of the base body, and m_1, m_2 are the masses of links concentrated at their ends. We assume that no external forces act on the space robot model. Let us select $\;= u_1, \theta_2 = u_2$ as controls and introduce a state vector $x \in R^3$ such that $x_1 = \;-\;_p, \; x_2 = \theta_1 - \theta_{1p}, \; x_3 = \theta_2 - \theta_{2p}$. It quantifies the errors between current values $(\;, \theta_1, \theta_2)$ and desired values $(\;_p, \theta_{1p}, \theta_{2p})$ of the coordinates. Then the control model (4.14) becomes

$$\frac{dx}{dt} = \begin{bmatrix} 0 \\ \overline{K}_o(x) \\ 0 \end{bmatrix} + \begin{bmatrix} 1 & 0 \\ -\overline{K}_1(x) & -\overline{K}_2(x) \\ 0 & 1 \end{bmatrix} \begin{bmatrix} u_1 \\ u_2 \end{bmatrix} \tag{4.15}$$

with

$$\overline{K}_o(x) = \frac{K_o(x)}{B_{1\theta1}(x)}, \overline{K}_1(x) = \frac{B_1\;(x)}{B_{1\theta1}(x)}, \overline{K}_2(x) = \frac{B_{1\theta2}(x)}{B_{1\theta1}(x)}.$$

When K_o is zero, it seems natural to formulate a control objective as to asymptotically stabilize the equilibrium $x = 0$. Then the system of Equation (4.15) is driftless, and the number of states $n = 3$ and $n-k = 2$. The equilibrium is not LAS. When K_o is not zero, the drift term never vanishes, and $x = 0$ is not an equilibrium. It implies that asymptotic stabilization of $x = 0$ is not an appropriate control objective. Instead, we can formulate a control problem as follows: make a system achieve $x(t_p) = 0$ for a given initial time $t = 0$ and some final time $t = t_p$. For this formulation of the control goal, we can apply the following time-varying transformation. Select $\xi \in R^3, \xi = (\xi_1, \xi_2, \xi_3)$ such that

$$\xi_1 = x_1,$$

$$\xi_2 = x_2 + \overline{K}_1(0)x_1 + \overline{K}_2(0)x_3 - \overline{K}_o(0)(t - t_p),$$

$$\xi_3 = x_3.$$

In the new coordinates, the control model (4.15) has the form

$$\frac{d\xi}{dt} = \begin{bmatrix} 0 \\ \overline{K}_o(\xi) - \overline{K}_o(0) \\ 0 \end{bmatrix} + \begin{bmatrix} 1 & 0 \\ -[\overline{K}_1(\xi) - \overline{K}_1(0)] & -[\overline{K}_2(\xi) - \overline{K}_2(0)] \\ 0 & 1 \end{bmatrix} \begin{bmatrix} u_1 \\ u_2 \end{bmatrix} \tag{4.16}$$

and the system has equilibrium at $\xi = 0$. It can be verified that the dimension of the equilibrium set is *2* and *n–k* = 2. Then the system is stabilizable by continuous static state feedbacks.

We may conclude that the control problem formulation is significant as well as the structure of the system model and constraints on it.

Problems

1. Prove that the smooth feedback law defined by Equation (4.5) in Example 4.2 makes the origin globally asymptotically stable equilibrium for the closed-loop system (4.3).
2. Prove that the feedback laws (4.7a) and (4.7b) make the closed-loop system (4.6) globally exponentially stable with respect to the homogenous norm $\rho(x)$.
3. Prove that the time derivative of the Lyapunov function (4.11) in Example 4.4 can be reduced to Equation (4.12). Use the dynamics (4.10) and the constraint equations in the proof.

References

Bloch, A.M., and S. Drakunov. 1994. Stabilization of a nonholonomic system via sliding modes. In *Proc. 33rd IEEE Conf. Decision and Contr.* 2961–2963.

Bloch, A.M. 2003. *Nonholonomic mechanics and control.* New York: Springer-Verlag.

Brockett, R.W. 1983. Asymptotic stability and feedback stabilization. In *Differential geometric control theory*, ed. R.W. Brockett, R.S. Millman, and H.J. Sussmann. 181–191. Boston, MA: Birkhauser.

Canadas-de-Wit, C., H. Khennouf, C. Samson, and O.J. Sordalen. 1993. Nonlinear control design for mobile robots. In *Advanced Mobile Robots—Theory and Applications*, ed. Y.F. Zheng. River Edge, NJ: World Scientific.

Coron, J.-M. 1992. Links between local controllability and local continuous stabilization. In *Proc. IFAC Nonl. Contr. Syst. Design Symp.* 477–482. Bordeaux: France.

De Luca, A., G. Oriolo, and C. Samson. 1998. Feedback control of a nonholonomic car-like robot. In *Robot Motion Planning and Control*, ed. J.-P. Laumond, 171–253. London: Springer.

Hui, R., and A.A. Goldenberg. 1991. Stability of nonholonomic systems. In *Proc. IEEE/ RSJ Int. Workshop on Intell. Rob. Syst. IROS* 260–265. Osaka: Japan.

Jarzębowska, E. 2007. Stabilizability and motion tracking conditions for nonholonomic control systems. *Math. Problems Eng.* Cairo: Hindawi.

Kolmanovsky, I., and N.H. McClamroch. 1995. Developments in nonholonomic control problems. *IEEE Contr. Syst. Mag.* 15:20–36.

Kolmanovsky, I., M. Reyhanoglu, and N.H. McClamroch. 1996. Switched mode feedback control laws for nonholonomic systems in extended power form. *Syst. Contr. Lett.* 27(1):29–36.

M'Closkey, R.T., and R.M. Murray. 1994. Exponential stabilization of driftless nonlinear control systems via time-varying, homogeneous feedback. In *Proc. 33rd Conf. Decision Contr.* 1317–1322.

Pomet, J.-B. 1992. Explicit design of time-varying stabilizing control laws for a class of controllable systems without drift. *Syst. Contr. Lett.* 18:147–158.

Samson, C. 1992. Time-varying feedback stabilization of car-like wheeled mobile controllable systems without drift. *Int. J. Rob. Res.* 12(1):55–64.

Sordalen, O.J., and O. Egeland. 1995. Exponential stabilization of nonholonomic chained systems. *IEEE Trans. Automat. Contr.* 40(1):35–49.

Szymański, M. 2011. Control of a mobile robot Pioneer 3-DX. B.S. diss., Warsaw University of Technology, Poland.

Walsh, G.O., and L.G. Bushnell. 1995. Stabilization of multiple input chained form control systems. *Syst. Contr. Lett.* 25:227–234.

5

Model-Based Tracking Control of Nonlinear Systems

Model-based tracking for nonlinear systems is a part of control where tracking is a goal and a controller is based upon a dynamic model. As stated in Section 3.7, model-based control surpasses kinematic-based control with regard to performance, control precision, ability to cope with external disturbances, friction, and noise. Also, this book focuses on two control design stages: modeling and controller design. As a consequence, we focus on constructive control procedures applicable to common classes of nonlinear systems significant in applications, and on modeling and control design tools. Of course, these constructive procedures and design tools available to control engineers are results of several theoretic research streams; these are discussed briefly in Chapter 3.

In this chapter, we present some model-based tracking control techniques and algorithms for holonomic, first order nonholonomic, and underactuated system models. Tracking control for systems subjected to high order nonholonomic constraints is presented in Chapter 7.

Based on examples, we recall the scope of applications of nonlinear control techniques, actually available and reported in the literature, to these three groups of nonlinear system models. This chapter is not a review of existing control techniques; these can be found in Arimoto (1990, 1995), Cortes (2002), Fradkov et al. (1999), Lewis et al. (2004), Nijmeijer and van der Shaft (1990), Slotine and Li (1996), Spong and Vidyasagar (1989), and references therein. The collection of control techniques and algorithms for nonlinear systems is rich due to the significance of these system models in applications.

5.1 A Unified Control-Oriented Model for Constrained Systems

In Section 3.5, we demonstrated that the generalized programmed motion equations (GPME) (2.199) for $p = 1$, i.e., Nielsen's equations in Maggi's form, are equivalent to the reduced-state Lagrange's equations. A dynamic control model required for tracking is developed for a system subjected to position or kinematic constraints of the form of Equation (2.22). Conclusions are then as follows:

1. Equations of motion of a system with material constraints can be obtained by the GPME for $p = 0$ or $p = 1$.
2. A dynamic model based on Lagrange's equations with multipliers can be replaced by a dynamic model based on the GPME for $p = 0$ or $p = 1$.

Then, to develop dynamic models for control applications, we use the GPME for $p = 1$ according to Algorithm 2.1. They yield

$$M(q)\ddot{q} + C(q,\dot{q})\dot{q} + D(q) = Q(t,q,\dot{q}),$$

$$B_1(q)\dot{q} = 0.$$

(5.1)

Equations (5.1) consist of $(n\text{-}k)$ equations of motion and k equations of constraints. The dimension of the matrix $M(q)$ is then $(n\text{-}k) \times n$, and $B_1(q)$ is a $(k \times n)$-dimensional matrix with $n > k$. We assume that the constraints are independent, i.e., the rank of $B_1(q)$ is k. In Equations (5.1) constraints are material, so comparing with (2.199), the term $V(q,\dot{q})$ is replaced by $C(q,\dot{q})\dot{q}$, which quantifies effects of Coriolis and centripetal forces. All external forces are denoted by $Q(t,q,\dot{q})$. If a dynamic model is generated for a system with holonomic constraints, they can be eliminated, at least locally, by appropriate selection of coordinates, and a dynamic model takes the form

$$M^*(q)\ddot{q} + C^*(q,\dot{q})\dot{q} + D^*(q) = Q^*(t,q,\dot{q}),$$ (5.2)

where now q is a m-vector, and $m = n\text{-}a$, with a being the number of holonomic constraint equations. When a system is subjected to both nonholonomic and holonomic constraints, they both can be presented in the form of Equation (2.22), and the dynamic model for the system is (5.1).

Definition 5.1: A dynamic model (5.1) for a system with material constraints is referred to as a unified control-oriented dynamic model.

Definition 5.2: The unified control-oriented dynamic model (5.1) with control inputs τ is referred to as a unified dynamic control model for a constrained system, and it is of the form

$$M(q)\ddot{q} + C(q,\dot{q})\dot{q} + D(q) = E(q)\tau,$$

$$B_1(q)\dot{q} = 0.$$

(5.3)

To transform Equations (5.3) to the state-space control form, the constraint equation may be presented as

$$\ddot{q} = G(q)\ddot{q}_1 + \dot{G}(q)\dot{q}_1,$$ (5.4)

where partition of the vector q is $q = (q_1, q_2)$ and $q_1 \in R^{n-k}, q_2 \in R^k$; q_1 is a vector of independent coordinates, and q_2 is a vector of dependent ones. Columns of the matrix $G(q)$ span the right null space of $B_1(q)$. It is a $(n \times m)$ matrix with $m = n-k$, and it has the form

$$G = \begin{bmatrix} I_{(m \times m)} \\ -B_{12}^{-1}(q)B_{11}(q) \end{bmatrix}, \tag{5.5}$$

where I is a $(m \times m)$ identity matrix, $B_{12}^{-1}(q)B_{11}(q)$ is a locally smooth $(k \times m)$ matrix function, the matrix $B_1(q)$ is expressed as $B_1 = [B_{11}(q), B_{12}(q)]$, $B_{11}(q)$ is a $k \times (n-k)$ matrix function, and $B_{12}(q)$ is a $(k \times k)$ locally nonsingular matrix function. Elimination of second order derivatives of dependent coordinates from the first equation in (5.3) yields

$$M(q)G(q)\ddot{q}_1 + [M(q)\dot{G}(q) + C(q,\dot{q})G(q)]\dot{q}_1 + D(q) = E(q)\tau, \tag{5.6}$$

$$\dot{q} = G(q)\dot{q}_1.$$

Equations (5.6) are the reduced-state dynamic model for a nonholonomic system.

Because we address control problems for nonholonomic systems, Equations (5.6) require more attention. First, they consist of a purely kinematic relationship $\dot{q} = G(q)\dot{q}_1$ and a reduced dynamics represented by the first $(n-k)$ equations (5.6). It can be proved through direct calculations that this reduced dynamics preserves the structure of the mechanical system model given by (3.93). The following hold for Equations (5.6):

1. The matrices $M(q)G(q) = \tilde{M}(q)$ and $D(q)$ are bounded functions of q, and their time derivatives $\dot{\tilde{M}}(q)$ and $\dot{D}(q)$ are also bounded in q and depend linearly on $\dot{q}_1$.
2. The matrix $\tilde{M}(q)$ is $(n-k) \times (n-k)$ symmetric and positive definite.
3. The matrices $\tilde{M}(q)$ and $[M(q)\dot{G}(q) + C(q,\dot{q})G(q)] = \tilde{C}(q,\dot{q})$ are related such that $\dot{\tilde{M}} = \tilde{C} + (\tilde{C})^T$.

We can formulate the following theorem for the dynamics (5.6).

Theorem 5.1

There exists a static state feedback $U(\dot{q}_1, q, u): R^m \times R^n \times R^m \rightarrow R^m$ such that the dynamics (5.6) can be transformed to the state-space control formulation (3.40).

Proof: First, introduce in Equations (5.6) a new state variable vector $x = (q, \dot{q}_1) = (x_1, x_2)$ such that $\dot{x}_1 = \dot{q} = (\dot{q}_1, \dot{q}_2)$, $\dot{x}_2 = \ddot{q}_1$ and $x_1 \in R^n, x_2 \in R^m$. Then,

(5.6) takes the form

$$M(x_1)G(x_1)\dot{x}_2 + [M(x_1)\dot{G}(x_1) + C(x_1, \dot{x}_1)G(x_1)]x_2 + D(x_1) = E(x_1)\tau,$$

$$\dot{x}_1 = G(x_1)x_2. \tag{5.7}$$

Now, select for the dynamics (5.7) a static state feedback $U(x_2, x_1, u): R^m \times R^n \times R^m \to R^m$ defined by the relation

$$M(x_1)G(x_1)u + [M(x_1)\dot{G}(x_1) + C(x_1, \dot{x}_1)G(x_1)]x_2 + D(x_1) = E(x_1)\tau.$$

Application of this state feedback to Equations (5.7) transforms them to the form

$$\dot{x}_1 = G(x_1)x_2,$$

$$\dot{x}_2 = u, \tag{5.8a}$$

which is a desirable state-space control formulation with $f(x) = (G(x_1), 0)$ and $g(x) = (0, e_i)$, and e_i is the standard basis vector in R^{n-k}.   q.e.d.

The first equation in (5.8a) is the constraint equation that must be satisfied during control. It is enough to control the independent coordinate vector x_2. The second equation in (5.8a) transforms immediately to the linear controllable dynamics.

The form (5.8a) can be transformed to the normal form equivalent to the one obtained, for instance, in Bloch (2003). Taking new state variables $z_1 = q_1$, $z_2 = q_2$, $z_3 = \dot{q}_1$, which are related to x_1 and x_2 such that $\dot{x}_1 = (\dot{z}_1, \dot{z}_2)$, $\dot{z}_3 = \dot{x}_2$, Equations (5.8a) can be written as

$$\dot{z}_1 = z_3,$$

$$\dot{z}_2 = G^*(z_1, z_2)z_3, \tag{5.8b}$$

$$\dot{z}_3 = u.$$

This form is equivalent to Equation (3.40), where $f(z) = (z_3, G^*(z_1, z_2)z_3, 0)$, $g_i = (0, 0, e_i)$, and e_i is the standard basis vector in R^{n-k}. The matrix G^* in (5.8b) is a ($k \times n-k$) submatrix of the matrix G defined in (5.5).

Based on Theorem 5.1, the computed torque controller can be applied to Equations (5.6). It transforms them to the form

$$\ddot{q}_1 = u,$$

$$\ddot{q}_2 = -B_{12}^{-1}(q)B_{11}(q)\ddot{q}_1 - \frac{d}{dt}\left[B_{12}^{-1}(q)B_{11}(q)\right]\dot{q}_1. \tag{5.9}$$

A vector of a new input is u, and it can be selected as

$$u = \ddot{q}_{1p} + 2\sigma\dot{\tilde{q}}_1 + \sigma^2\tilde{q}_1, \tag{5.10}$$

where the subscript p stands for "program," $\tilde{q}_1 = q_{1p} - q_1$ is a position tracking error, and σ is a convergence rate diagonal matrix. The tracking error $\tilde{q}_1$ satisfies the equation $\ddot{\tilde{q}}_1 + 2\sigma\dot{\tilde{q}}_1 + \sigma^2\tilde{q}_1 = 0$ and converges to zero exponentially, i.e., the system motion converges to motion prespecified by a designer.

It is often more convenient to introduce an extended torque vector that we refer to as a virtual torque vector. We introduce it by rearranging Equations (5.6) to the form

$$
\begin{bmatrix} M(q)G(q) & 0_{(n-k)\times k} \\ B_{12}^{-1}(q)B_{11}(q) & I_{k\times k} \end{bmatrix} \begin{bmatrix} \ddot{q}_1 \\ \ddot{q}_2 \end{bmatrix} + \begin{bmatrix} M(q)\dot{G}(q) + C(q,\dot{q})G(q) & 0_{(n-k)\times k} \\ \dfrac{d}{dt}\left[B_{12}^{-1}(q)B_{11}(q)\right] & 0_{k\times k} \end{bmatrix} \begin{bmatrix} \dot{q}_1 \\ \dot{q}_2 \end{bmatrix}
$$

$$
+ \begin{bmatrix} D(q) \\ 0 \end{bmatrix} = \begin{bmatrix} E(q) & 0_{(n-k)\times k} \\ 0_{k\times(n-k)} & 0_{k\times k} \end{bmatrix} \begin{bmatrix} \tau_1 \\ \tau_2 \end{bmatrix}, \tag{5.11}
$$

where now τ is a n-dimensional vector of the virtual torque. The number of components τ_i in Equation (5.11) is equal to the number of coordinates that describe motion. This form of the unified dynamic control model is useful for constrained, specifically programmed, motion tracking. It means that in a programmed motion tracking, each "direction" is controlled, and a virtual torque is also assigned to a "material constraint direction." As it is demonstrated in examples in Chapter 7, this torque is equal to zero because the constraint must be satisfied. Also, if a system is nonholonomic in nature, it may happen that not all τ_i are independent, and real torques T_j that are applied to a real system have to be determined, and j is equal to the number of degrees of freedom. Also, the virtual torque vector is helpful in designing a controller for an underactuated system. Details are provided in Chapter 7.

We demonstrated that for a system subjected to both holonomic and nonholonomic material constraints, its dynamic control model derived by the GPME can be presented in the standard state-space form (5.8a). This allows us to reformulate for the dynamics (5.3) all theoretical control results obtained for nonlinear dynamic control models using classical methods (see, e.g., Bloch 2003; Isidori 1989; Sontag 1990). Specifically, we reformulate the following.

Definition 5.3: For a system subjected to constraints $B_1(q)\dot{q} = 0$, assumed to be completely nonholonomic, a $(2n-k)$-dimensional smooth manifold of the phase space $M = \{(q,\dot{q}) \in R^n \times R^n : B_1(q)\dot{q} = 0\}$ is referred to as a constraint manifold.

Definition 5.4: A solution $(q_e, 0)$ of Equations (5.3) or, equivalently, a solution $(z_1, z_2, 0)$ of Equations (5.8b) is referred to as an equilibrium solution if it is a constant solution.

Theorem 5.2

Let $k \geq 1$ and let $(q_e, 0)$ denote an equilibrium solution in M. The nonholonomic control system (5.3) is not asymptotically stabilizable using C^1 feedback to $(q_e, 0)$.

Theorem 5.3

Let $k \geq 1$ and let $(q_e, 0)$ denote an equilibrium solution in M. The nonholonomic control system (5.3) is strongly accessible at $(q_e, 0)$.

If the accessibility rank condition holds for any $q \in M$, we can say that the system (5.3) is strongly accessible.

Theorem 5.4

Let $k \geq 1$ and let $(q_e, 0)$ denote an equilibrium solution in M. The nonholonomic control system (5.3) is small-time locally controllable at $(q_e, 0)$.

The theorems above restated to dynamics (5.3) are selected with regard to the scope of the book. The proofs can be found in Bloch (2003) and Isidori (1989).

The unified dynamic control model (5.3) admits properties useful in a controller design process. This is due to the GPME that yield equations with constraint reaction forces eliminated and enable selection of independent coordinates according to the selection of control inputs. These properties are the generic features of the GPME.

5.2 Tracking Control of Holonomic Systems

Tracking control of holonomic systems can be considered a solved problem, at least theoretically. We present an example to provide a reader with comparisons between tracking control designs and accompanying challenges for holonomic and nonholonomic systems. Readers interested in control of holonomic systems may consult, for example, Arimoto (1990), Kelly et al. (2005), Lewis et al. (2004), and references therein.

Example 5.1

As explained in Section 3.9, the backstepping technique is suitable for some flight control problems. Consider a simplified example of the flight path angle control. The aircraft dynamics can be written as

$$\dot{V} = \frac{1}{m}(-D + T \cos\alpha - mg \sin\gamma),$$

$$\dot{\gamma} = \frac{1}{mV}(L + T \sin\alpha - mg \cos\gamma),$$

$$\dot{\theta} = p,$$

$$\dot{p} = \frac{1}{I_y}(M + TZ_{TP}).$$

(5.12)

The simplification of the model (5.12) relies on the consideration of control about the longitudinal aircraft axis. The state variables are as follows: V is air velocity, γ is the flight path angle, θ is the pitch angle, p is the pitch rate, $\alpha = (\theta\text{-}\gamma)$ is the attack angle, T-engine trust force that contributes to the pitching moment due to a thrust point offset, Z_{TP}. The aerodynamics effects are described by the lift force L, drag force D, and the pitching moment M. The controlled variables are V and γ. The control input is the elevator angle δ. More details about aircraft control can be found in Stevens and Lewis (1992).

To illustrate the backstepping controller design, we neglect in the aircraft dynamics the dependence of lift on the pitch and elevator angle rates. The change of δ produces changes of the pitching moment. This will affect the angle of attack to which the lift force is related. This assumption is the same as that needed to apply dynamic inversion to flight control (compare to Lane and Stengel 1988).

The lift force and the pitching moment are functions of aerodynamic coefficients and wing parameters:

$$L = \frac{1}{2}\rho V^2 S c_L(\alpha), \quad M = \frac{1}{2}\rho V^2 S \bar{c} c_m(\alpha, q, \delta),$$

where ρ is the air density, S is the wing area, and $\bar{c}$ is the mean aerodynamic chord.

The control goal is to control γ to some reference γ_{ref}. The angle of attack at steady state, α_0, is specified through $\dot{\gamma} = 0$, and it is computed online and updated at each time step, using current values of ρ, V^2, δ, and T. In Equations (5.12), the angle γ depends upon the gravity, but this term is not significant and will be taken into account at equilibrium, so the equation for the changes of γ may be written as

$$\dot{\gamma} = \frac{1}{mV}(L(\alpha) + T\sin\alpha - mg\cos\gamma_{ref}) \equiv \phi(\alpha - \alpha_0), \tag{5.13}$$

where $\phi(0) = 0$. Also, taking into account the way in which lift depends on α, i.e., $L(\alpha)$ function and $T\sin\alpha$, it may be concluded that $\alpha\phi(\alpha) > 0$ for $\alpha \neq 0$. This property is significant for the backstepping controller design. Also, it reflects some range of α, for which Equations (5.12) are still accurate.

The backstepping controller is designed in three steps (see Lemma 3.4).

Step 1: Introduce the control error $z_1 = \gamma - \gamma_{ref}$, whose dynamics is

$$\dot{z}_1 = \phi(\alpha - \alpha_0). \tag{5.14}$$

Select a Lyapunov function candidate, in control context often referred to as a control Lyapunov function, as

$$V_1 = \frac{1}{2}z_1^2.$$

To determine the stabilizing function θ_s for the control input θ in Equation (5.14), time derivative of V_1 can be written as

$$\dot{V}_1 = z_1\phi(\alpha - \alpha_0) = z_1\phi(\theta - z_1 - \gamma_{ref} - \alpha_0) = z_1\phi(-(1+c_1)z_1 + \theta - (-c_1 z_1 + \gamma_{ref} + \alpha_0))$$

$$= z_1\phi(-(1+c_1)z_1 + \theta - \theta_s).$$

With $\theta_s = (-c_1 z_1 + \gamma_{ref} + \alpha_0)$, the time derivative of V_1 satisfies

$$\dot{V}_1 \leq z_1 \phi(-(1+c_1)z_1). \qquad c_1 > -1$$

Step 2: For the virtual control law,

$$\theta_s = (-c_1 z_1 + \gamma_{ref} + \alpha_0), \qquad (5.15)$$

select the control error $z_2 = \theta - \theta_s = \theta + c_1 z_1 - \gamma_{ref} - \alpha_0$ and a new state $\xi = -(1+c_1)z_1 + z_2$. Then, the new dynamics is

$$\dot{z}_1 = \phi(\xi),$$

$$\dot{z}_2 = \dot{\theta} + c_1 \phi(\xi).$$

It can be verified that $\dot{\xi} = -\phi(\xi) + \dot{\theta} = -\phi(\xi) + p$. Select a control Lyapunov function V_2 as

$$V_2 = \frac{c_2}{2} z_1^2 + \frac{1}{2} z_2^2 + F(\xi), \qquad c_2 > 0$$

where an extra term $F(\xi)$ was added according to Krstic and Kokotovic (1995). It is required to be positive definite and radially growing. Computing the time derivative of V_2 enables finding a new stabilizing function for p, i.e., p_s.

$$\dot{V}_2 = c_2 z_1 \phi(\xi) + z_2 (p + c_1 \phi(\xi)) + \dot{F}(\xi)(-\phi(\xi) + p) = (c_2 z_1 + c_1 z_2 - \dot{F}(\xi))\phi(\xi) + (z_2 + \dot{F}(\xi))p.$$

The time derivative of V_2 may be made negative if to select

$$p_s = -c_3 z_2, \qquad c_3 > 0 \qquad (5.16)$$

$$\dot{F}(\xi) = c_4 \phi(\xi), \qquad F(0) = 0, \qquad c_4 > 0 \qquad (5.17)$$

Then

$$\dot{V}_2 = [c_2 z_1 + (c_1 - c_3 c_4)z_2]\phi(\xi) - c_4 \phi(\xi)^2 - c_3 z_2^2.$$

Selecting $c_2 = -(1+c_1)(c_1 - c_3 c_4)$ and $c_3 c_4 > c_1$ and using the condition $\alpha\phi(\alpha) > 0$ for $\alpha \neq 0$, we obtain

$$\dot{V}_2 = (c_1 - c_3 c_4)\xi\phi(\xi) - c_4 \phi(\xi)^2 - c_3 z_2^2 < 0.$$

Step 3: The final iteration requires introduction of the third control error $z_3 = \dot{\theta} - \dot{\theta}_s = p - p_s = p + c_3 z_2$. Then, the new dynamics is

$$\dot{z}_1 = \phi(\xi),$$

$$\dot{z}_2 = z_3 - c_3 z_2 + c_1 \phi(\xi), \qquad (5.18)$$

$$\dot{z}_3 = u + c_3 (z_3 - c_3 z_2 + c_1 \phi(\xi))$$

with

$$u = \frac{1}{I_y}(M(\alpha, q, \delta) + TZ_{TP})$$

being the control variable. Also, $\dot{\xi} = -\phi(\xi) + z_3 - c_3 z_2$. Select a control Lyapunov function V_3, which penalizes z_3 as

$$V_3 = c_5 V_2 + \frac{1}{2} z_3^2, \quad c_5 > 0.$$

We can demonstrate that the time derivative of V_3 may be made negative with the use of the condition $\alpha \phi(\alpha) > 0$ for $\alpha \neq 0$, i.e.,

$$\dot{V}_3 = c_5 \left[(c_1 - c_3 c_4) \xi \phi(\xi) - c_4 \phi^2(\xi) - c_3 z_2^2 + z_3(z_2 + c_4 \phi(\xi)) \right] + z_3 [u + c_3(z_3 - c_3 z_2 + c_1 \phi(\xi))] \leq$$

$$-c_4 c_5 \phi^2(\xi) - c_3 c_5 z_2^2 + z_3 \left[u + c_3 z_3 + z_2 (c_5 - c_3^2) + (c_1 c_3 + c_4 c_5) \phi(\xi) \right].$$

Now, select the parameters and u such that $c_5 = -c_3^2$,

$$u = -c_6 z_3, \quad \text{with} \quad c_6 > c_3. \tag{5.19}$$

Then,

$$\dot{V}_3 \leq -c_3^2 c_4 \phi^2(\xi) - c_3^3 z_2^2 - (c_6 - c_3) z_3^2 + z_3 \left(c_1 c_3 + c_3^2 c_4 \right) \phi(\xi).$$

The right-hand side of the relation above can be written in the form

$$\dot{V}_3 \leq -c_3^3 z_2^2 - (c_6 - c_3) \left[z_3 - \frac{c_1 c_3 + c_3^2 c_4}{2(c_6 - c_3)} \phi(\xi) \right]^2 - \left[c_3^2 c_4 - \frac{\left(c_1 c_3 + c_3^2 c_4 \right)^2}{4(c_6 - c_3)} \right] \phi^2(\xi).$$

It can be seen that $\dot{V}_3$ is negative definite if the coefficient at $\phi^2(\xi)$ is negative. It may be made negative for

$$c_6 > c_3 \left[1 + \frac{(c_1 + c_3 c_4)^2}{4 c_3 c_4} \right]. \tag{5.20}$$

The left-hand side of (5.20) reaches minimum, with respect to c_4, which satisfies $c_4 > 0$, for $c_4 = c_1/c_3$. Then, the relation (5.20) may be replaced by

$$c_6 > c_3(1 + c_1).$$

Getting back to the original state variables, the control law (5.19) can be written as

$$u = -c_6 [p + c_3(\theta + c_1(\gamma - \gamma_{ref}) - \gamma_{ref} - \alpha_0)] \tag{5.21}$$

with the design parameters related as

$$c_1 > -1, \quad c_3 > 0,$$

$$c_6 > \begin{cases} c_3 & c_1 \leq 0 \\ c_3(1 + c_1) & c_1 > 0 \end{cases} \tag{5.22}$$

This ends the backstepping procedure of designing a feedback controller for the simplified aircraft dynamics (5.12). It is a globally stabilizing controller. The truth is, however, that the control law (5.21) respects the angular pitch acceleration $\dot{p}$ as the control variable. The next part of the controller design requires the relation of the elevator angle with u. This stage of the specific controller design can be found in Harkegard and Glad (2000).

The backstepping procedure that results in the control law (5.21) may be compared to feedback linearization design. Feedback linearization design requires the following steps:

Select $z_1 = \gamma - \gamma_{ref}$, then

$$\dot{z}_1 = \phi(\alpha - \alpha_0) = z_2,$$

$$\dot{z}_2 = \dot{\phi}(\alpha - \alpha_0)(p - z_2) = z_3, \tag{5.23}$$

$$\dot{z}_3 = \ddot{\phi}(\alpha - \alpha_0)(p - z_2)^2 + \dot{\phi}(\alpha - \alpha_0)(u - z_3) = v.$$

Selecting $v = -k_1 z_1 - k_2 z_2 - k_3 z_3$, we can verify that it renders the closed-loop system. Solving Equations (5.23) for u yields

$$u = z_3 + \frac{v - \ddot{\phi}(\alpha - \alpha_0)(q - z_2)^2}{\dot{\phi}(\alpha - \alpha_0)}. \tag{5.24}$$

A couple of conclusions can be drawn from the controller form (5.24). First, it depends on ϕ and its first and second derivative. Practically, it means that the lift force has to be well known to compute u accurately. Second, the derivative of ϕ is in the denominator in Equation (5.24). The singularity may occur then, if $\dot{\phi}$ approaches zero; this may happen around the stall angle, where the lift force does not increase as α increases. The comparison of Equations (5.21) and (5.24) reveals potential benefits of the backstepping design.

5.3 Tracking Control of First Order Nonholonomic Systems

In this section, we present an example of control design for a trajectory tracking of a nonholonomic system. This classical tracking control problem is approached using the GPME and the concept of the programmed constraint.

Example 5.2

Consider the unicycle model, for which the material and programmed constraints are specified by Equations (2.208) in Example 2.15. Its unified dynamic model is governed by Equations (2.210), and the unified dynamic control model in the extended form (5.11) is

$$
\begin{bmatrix}
1 & 0 & -r\cos & 0 \\
0 & 1 & -r\sin & 0 \\
0 & 0 & mr^2 + I_\theta & 0 \\
0 & 0 & 0 & I
\end{bmatrix}
\begin{bmatrix}
\ddot{x} \\
\ddot{y} \\
\ddot{\theta} \\
\ddot{}
\end{bmatrix}
+
\begin{bmatrix}
r\dot{\theta}\sin + \sigma_x \\
-r\dot{\theta}\cos + \sigma_y \\
0 \\
0
\end{bmatrix}
=
\begin{bmatrix}
\tau_1 \\
\tau_2 \\
\tau_3 \\
\tau_4
\end{bmatrix}. \tag{5.25}
$$

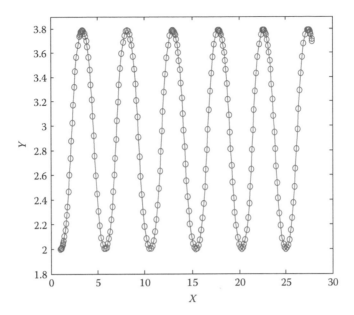

FIGURE 5.1
Programmed motion tracking by the computed torque controller (ooo).

For simulation purposes, the nonholonomic constraint equations are differentiated. To eliminate drift from solutions of Equations (5.25), the differentiated constraint equations are stabilized using Baumgarte's stabilization method (see Jarzębowska 2007 for details). MATLAB with *ODE15s* is applied to the simulation studies. All dimensions in figures are in SI units unless stated otherwise.

The programmed trajectory tracking by the computed torque controller (5.10) is presented in Figure 5.1. Figure 5.2 presents torques $\tau_1 - \tau_4$ required to track the programmed motion.

Examining Figure 5.2, the meaning of the name "virtual" given to control torques is clear. As expected, τ_1 (marked with "∇") and τ_2 (marked with "o") are equal to zero; these are the virtual torques related to the nonholonomic constraint equation. The nonzero virtual torques are τ_3 and τ_4 marked with "$\square$" and "+", respectively. However, referring to implementation problems, e.g., for a two-wheeled robot, we have to determine real torques T_j from the virtual torques τ_3 and τ_4. The torques T_j, $j = 1,2$, are generated by motors by which wheels are powered, and they control the change in wheel angular velocities ω_1 and ω_2. The mass center velocity of the robot is $v_c = r(\omega_1 + \omega_2)/2$. The change in the heading angle is equal to $\dot{} = r(\omega_1 - \omega_2)/2b$, with $2b$ being the distance between the wheels. Then, relations between the virtual and real torques are

$$\tau_3 = \frac{T_1 + T_2}{2}, \quad \tau_4 = \frac{r}{2b}(T_1 - T_2). \tag{5.26}$$

Programmed motion tracking using the computed torque controller is conceptually simple and easy in simulation, but it requires measuring the acceleration

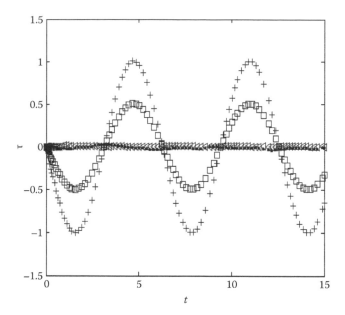

FIGURE 5.2
Computed torques for tracking the unicycle motion.

and ensuring that the inverse of the mass matrix exists. Both requirements may be difficult to preserve. Also, most industrial robots have only position and velocity sensors. For programmed motions, we can reduce much online computation introducing a control law that takes advantage of system dynamics along these motions. We present the control law of Wen and Bayard (1988), which is used in Jarzębowska (2007) for the first time to track motion of a nonholonomic system. It is of the form

$$\tau = M(q_p)\ddot{q}_{1p} + C(q_p, \dot{q}_{1p})\dot{q}_{1p} + D(q_p) + k_d\dot{e} + k_s e, \tag{5.27}$$

where $e = q_{1p} - q_1$. This control law is selected based on results of comparative simulations that proved its better performance versus complexity. Control gains are selected to be $k_d = 24$, $k_s = 15$. Figures 5.3a,b present comparisons of programmed motion tracking executed by the computed torque (•••) and the Wen–Bayard controller (ooo). Figures 5.4a,b present torques $\tau_1 - \tau_4$ generated by the Wen–Bayard controller. They are marked with the same kind of lines as in Figure 5.2. It can be seen that the torque τ_4 needed to track   is about three times smaller than that generated by the computed torque.

Finally, Figure 5.5 shows the comparison of satisfaction of the constraint equations along y direction for the two controllers. For the Wen–Bayard controller with the gains selected as above, the constraint equations are worse satisfied. Figure 5.6 presents simulation results when the nonholonomic constraint equation in y direction is violated. The controlled system gets "a kick" and some short time is required to get the constraints satisfied again. This kick is undesirable in

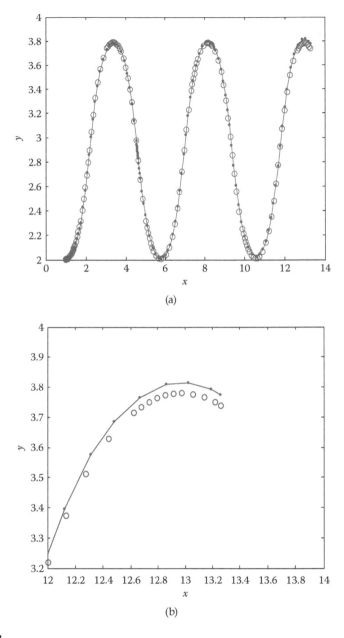

FIGURE 5.3
(a) Programmed motion tracking: computed torque (●●●) and the Wen–Bayard (ooo) controllers.
(b) Performance comparisons of the Wen–Bayard (ooo) and computed torque (●●●) controllers.

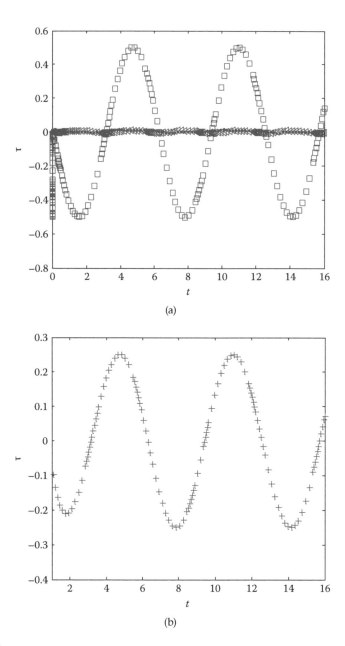

FIGURE 5.4
(a) Torques in the Wen–Bayard control law versus time: τ_1 (∇), τ_2 (o), τ_3 ($\square$). (b) Torque τ_4 in the Wen–Bayard control law versus time.

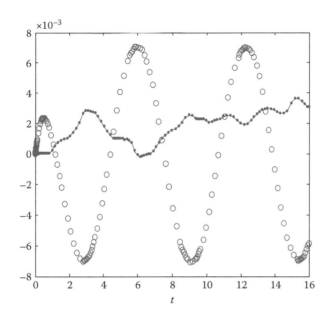

FIGURE 5.5
Constraint equations satisfaction in time: computed torque (···), the Wen–Bayard controller (ooo).

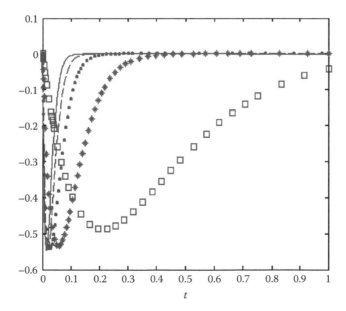

FIGURE 5.6
Constraint violation in tracking for σ: 80(—), 60(---), 40(•••), 20(***), 5(□□□).

motion of a real system, because it means that the wheels just slip. We can conclude that when moderate accuracy is satisfactory, the Wen–Bayard controller is recommended. Control torques are smaller in magnitudes and online computation is reduced because it utilizes information about system dynamics along the programmed motion.

5.4 Tracking Control of Underactuated Systems

A system is referred to as underactuated when it possesses fewer inputs than degrees of freedom, that is, the dimension of the configuration space exceeds the dimension of the control input space. Dynamic models of underactuated systems are classified as second order nonholonomic system models. This is due to the equation of an underactuated degree of freedom, which is not integrated in general and is regarded as a nonholonomic constraint equation of the second order. Sometimes, underactuated systems are referred to as systems with dynamic or design constraints, because their underactuation may originate from design or operation requirements.

Underactuated systems are wheeled mobile robots, manipulators with unactuated joints, or space robots without jets or momentum wheels (Reyhanoglu et al. 1996; Shiroma et al. 1997). Sometimes, specific properties of these systems are utilized to facilitate control design. For example, utilizing breaking mechanisms with which unactuated joints are equipped or including gravity terms, make system models linearization about equilibrium controllable (Arai and Tachi 1990; Shiroma et al. 1997). In our development, the gravity term is absent.

Underactuated systems equipped with wheels form a group of systems of interesting properties. Examples are a snake-board (Ostrowski et al. 1997), a roller-racer (Krishnaprasad and Tsakiris 1998, 2001) or its version trikke, a roller-blader (Chitta et al. 2004), a roller-walker (Hirose and Takeuchi 1996), or snake-like robots (Date et al. 2000; Hirose and Morishima 1990). They can move their bodies due to relative motion of their joints. Their joint variables are moved in cyclic patterns that generate periodic shape variations called gaits. This motion is referred to as undulatory motion. Additionally, for some of these systems, their motions may be amplified by an appropriate weight transfer of a rider. All are nonholonomic, and the constraints arise from a nonslip rolling of the wheels.

Underactuated systems are challenging from dynamic and control design perspectives. One of the challenges is a drift term that is present in the state-space control formulation for them. As a consequence, it is difficult to develop intuitive control strategies, as it was possible for driftless control models. Another challenge is the second order nonholonomic constraint equation that complicates direct applications of control techniques developed for first order nonholonomic systems.

The first observation about underactuated systems is that until we are interested in dynamic models for them, i.e., no control design is addressed, the problem of underactuation does not exist. For example, for a manipulator model, its dynamics can be developed using the Lagrange approach. For the roller-racer, its dynamics can be developed using either the Lagrange or Boltzmann–Hamel approaches. Underactuation shows up in the control setting. A control design starts from an evaluation of a system design and selection of available control input. If there are less control inputs than degrees of freedom, a system is underactuated.

In this section, we are interested in the transformation of dynamics of underactuated systems to the form that enables using existing strategies and tracking control algorithms at nonlinear control disposal. We demonstrate that underactuated systems can be considered as nonholonomic systems with second order nonholonomic constraints. Constrained dynamics and control dynamics for them can be obtained by the Lagrange approach or using the GPME presented in Chapter 2.

We start the formulation of control dynamics for an underactuated system from Equations (5.2) for a holonomic system with control inputs added (Jarzębowska 2007), i.e.,

$$M(q)\ddot{q} + C(q,\dot{q}) = E(q)\tau. \tag{5.28}$$

We assume that q is a $(n \times 1)$-dimensional vector of generalized coordinates that belongs to some configuration manifold Ω, τ are independent control vectors, and $\tau \in R^m$, $m < n$. The matrix $E(q) \in R^{n \times m}$ means that there are fewer control inputs than degrees of freedom. Now, let us assume that actuated degrees of freedom are represented by elements $q_1 \in R^m$ and unactuated by elements $q_2 \in R^{n-m}$. Partition of the generalized coordinate vector is then $q = (q_1, q_2)$. Some rearrangement of coordinates may be needed to obtain vectors q_1 and q_2. The control dynamics of Equation (5.28) can be written as

$$M_{11}(q)\ddot{q}_1 + M_{12}(q)\ddot{q}_2 + C_1(q,\dot{q}) = E_1(q)\tau,$$
$$M_{21}(q)\ddot{q}_1 + M_{22}(q)\ddot{q}_2 + C_2(q,\dot{q}) = 0, \tag{5.29}$$

where $C_1(q,\dot{q}) \in R^m$, $C_2(q,\dot{q}) \in R^{n-m}$, M_{ij}, $i, j = 1, 2$, are components of the $(n \times n)$ inertia matrix $M(q)$. The matrix $E_1(q) \in R^{m \times m}$ is invertible for all $q \in \Omega$, and it is obtained from $E(q)$ partition, that is, $E(q) = [E_1(q), 0]^T$. The second equation in (5.29) can be solved for $\ddot{q}_2$ such that

$$\ddot{q}_2 = -M_{22}^{-1}(q)[M_{21}(q)\ddot{q}_1 + C_2(q,\dot{q})]$$

and substituted to the first one yields

$$\hat{M}(q)\ddot{q}_1 + \hat{C}(q,\dot{q}) = E_1(q)\tau, \tag{5.30}$$

where $\hat{M}(q) = M_{11}(q) - M_{12}(q)M_{22}^{-1}(q)M_{21}(q)$, $\hat{C}(q,\dot{q}) = C_1(q,\dot{q}) - M_{12}(q)M_{22}^{-1}(q)C_2(q,\dot{q})$.
Application of the partial feedback linearizing controller of the form

$$\tau = E_1^{-1}(q)[\hat{M}(q)u + \hat{C}(q,\dot{q})] \tag{5.31}$$

transforms Equation (5.30) to

$$\ddot{q}_1 = u,$$
$$\ddot{q}_2 = R(q)\ddot{q}_1 + H(q,\dot{q}), \tag{5.32}$$

with $R(q) = -M_{22}^{-1}(q)M_{21}(q)$, $H(q,\dot{q}) = -M_{22}^{-1}(q)C_2(q,\dot{q})$.
The system of Equations (5.32) can be presented in the canonical control form (3.40), i.e.,

$$\dot{x} = f(x) + \sum_{i=1}^{m} g_i(x)u_i, \tag{5.33}$$

where $x = (q_1, q_2, \dot{q}_1, \dot{q}_2) = (x_1, x_2, x_3, x_4)$, $f(x) = (x_3, x_4, 0, H_i)$, $g_i = (0, 0, e_i, R_i)$, H_i is the i-th component of the $(n-m) \times 1$ vector, and R_i is the i-th column of the matrix function $R(x_1, x_2)$.

Let us now show that a tracking controller for underactuated systems can be designed in the same way as for fully actuated nonholonomic systems. To this end, we start from the GPME for $p = 1$ (2.199) or, equivalently, (5.3) and write it as

$$M(q)\ddot{q}_1 + C(q,\dot{q}_1) = E(q)\tau,$$
$$\ddot{q} = G(q)\ddot{q}_1 + \dot{G}(q)\dot{q}_1. \tag{5.34}$$

Because the constraint equation must be satisfied, it is enough to choose a control law such that $q_1 \to q_p$ for $q_1 \in R^m$, $q_2 \in R^k$, where the number of degrees of freedom is $m = n-k$. Then $E(q) \in R^{m \times m}$ and $\tau \in R^m$, $m < n$. Assume now that we have an underactuated system, for which s, $s < m$, is the number of inputs. We partition q_1 as $q_1 = (q_{1a}, q_{1f})$, where $q_{1a} \in R^s$, $q_{1f} \in R^{m-s}$, and subscripts a and f stand for "actuated" and "free," respectively. Then, Equations (5.34) can be presented as

$$\bar{M}(q)\ddot{q} + \bar{C}(q,\dot{q}) = \bar{E}(q)\tau, \tag{5.35}$$

where now q is of the form $q = (q_{1a}, q_{1f}, q_2)$, $\bar{M}$ is a $(n \times n)$ kind of inertia matrix extended to include terms from the constraint equation, and $\bar{C}$ and $\bar{E}$ are

$$\bar{M}(q) = \begin{bmatrix} M_{11}(q) & M_{12}(q) & 0_{s \times k} \\ M_{21}(q) & M_{22}(q) & 0_{(m-s) \times k} \\ G_{3a}(q) & G_{3f}(q) & I_{k \times k} \end{bmatrix}, \quad \bar{C}(q, \dot{q}) = \begin{bmatrix} C_1(q, \dot{q}) \\ C_2(q, \dot{q}) \\ 0 \end{bmatrix}, \quad \bar{E}(q) = \begin{bmatrix} E_1(q) \\ 0 \\ 0 \end{bmatrix}.$$

(5.36)

Dimensions of submatrices in Equations (5.36) are as follows: $M_{11}(q)$ is a $(s \times s)$ matrix, $M_{12}(q)$ is $(s \times m\text{-}s)$, $M_{21}(q)$ is $(m\text{-}s \times s)$, and $M_{22}(q)$ is $(m\text{-}s \times m\text{-}s)$. The last row in $\bar{M}$ consists of submatrices that result from the partition of $G(q)$:

$$G(q) = \begin{bmatrix} I_{s \times s} & 0 \\ 0 & I_{(m-s) \times (m-s)} \\ G_{3a} & G_{3f} \end{bmatrix}, \quad G_{3a} \text{ is } (k \times s), \text{ and } G_{3f} \text{ is a } k \times (m-s) \text{ matrix.}$$

The vector $\bar{C}(q, \dot{q})$ components are $C_1 \in R^s, C_2 \in R^{m-s}$ and $C_3 \in R^k$, which is a zero vector. Let us now reorder and rename the coordinates, that is, $q_{1a} = q_1$, $q_1 \in R^s$ and $(q_{1f}, q_2) = q_2$, $q_2 \in R^{n-s}$. Then, Equations (5.35) with the coordinate vector $q = (q_1, q_2)$ are equivalent to Equations (5.29). Consequently, we can apply the partial feedback linearizing controller in the form

$$\tau = E_1^{-1}(q)[\bar{M}(q)u + \bar{C}(q, \dot{q})] \tag{5.37}$$

which transforms (5.35) into (5.32).

We demonstrated that the dynamic control model (5.3) obtained by the GPME can be accommodated to underactuated systems with holonomic and nonholonomic constraints. The results (5.35) and (5.37) enable applying all theoretical control results obtained for underactuated systems (Fantoni and Lozano 2001; Reyhanoglu et al. 1996). Specifically, the system (5.32) is not asymptotically stabilizable to $(q_e, 0)$ by the time invariant continuous state feedback law. However, different control objectives can be pursued for it, for example, stabilization to a manifold of equilibrium or to a trajectory that does not converge to a point.

Example 5.3

Consider the two-link planar manipulator model presented in Example 1.5, which is now equipped with one actuator. We formulate the following tracking control objective for it. Its end-effector is to move according to a specified programmed motion, and it is the only constraint put upon the manipulator. The first joint is actuated and the second is not.

The manipulator dynamic control model with the first joint actuated is

$$
\begin{bmatrix} M_{11} & M_{12} \\ M_{21} & M_{22} \end{bmatrix} \begin{bmatrix} \ddot{\Theta}_1 \\ \ddot{\Theta}_2 \end{bmatrix} + \begin{bmatrix} C_{11} & C_{12} \\ C_{21} & C_{22} \end{bmatrix} \begin{bmatrix} \dot{\Theta}_1 \\ \dot{\Theta}_2 \end{bmatrix} = \begin{bmatrix} \tau_1 \\ 0 \end{bmatrix},
\tag{5.38}
$$

or, equivalently,

$$
\begin{bmatrix} \alpha + 2\beta\cos\Theta_2 & \delta + \beta\cos\Theta_2 \\ \delta + \beta\cos\Theta_2 & \delta \end{bmatrix} \begin{bmatrix} \ddot{\Theta}_1 \\ \ddot{\Theta}_2 \end{bmatrix}
$$

$$
+ \begin{bmatrix} -\dot{\Theta}_2\beta\sin\Theta_2 & -\beta\sin\Theta_2(\dot{\Theta}_1 + \dot{\Theta}_2) \\ \dot{\Theta}_1\beta\sin\Theta_2 & 0 \end{bmatrix} \begin{bmatrix} \dot{\Theta}_1 \\ \dot{\Theta}_2 \end{bmatrix} = \begin{bmatrix} \tau_1 \\ 0 \end{bmatrix},
$$

where $\alpha = I_{z1} + I_{z2} + m_1 r_1^2 + m_2(l_1^2 + r_2^2), \beta = m_2 l_1 r_2, \delta = I_{z2} + m_2 r_2^2$.

The equation for the unactuated joint is a second order nonholonomic constraint equation, because the variable Θ_2 is present in the inertia matrix. Following the transformations leading to Equations (5.30), we obtain

$$
\ddot{\Theta}_2 = -\frac{1}{\delta} \left[(\delta + \beta\cos\Theta_2)\ddot{\Theta}_1 + \beta\sin\Theta_2\dot{\Theta}_1^2 \right]
$$

and then

$$
\left[(\alpha + 2\beta\cos\Theta_2) - \frac{1}{\delta}(\delta + \beta\cos\Theta_2)^2 \right] \ddot{\Theta}_1 - \dot{\Theta}_2\beta\sin\Theta_2(2\dot{\Theta}_1 + \dot{\Theta}_2)
$$

$$
- \frac{\dot{\Theta}_1^2\beta\sin\Theta_2}{\delta}(\delta + \beta\cos\Theta_2) = \tau_1.
$$

Using the partial feedback linearizing controller,

$$
\tau_1 = \left[(\alpha + 2\beta\cos\Theta_2) - \frac{1}{\delta}(\delta + \beta\cos\Theta_2)^2 \right] u
$$

$$
- \left[\dot{\Theta}_2\beta\sin\Theta_2(2\dot{\Theta}_1 + \dot{\Theta}_2) + \frac{\dot{\Theta}_1^2\beta\sin\Theta_2}{\delta}(\delta + \beta\cos\Theta_2) \right],
$$

Equations (5.32) for the manipulator model become

$$
\ddot{\Theta}_1 = u,
$$

$$
\ddot{\Theta}_2 = -\frac{1}{\delta}(\delta + \beta\cos\Theta_2)\ddot{\Theta}_1 - \frac{1}{\delta}\beta\sin\Theta_2\dot{\Theta}_1^2.
\tag{5.39}
$$

Equations (5.39) can be expressed in the state-space control form by defining the state variables as $x_1 = \Theta_1, x_2 = \Theta_2, x_3 = \dot{\Theta}_1, x_4 = \dot{\Theta}_2$. Then, we obtain

$$\dot{x}_1 = x_3,$$

$$\dot{x}_2 = x_4,$$

$$\dot{x}_3 = u,$$ (5.40)

$$\dot{x}_4 = -\frac{x_3^2 \beta \sin x_2}{\delta} - \frac{1}{\delta}(\delta + \beta \cos x_2)u$$

or

$$\dot{x} = f(x) + g(x)u,$$

where $f(x) = (x_3, x_4, 0, -x_3^2\beta \sin x_2/\delta)$, $g(x) = (0, 0, e_1, -(\delta + \beta \cos x_2)/\delta)$ with e_1 as the standard basis vector in R^1, are the drift and control vector fields on $\Omega = (-\pi/2, \pi/2) \times (-\pi/2, \pi/2) \times R^2$.

It can be verified that the system (5.40) is strongly accessible because the space spanned by the vectors g, $g_1 = [f, g]$, $g_2 = [g_1, [g, g_1]]$, $g_3 = [f, g_2]$ is of dimension 4 at any $x \in \Omega$. Also, the system is small time locally controllable (STLC) at any equilibrium in Ω. Equilibrium solutions are $x_3 = x_4 = 0$. The function f vanishes at equilibrium. Brackets g, g_1, g_2, g_3 are the "good" brackets, and they constitute a spanning set when evaluated at any equilibrium in Ω.

The conclusion is that the constrained control problem always has a solution. That is, in the neighborhood of any equilibrium state in Ω, any initial state can be transferred to any final state in an arbitrarily short time interval, while maintaining satisfaction of the constraint. Based on the above results, we formulate the following.

Proposition 5.1 (Jarzębowska 2007)

Let $\Omega_e = \{x \in \Omega \mid x_3 = x_4 = 0\}$ denote the equilibrium manifold and $x_e \in \Omega_e$ be an equilibrium solution. The following properties hold for the two-link manipulator dynamics described by Equations (5.39):

1. The system is strongly accessible because the space spanned by (g, g_1, g_2, g_3), where $g_1 = [f, g], g_2 = [g_1, [g, g_1]], g_3 = [f, g_2]$ is of dimension 4 at any $x \in \Omega$.

2. The system is small time locally controllable at x_e because the brackets satisfy sufficient conditions for small time local controllability.

3. There is no time invariant continuous feedback law, which asymptotically stabilizes the closed loop to x_e.

The controllability properties formulated in Proposition 5.1 guarantee that there exist control problem solutions with no excitation of the unactuated

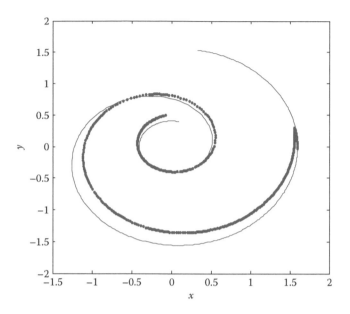

FIGURE 5.7
Programmed motion tracking by the PD controller.

joint 2. Figure 5.9 illustrates this theoretical result. Let us select the programmed constraint (1.7) and apply the PD controller with gains $k_s = 20, k_d = 10$. Initial conditions for the reference motion are $y_{0r} = [1,2;0.43;-0.9;-0.5;0.35;0.35]$ and for controlled $y_{0c} = [0;0;-0.9;-0.5]$. The details about the tracking strategy are discussed in Chapter 7. Tracking results, time histories of position tracking errors, and control torques are presented in Figures 5.7, 5.8, and 5.10.

Other examples of underactuated systems are acrobot (Spong 1995) or pendubot (Spong and Block 1995; Zhang and Tarn 2002). More about underactuated systems control can be found in (Spong 1996; Fantoni and Lozano 2001).

5.5 Tracking Control Algorithms Specified in Quasi-Coordinates

Quasi-coordinates are used rather rarely to model system dynamics and design model-based controllers. However, they may be convenient in applications to some control problems and in this regard be competitive to the generalized coordinates.

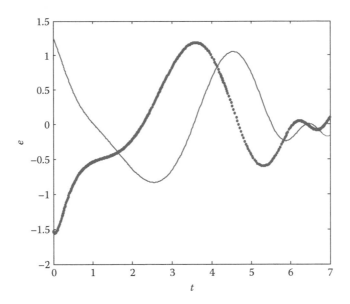

FIGURE 5.8
End-effector position tracking errors e_x (—) and e_y (ooo) vs. time.

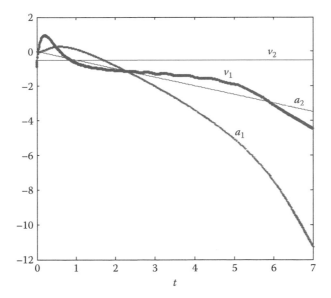

FIGURE 5.9
Controlled joint positions (a_1, a_2) and velocities (v_1, v_2) vs. time.

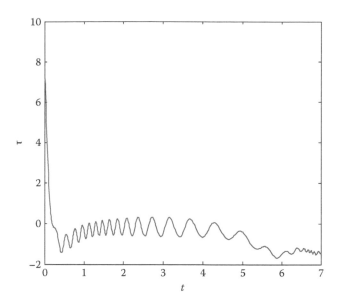

FIGURE 5.10
The control torque applied to the actuated link vs. time.

Example 5.4

In Example 2.10, equations of motion of a roller-racer are derived—Equations (2.154). They are derived using the Boltzmann–Hamel equations. The roller-racer is presented in Figure 1.4.

A roller-racer is one of these fascinating toys that perform undulatory motion. The fundamental means of its propulsion is the pivoting of the steering handlebar around the joint axis and the nonholonomic constraints. The purely kinematic analysis of the roller-racer is not allowed. We cannot determine its global motion by just the shape variation, because it does not possess a sufficient number of nonholonomic constraint equations for this. Kinematics must be complemented with the system dynamics. Research on the roller-racer has been concentrated on kinematic controllability and motion planning (Bullo and Lewis 2003), optimal control (Ostrowski et al. 1997), and dynamics and properties of control (Krishnaprasad and Tsakiris 1998). All studies that concern constrained dynamics and control are based on Lagrange's equations with multipliers followed by the reduction procedure. In Jarzębowska and Lewandowski (2006), motivated by specific features of the roller-racer and available tools from analytical mechanics, the use of the Boltzmann–Hamel equations to develop a control-oriented dynamic model and a control algorithm for the roller-racer are presented.

The Boltzmann–Hamel equations are used for robotic system dynamics rather rarely, for example, in Cameron and Book (1997) the authors derive these equations for simple wheeled vehicles. We demonstrate that the Boltzmann–Hamel equations offer a competitive tool to develop control-oriented dynamic models for constrained systems, and they may facilitate subsequent controller design.

Let us recall that the propulsion and steering come from a rotary motion at the joint connecting two segments of the toy. A torque τ applied at this joint is the only control of the device. Motion of the roller-racer is described by $q = (x, y, \theta, \psi) \in \Omega$, $\Omega = SE(2) \times S^1$, where (x, y, θ) describe a position, and ψ is a shape variable. Equations of nonholonomic constraints as in Example 2.10 are

$$\dot{x}\sin\theta - \dot{y}\cos\theta = 0,$$

$$-\dot{x}\sin\psi + \dot{y}\cos\psi + \dot{\theta}l_1\cos(\theta - \psi) + l_2\dot{\psi} = 0. \tag{5.41}$$

A (4×1) vector of quasi-velocities, in which the Boltzmann–Hamel equations are derived is

$$\omega_1 = v = \dot{x}\cos\theta + \dot{y}\sin\theta,$$

$$\omega_2 = \dot{\theta},$$

$$\omega_3 = \dot{x}\sin\theta - \dot{y}\cos\theta = 0, \tag{5.42}$$

$$\omega_4 = -\dot{x}\sin\psi + \dot{y}\cos\psi + l_1\dot{\theta}\cos(\psi - \theta) + l_2\dot{\psi} = 0.$$

The Boltzmann–Hamel equations for the roller-racer have the form

$$M(\ )\dot{\omega} + C(\ ,\ ^{\cdot},\omega)\omega + D\omega = E(\ )\tau,$$

$$\dot{q} = B(q)\omega, \tag{5.43}$$

$$^{\cdot} = \omega_1 e_1 + \omega_2(e_2 - 1). \tag{5.43a}$$

where $\omega = (\omega_1, \omega_2)$ is a (2×1) vector, and ω_3, ω_4 are set equal to zero because they satisfy the constraints (5.42).

To design a controller that tracks a desired maneuver of the roller-racer, we apply a static state feedback linearization to Equations (5.43) and use it to design a computed torque controller. The maneuver consists of driving a circular trajectory with some desired forward velocity. A rider learns what value of this velocity is enough to start a "smooth" maneuver and changes his or her orientation with respect to the world coordinates so we parameterize the desired trajectory in terms of θ. For illustrative purposes, the desired θ is given as $\theta_d = 0.2(t - t_1)$, where t_1 is the time at which a maneuver may start. We design a controller with respect to the quasi-velocity ω_2. To this end, we take the first of Equations (5.43) and (5.43a). We may solve the first of Equations (5.43) for $\dot{\omega}_1$ and substitute into the second to obtain

$$\dot{\omega}_2 = f_2(\ ,\omega) + b_2(\ )\tau \tag{5.44}$$

where τ is a single input and

$$f_2 = \frac{(m_{11} + d_1)a_2 - m_{21}a_1}{m_{12}m_{21} - (m_{11} + d_1)m_{22}}, \qquad b_2 = \frac{e_1 m_{21} - e_2(m_{11} + d_1)}{m_{12}m_{21} - (m_{11} + d_1)m_{22}},$$

with m_{ij} being the components of the inertia matrix, for which

$$\det(M) = m J_1 + m J_2 \left(\frac{I_1}{I_2}\right)^2 \cos^2 \quad + \frac{J_1 J_2}{I_2^2} \sin^2 \quad , \det(M) \neq 0 \text{ for all} \quad ,$$

and $a_1 = C_{11}\omega_1 + C_{12}\omega_2, a_2 = C_{21}\omega_1 + C_{22}\omega_2$.
Using a partial feedback linearizing controller,

$$\tau = \frac{1}{b_2(\ )}[u - f_2(\ ,\omega)] \tag{5.45}$$

Equation (5.44) becomes

$$\dot{\omega}_2 = u. \tag{5.46}$$

The control algorithm u is designed as

$$u = u_{\omega 2} + u_{\omega 1} \tag{5.47}$$

with

$$u_{\omega 2} = \dot{\omega}_{2d} - k_{p2}(\omega_2 - \omega_{2d}) - k_{i2}\int_0^t (\omega_2 - \omega_{2d})\,dt,$$

$$u_{\omega 1} = [-k_{p1}(\omega_1 - \omega_{1d}) - k_{i1}\int_0^t (\omega_1 - \omega_{1d})\,dt]\sin wt.$$

The controller (5.47) consists of two parts. The first is for tracking a desired maneuver, a trajectory in our case; and the second accelerates the system to a desired forward velocity, which is enough to start maneuvering, and it does not allow the velocity to drop. The higher is w, the faster the forward velocity grows. When friction is ignored in the system model, then $u_{\omega 1}$ can be switched off when the system reaches a desired forward velocity. For the controller (5.47) we formulate the following theorem.

Theorem 5.5

The computed torque controller u designed according to Equation (5.47) provides a desired trajectory tracking by an undulatory motion-like pattern with a bounded tracking error.

Proof: Let us denote by

$$e_1 = \omega_1 - \omega_{1d}, \quad \varepsilon_1 = \int_0^t e_1\,dt, \quad e_2 = \theta - \theta_d, \quad \varepsilon_2 = \int_0^t e_2\,dt,$$

the tracking errors and assume that $u_{\omega 1}$ acts for some time t_1, which is enough to accelerate the roller-racer to the desired velocity value. Then, the closed-loop error dynamics can be written as

$$\ddot{e}_2 + k_{p2}\dot{e}_2 + k_{i2}e_2 + \xi(t) = 0, \tag{5.48}$$

where $\xi(t)$ is bounded and treated as a bounded disturbance. As long as it is bounded, so is the error e_2. If it switches off $u_{\omega 1}$ in the case when friction is neglected and at the moment when the forward velocity is large enough, the controller (5.47) ensures asymptotic tracking of a desired trajectory. This can be verified by direct calculation of the closed-loop characteristic polynomial for Equation (5.48). When, in the presence of friction, $u_{\omega 1}$ is on all the time, the roller-racer performs its undulatory motion about a desired trajectory. q.e.d.

The simulation results are presented for two cases: one is motion without friction for $d_1 = 0, d_2 = 0$, and the second is motion including friction for $d_1 = 0.5, d_2 = 0$. In both cases, initial motion conditions are the same, as in Example 2.10, and the controller gains are set equal to $k_{p2} = 20$, $k_{i2} = 100$ and $k_{p1} = 200$, $k_{i1} = 8$, and $w = 2$. Figures 5.11 through 5.15 present simulation results for the first case. The controller $u_{\omega 1}$ was switched off when the forward velocity was enough to start maneuvering. Figure 5.14 demonstrates actions of both $u_{\omega 2}$ and $u_{\omega 1}$ (speed up), and then $u_{\omega 2}$ separately (tracking a circle). In this simulation study, however, there is a problem with the

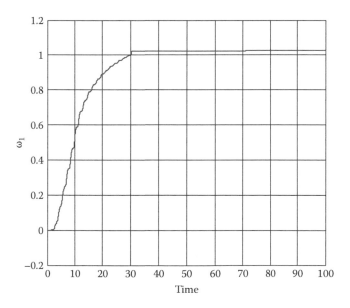

FIGURE 5.11
Increase of ω_1 to a desired value (no friction), $w = 2$.

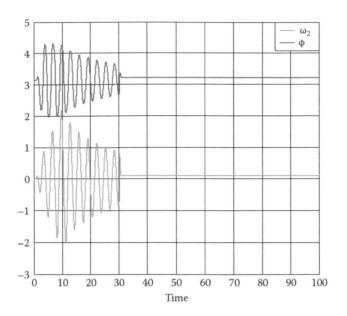

FIGURE 5.12
Angular velocity ω_2 and   versus time (no friction).

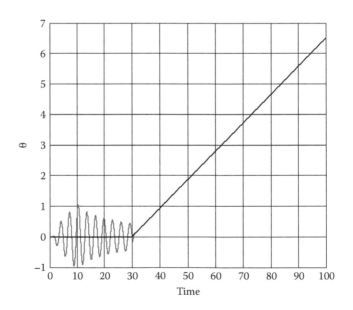

FIGURE 5.13
Control of the angle θ (no friction).

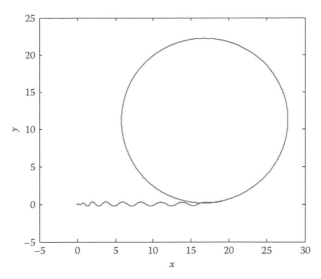

FIGURE 5.14
Desired trajectory tracking (no friction).

torque that exhibit jump after switching off $u_{\omega 1}$ as seen in Figure 5.15. For the second case, $u_{\omega 1}$ corrects forward velocity, and its action is clearly seen in Figures 5.18 and 5.19. The computed torque in this case exhibits no jump (Figure 5.20). In Figures 5.16 and 5.17 the forward and angular velocities of the roller-racer tracking a circle are presented.

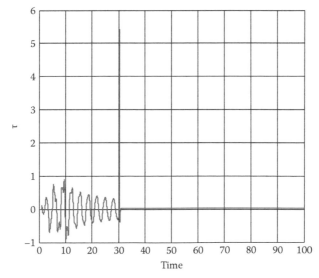

FIGURE 5.15
Control torque versus time (no friction).

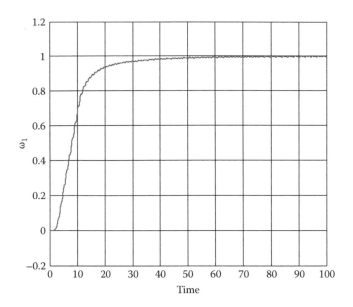

FIGURE 5.16
Increase of ω_1 to a desired value (with friction), $w = 2$.

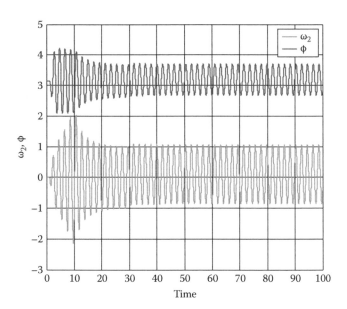

FIGURE 5.17
Angular velocity ω_2 and the angle ϕ vs. time (with friction).

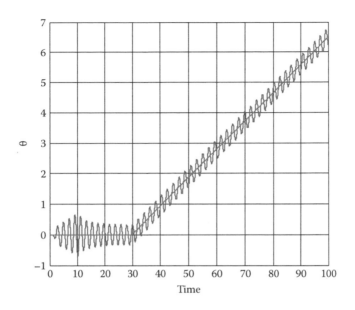

FIGURE 5.18
Control of the angle θ (with friction).

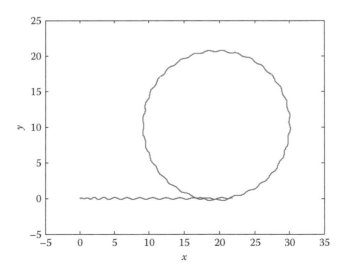

FIGURE 5.19
Desired trajectory tracking (with friction).

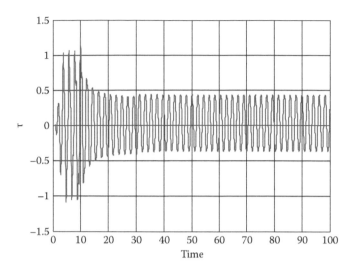

FIGURE 5.20
Control torque versus time (with friction).

To conclude, we presented a competitive method of derivation of a dynamic model for a nonholonomic system based on the Boltzmann–Hamel equations and the subsequent tracking controller design. The Boltzmann–Hamel equations yield a dynamic model that facilitates a controller design.

Problems

1. For the unicycle model from Example 5.2, take the same desired trajectory as in the example, and design a controller based upon the Lagrange approach. Apply the PD controller to the reduced state equations; compare the results and the amount of work needed to obtain the dynamic control model.
2. For the roller-racer model from Example 5.4, derive equations of motion using the Lagrange equations with multipliers and transform it to the reduced state form. Design a tracking controller, to track the circular trajectory as in Example 5.4.

References

Arai, H., and S. Tachi. 1990. Dynamic control of a manipulator with passive joints— position control experiments by a prototype manipulator. In *Proc. IEEE Int. Workshop on Intell. Rob. Syst.* 935–949. Tsukuba: Japan.
Arimoto, S. 1990. Design of robot control systems. *Advanced Robotics* 4:79–91.

Arimoto, S. 1995. Fundamental problems of robot control: Part II a nonlinear circuit theory towards an understanding of dexterous motions. *Robotica* 13:111–122.

Bloch, A.M. 2003. *Nonholonomic mechanics and control.* vol. 24: Interdisciplinary applied mathematics. New York: Springer.

Bullo, F., and A.D. Lewis. 2003. Kinematic controllability and motion planning for the snake-board. *IEEE Trans. Robot. Automat.* 19(3):494–498.

Cameron, J.M., and W.J. Book. 1997. Modeling mechanisms with nonholonomic joints using the Boltzmann-Hamel equations. *Int. J. Robot. Res.* 16(1):47–59.

Chitta, S., F. Heger, and V. Kumar. 2004. Design and gait control of a rollerblading robot. In *Proc. 2004 IEEE Int. Conf. Robot. Automat.* 386–391.

Cortes, J. Monforte. 2002. *Geometric, control and numerical aspects of nonholonomic systems.* Lecture Notes in Mathematics. New York: Springer.

Date, H., Y. Hoshi, and M. Sampei. 2000. Locomotion control of a snake-like robot based on dynamic manipulability. *Proc. IEEE/RSJ Intl. Conf. Intell. Robot. Syst.* 2236–2241.

Fantoni, I., and R. Lozano. 2001. *Non-linear control for underactuated mechanical systems.* New York: Springer.

Fradkov, A.L., I.V. Miroshnik, and V.O. Nikifor. 1999. *Nonlinear and adaptive control of complex systems.* Dordrecht: Kluwer.

Harkegard, O., and S.T. Glad. 2000. A backstepping design for flight path angle control. In *Proc. 39th IEEE Conf. Decision Contr.* 3570–3575. Sydney: Australia.

Hirose, S., and H. Takeuchi. 1996. Study on roller-walker (basic characteristics and its control). In *Proc. 1996 IEEE Int. Conf. Robot. Automat.* 3265–3270.

Hirose, S., and A. Morishima. 1990. Design and control of a mobile robot with an articulated body. *Int. J. Robot. Res.* 9(2):99–114.

Isidori, A. 1989. *Nonlinear control systems.* Berlin: Springer.

Jarzębowska, E., and R. Lewandowski. 2006. Modeling and control design using the Boltzmann-Hamel equations: A roller-racer example. In *Proc. Symp. Robot Control, SYROCO.* Bologna. Italy.

Jarzębowska, E. 2007. *Model-based tracking control strategies for constrained mechanical systems.* Palermo: International Society for Advance Research.

Kelly, L., V. Santibanez, and A. Loria. 2005. *Control of robot manipulators in joint space.* London: Springer-Verlag.

Khalil, H.K. 2001. *Nonlinear systems.* Englewood Cliffs, NJ: Prentice Hall.

Krishnaprasad, P.S., and D.P. Tsakiris. 1998. SE(2)-snakes and motion control: A study of the roller-racer. *Technical report, Center for Dynamics and Control of Smart Structures,* University of Maryland, College Park.

Krishnaprasad, P.S., and D.P. Tsakiris. 2001. Oscillations, SE(2)-snakes and motion control: A study of the roller-racer. *Dyn. Syst.* 16(4):347–397.

Krstic, M., and P.V. Kokotovic. 1995. Lean backstepping design for a jet engine compressor model. In *Proc. Fourth IEEE Conf. Control Appl.* 1047–1052.

Lane, S.H., and R.F. Stengel. 1988. Flight control design using non-linear inverse dynamics. *Automatica* 24(4):471–483.

Lewis, F., D.M. Dowson, and Ch.T. Abdallach. 2004. *Robot manipulator control, theory and practice.* New York: Marcel Dekker.

Nijmeijer, H., and A. van der Shaft. 1990. *Nonlinear dynamical control systems.* New York: Springer-Verlag.

Ostrowski, J., J.P. Desai, and V. Kumar. 1997. Optimal gait selection for nonholonomic locomotion systems. In *Proc. 1997 IEEE Int. Conf. Robot. Automat.* 786–791.

Reyhanoglu, M., A. van der Schaft, N.H. McClamroch, and I. Kolmanovsky. 1996. Nonlinear control of a class of underactuated systems. In *Proc. 35th Conf. Decision Contr.* 1682–1687. Kobe: Japan.

Shiroma, N., H. Arai, and K. Tanie. 1997. Nonlinear control of a planar free-link under a nonholonomic constraint. In *Proc. ICA* 103–109. Monterey, CA.

Slotine, J., and W. Li. 1996. *Applied nonlinear control*. Englewood Cliffs, NJ: Prentice Hall.

Sontag, D. 1990. *Mathematical control theory*. New York: Springer.

Spong, M.W., and M. Vidyasagar. 1989. *Robot control and dynamics*. New York: Wiley.

Spong, M.W. 1995. The swing up control problem for the acrobat. *IEEE Contr. Syst.* 15(1):49–55.

Spong, M.W., and D. Block. 1995. The pendubot: A mechatronic systems for control research and education. In *Proc. IEEE Conf. Decision Contr.* 555–557. New Orleans, LA.

Stevens, B.L., and F. Lewis. 1992. *Aircraft control and simulation*. New York: John Wiley & Sons.

Vidyasagar, M. 1992. *Nonlinear systems analysis*. Englewood Cliffs, NJ: Prentice Hall.

Wen, J., and D.S. Bayard. 1988. A new class of control laws for robotic manipulators. Part 1: Non-adaptive case. *Int. J. Contr.* 47(8):1361–1385.

Zhang, M., and T.J. Tarn. 2002. Hybrid control of the pendubot. *IEEE Trans. Mechatronics* 7:79–86.

6

Path Following Strategies for Nonlinear Systems

Path following is a control task that consists of following a prespecified path usually parameterized by its arc length. Usually it is assumed that a vehicle that follows the path moves forward, and the time information, comparing to tracking, is not a control demand, i.e., there are no temporal specifications.

Design of feedback controllers for the path following may be considered from two points of view. First, by separating the geometric and time information of a path, path following may be seen as a subproblem of trajectory tracking (Coelho and Nunes 2005). Second, looking at the controlled states, path following may be a part of stabilization (Canudas de Wit et al. 1996).

Path following control problems are addressed in a lot of works (see, e.g., Micaelli and Samson 1993, Samson and Ait-Abderrahim 1991 for the first results, Jiang and Nijmeijer 1996 and Canudas de Wit et al. 1993 for references). The path following problem may be addressed at the kinematic control level, as in Micaelli and Samson (1993), Samson and Ait-Abderrahim (1991), Seo and Lee (2006), or at the dynamic control level as in Bakker et al. (2010), Coelho and Nunes (2005), and Soetanto et al. (2003). Significant progress has been made in path following, in both the model formulation and controller design, from the time of publishing the pioneering works. In Soetanto et al. (2003), results obtained in Micaelli and Samson (1993) were extended to the dynamic level with parameter uncertainties, and a global convergence of following is achieved. In Coelho and Nunes (2005), a discrete state-space controller is applied. It guarantees overall system stability in the presence of external disturbances, modeling errors, and noise. In Seo and Lee (2006), bounded angular velocity errors are taken into account, and a sliding mode-type controller is designed to obtain a robust vehicle steering. The robustness applies to the input disturbance and uncertain dynamics. A classical presentation of the path following problem in the context of other control problems can be found in Canudas de Wit et al. (1996).

6.1 Path Following Strategies Based on Kinematic Control Models

Path following at the kinematic level is formulated as follows: Given a car-like vehicle of a specified kinematics, design a feedback controller that enables path following such that a distance to the path and the orientation error tend to zero.

Consider then a mobile robot whose kinematics is equivalent to that of a unicycle. The robot is equipped with two differential-drive wheels in the same axis and one castor wheel that prevents the robot platform from falling. Denote by (x,y) the position of the center of the rear axis and by φ the heading angle of the robot (see Figure 6.1).

The kinematic constraints on the robot motion have the form as presented in Example 3.2:

$$\dot{x} = v\cos\varphi$$

$$\dot{y} = v\sin\varphi, \tag{6.1}$$

$$\dot{\varphi} = \omega,$$

where control inputs are the forward velocity $v = \dot{x}\cos\varphi + \dot{y}\sin\varphi$, and the robot angular velocity is ω. Also, it is usually assumed that motors of the robot are capable of delivering larger velocities than the ones desirable during normal operation.

Let us denote a control input (2×1) vector by $u = (v, \omega)$ and a velocity (2×1) vector of robot wheels by $u_r = (v_r, v_l)$. Then, the forward and angular

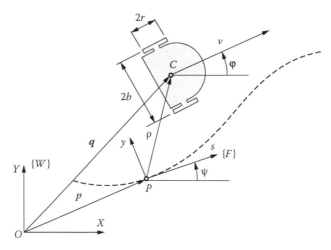

FIGURE 6.1
Mobile robot parameters.

velocities of the robot are related to the right and left wheel velocities as follows:

$$u = \begin{bmatrix} v \\ \omega \end{bmatrix} = \begin{bmatrix} 1/2 & 1/2 \\ 1/2b & -1/2b \end{bmatrix} \begin{bmatrix} v_r \\ v_l \end{bmatrix} \tag{6.2}$$

with $2b$ being the distance between the wheels.

The path following problem is parameterized as follows. The kinematic model of the robot is transformed to the Frenet frame {F} that may move along the path. The axes of the Frenet frame are the body axes of a virtual vehicle that should be followed by a real one. In what follows, we use the notation that is now standard in textbooks; for more details, see Micaelli and Samson (1993), Samson and Ait-Abderrahim (1991), and Soetanto et al. (2003).

The origin of the Frenet frame {F} enables evolving according to a conveniently defined time function. It yields an extra controller design parameter. According to Figure 6.1, P is an arbitrary point on the path to be followed, and C is the mass center of the robot. The frame {F} origin is located at P, and its axes are (s,y). Coordinates of C may be expressed by two vectors, either as $q = (X,Y,0)$ or $\rho = (s_1,y_1,0)$. The position vector for P can be denoted by p. The rotation matrix that describes transformation from the world coordinate frame {W} coordinates (X,Y) to the frame {F} coordinates (s,y) is $R = R(\psi)$, where ψ is the angle between the horizontal axis parallel to X and s. The following relations hold:

$$\dot{\psi} = c_c(s)\dot{s},$$
$$\dot{c}_c(s) = g_c(s)\dot{s}, \tag{6.3}$$

where $c_c(s)$ is the path curvature,

$$g_c(s) = \frac{dc_c(s)}{ds}.$$

The velocity of P in {W} expressed in {F} becomes

$$\left(\frac{dp}{dt}\right)_F = [\dot{s},0,0]^T,$$

and the velocity of C in {W} is

$$\left(\frac{dq}{dt}\right)_W = \left(\frac{dp}{dt}\right)_W + R^{-1}\left(\frac{d\rho}{dt}\right)_F + R^{-1}(\dot{\psi} \times r). \tag{6.4}$$

Multiplying Equation (6.4) by R from the left yields

$$R\left(\frac{dq}{dt}\right)_W = \left(\frac{dp}{dt}\right)_F + \left(\frac{d\rho}{dt}\right)_F + \dot{\psi} \times r. \tag{6.5}$$

Using (6.3) and the relations

$$
\left(\frac{dq}{dt}\right)_W = \begin{bmatrix} \dot{X} \\ \dot{Y} \\ 0 \end{bmatrix}, \quad \left(\frac{d\rho}{dt}\right)_F = \begin{bmatrix} \dot{s}_1 \\ \dot{y}_1 \\ 0 \end{bmatrix}, \quad \dot{\psi} \times r = \begin{bmatrix} -c_c(s)\dot{s}y_1 \\ c_c(s)\dot{s}s_1 \\ 0 \end{bmatrix},
$$

Equation (6.5) takes the form

$$
R\begin{bmatrix} \dot{X} \\ \dot{Y} \\ 0 \end{bmatrix} = \begin{bmatrix} \dot{s}(1 - c_c(s)y_1) + \dot{s}_1 \\ \dot{y}_1 + c_c(s)\dot{s}s_1 \\ 0 \end{bmatrix}.
\tag{6.6}
$$

Solving Equation (6.6) for $\dot{s}_1, \dot{y}_1$ yields

$$
\dot{s}_1 = [\cos\psi \quad \sin\psi]\begin{bmatrix} \dot{X} \\ \dot{Y} \end{bmatrix} - \dot{s}(1 - c_c(s)y_1),
$$

$$
\tag{6.7}
$$

$$
\dot{y}_1 = [-\sin\psi \quad \cos\psi]\begin{bmatrix} \dot{X} \\ \dot{Y} \end{bmatrix} - c_c(s)\dot{s}s_1.
$$

The relations in Equation (6.7) are different than the ones obtained in Micaelli and Samson (1993) due to the translation of the {F} origin. There, as well as in Seo and Lee (2006) and Canudas de Wit et al. (1996), $s_1 = 0$, because P is defined as the projection of C on the path. What follows is that Equations (6.7) have to be solved for $\dot{s}$, and the singularity shows up at $y_1 = 1/c_c$. It results in the requirement on an initial location of C, which must remain in a tube around the path. The tube radius must be less than $1/c_{cmax}$, where c_{cmax} is the maximum path curvature. When c_{cmax} is large, even in a small section of the path, this restriction constrains the robot initial position no matter where it starts its motion with regard to the path. When $s_1 \neq 0$, an extra degree of freedom is introduced to a controller design, but the singularity can be avoided. A new virtual target is assigned to the variable s_1.

The robot kinematics in {W} is described by Equations (6.1). Substituting (6.1) into (6.7) and denoting by $\theta_r = -\psi$, we may obtain the kinematic model in {F} coordinates as

$$
\dot{s}_1 = -\dot{s}(1 - c_c(s)y_1) + v\cos\theta_r,
$$

$$
\dot{y}_1 = -c_c(s)\dot{s}s_1 + v\sin\theta_r,
\tag{6.8}
$$

$$
\dot{\theta}_r = \dot{\,} - c_c(s)\dot{s}.
$$

For the kinematic model (6.8), a controller may be designed. The control goal is to drive y_1, s_1 and θ_r to zero. Following Samson and Ait-Abderrahim (1991) and Soetanto et al. (2003), the control laws may be selected as

$$\dot{s} = v\cos\theta_r + k_1 s_1,$$

$$\dot{\theta}_r = \dot{\delta} - \gamma y_1 v \frac{\sin\theta_r - \sin\delta}{\theta_r - \delta} - k_2(\theta_r - \delta), \tag{6.9}$$

where it is assumed that for some design parameter γ, the following hold:

- $\lim_{t\to\infty} v(t) \neq 0,$
- $\delta(0, v) = 0,$
- $y_1 v \sin\delta(y_1, v) \leq 0 \quad \forall y \quad \forall v.$

The function δ is significant in the transient maneuvers during the path approach phase (see Micaelli and Samson 1993 for details). To prove the convergence of the closed-loop Equations (6.8) and (6.9), the following Lyapunov function candidate can be selected as

$$V_1 = \frac{1}{2}\left(s_1^2 + y_1^2\right) + \frac{1}{2\gamma}[\theta_r - \delta(y_1, v)]^2. \tag{6.10}$$

It can be verified that for controls (6.9), $\dot{V}_1 \leq 0$.

In the kinematic control design, the robot velocity $v(t)$ follows a desired velocity function $v_d(t)$. The angular velocity $\dot{\theta}_r$ is assumed to be a control input.

6.2 Path Following Strategies Based on Dynamic Control Models

Let us now, using the backstepping approach, extend the kinematic controller (6.9) to the dynamic one. A dynamic model of the two-wheeled mobile robot, specified in {F} coordinate frame, is of the form

$$\dot{v} = \frac{\tau_1 + \tau_2}{mr},$$

$$\dot{\omega} = \frac{b(\tau_1 - \tau_2)}{Ir} - c_c\ddot{s} - g_c\dot{s}^2, \tag{6.11}$$

where m is the robot mass, I is its moment of inertia, r is the wheel radius, $2b$ is the distance between the wheels, and τ_1, τ_2 are control inputs.

The path following task at the dynamic level is formulated as follows: Given a desired velocity profile $v_d(t) > v_{min} > 0$, design feedback control laws for τ_1 and τ_2 to drive y_1, θ_r and $v(t) - v_d(t)$ asymptotically to zero.

Notice the following differences between the kinematic and dynamic controller designs:

- In the dynamic design, a feedback controller has to be designed such that the tracking error $v(t) - v_d(t)$ approaches zero.

- The robot angular velocity is not a control input.

The virtual control law for the desired behavior of $\dot{\theta}_r$ in Equations (6.9) can be designed as follows (Krstic et al. 1995; Soetanto et al. 2003):

$$\dot{\theta} = \dot{\delta} - \gamma y_1 v \frac{\sin \theta_r - \sin \delta}{\theta_r - \delta} - k_2 (\theta_r - \delta). \tag{6.12}$$

Let $e = \dot{\theta}_r - \dot{\theta}$ be the following error. Replacing $\dot{\theta}_r = e + \dot{\theta}$ in the derivative of the Lyapunov function candidate V_1 selected as in Equation (6.10), we may show that

$$\dot{V}_1 = -k_1 s_1^2 + y_1 v \sin \delta - \frac{(\theta_r - \delta)^2}{\gamma} + \frac{(\theta_r - \delta)}{\gamma} e. \tag{6.13}$$

Select now a new Lyapunov function candidate V_2 as

$$V_2 = V_1 + \frac{1}{2} \left[e^2 + (v - v_d)^2 \right] \tag{6.14}$$

It can be verified that for

$$\dot{e} = -\frac{(\theta_r - \delta)}{\gamma} - \frac{k_3}{I} e, \qquad \dot{v} = \dot{v}_d - \frac{k_4}{m} (v - v_d), \tag{6.15}$$

the derivative of V_2 satisfies $\dot{V}_2 \leq 0$.

Solving the robot dynamics (6.11) for τ_1 and τ_2, we obtain that

$$\tau_1 = \frac{r}{2} \left[mf_2 - k_4(v - v_d) + \frac{I}{b} f_1 - \frac{k_3}{b} e \right],$$

$$\tau_2 = \frac{r}{2} \left[mf_2 - k_4(v - v_d) - \frac{I}{b} f_1 + \frac{k_3}{b} e \right] \tag{6.16}$$

with $f_1 = \ddot{\theta} - \frac{(\theta_r - \delta)}{\gamma} + c_c \ddot{s} + g_c \dot{s}^2$, $f_2 = \dot{v}_d$.

Details about the simulation study and the controller (6.16) performance with the additional adaptation for uncertainties of parameters m, I, r, b, can be found in Soetanto et al. (2003) and references therein.

Problems

1. Prove that the time derivative of the Lyapunov function candidate V_1 defined in Equation (6.10) is negative semi-definite for control inputs (6.9).
2. Derive the equations of motion for the mobile robot and transform them to the dynamic control model (6.11) in {F} frame.
3. Prove that the time derivative of the Lyapunov function candidate V_2 defined in Equation (6.14) is negative semi-definite for control inputs (6.16).

References

Bakker, T., K. Asselt, and J. Bontsema. 2010. A path following algorithm for mobile robots. *Auton. Robot.* 29:85–97.

Coelho, P., and U. Nunes. 2005. Path-following control of mobile robots in presence of uncertainties. *IEEE Trans. Robot.* 21(2):252–261.

Canudas de Wit, C., H. Khennouf, C. Samson, and O. Sordalen. 1993. Nonlinear control design for mobile robots. In *Recent trends in mobile robotics*, ed. Y.F. Zheng, World science series in robotics and automated systems 11:121–156.

Canudas-de-Wit, C., B. Siciliano, and G. Bastin, ed. 1996. *Theory of robot control*. Communications and Control Engineering Series. New York: Springer.

Jiang, Z., and H. Nijmeijer. 1999. A recursive technique for tracking control of nonholonomic systems in chained form. *IEEE Trans. Robot. Automat.* 44(2):265–279.

Krstic, M., I. Kanellakopoulos, and P. Kokotovic. 1995. *Nonlinear and adaptive control design*. New York: John Wiley & Sons.

Micaelli, A., and C. Samson. 1993. Trajectory tracking for unicycle type and two-steering wheels mobile robots. *Technical Report No. 2097, INRIA*, Sophia Antipolis, France.

Samson, C., and K. Ait-Abderrahim. 1991. Mobile robot control part 1: Feedback control of a nonholonomic mobile robot. *Technical Report No. 1281, INRIA*, Sophia Antipolis, France.

Seo, K., and J.S. Lee. 2006. Kinematic path-following control of a mobile robot under bounded angular velocity error. *Adv. Robot.* 20(1):1–23.

Soetanto, D., L. Lapierre, and A. Pascoal. 2003. Adaptive, non-singular path-following control of dynamic wheeled robots. In *Proc. 42nd IEEE Conf. Decision Contr.* 1765–1770. Maui, HI.

7

Model Reference Tracking Control of High Order Nonholonomic Systems

In Chapters 2 and 5, we present a unified dynamic model (2.199) and a unified dynamic control model (5.3) for a constrained system. Both models are derived based on the generalized programmed motion equations (GPME). The unified dynamics (2.199) can incorporate both material and non-material constraints of an arbitrary order. The two results contribute to theoretical development in both mechanics and control of constrained systems. The first contribution is an extension of the class of constraints that are merged into a dynamic model. The essential difference between our constrained dynamic model and models presented in the literature is that only material position and velocity constraints are merged into the latter models. Second, Equations (2.199) are in the reduced-state form. The third contribution relates to control, and it is the introduction of the unified formulation of constraints (3.78). Also, (5.3) is competitive to dynamic control models control theory uses, because it is free of the constraint reaction forces.

These results in modeling and control of constrained systems conform to the latest trends of integrating mechanics and control tools in the design of controllers.

Based on these results, a new tracking control strategy is presented in this chapter. It is a model reference tracking control strategy for programmed motion, which we call strategy for programmed motion tracking for brevity. First, we present architecture of the strategy. We demonstrate that it enables tracking any motion specified by equations of constraints, and it is simpler in the implementation in comparison with traditional tracking strategies designed for nonholonomic systems. Second, we demonstrate the effectiveness of programmed motion tracking through simulations. Several tasks formulated by other researchers are adopted and executed by our tracking strategy. Also, we present tracking tasks that can be achieved using the strategy for programmed motion tracking only. To help comparisons, we consider two types of examples that are popular illustrative examples in control references, i.e., a unicycle-type robot model and a two-link planar manipulator model.

7.1 Model Reference Tracking Control Strategy for Programmed Motion

7.1.1 A Reference Dynamic Model for Programmed Motion

Solutions of the constrained dynamics (2.199) can predefine system's positions, velocities, and their time derivatives in the programmed motion, i.e., they can plan the programmed motion. Based on the preplanned motion a controller can be designed. Hence, we formulate the following definition.

Definition 7.1: The dynamic model (2.199) for a system subjected to both material and programmed constraints

$$M(q)\ddot{q} + V(q,\dot{q}) + D(q) = Q(t,q,\dot{q}),$$

$$B(t,q,\dot{q},...,q^{(p-1)})q^{(p)} + s(t,q,\dot{q},...,q^{(p-1)}) = 0, \tag{7.1}$$

is referred to as a reference dynamic model for programmed motion.

The reference dynamic model for programmed motion is developed in three steps:

1. Formulation of the constraint equations (3.78)
2. Generation of equations of the reference dynamics (7.1)
3. Solution of the reference dynamics (7.1)

The introduction of this model to control is the first step in applying the latest theoretical results from analytical dynamics to control theory. From the point of view of control applications, the reference dynamic model for programmed motion (7.1) is an extension of the model reported in You and Chen (1993) and Yun and Sarkar (1998), where the authors state that they present a unified approach to dynamic modeling and control of holonomic and nonholonomic systems. However, they consider systems with constraint equations of first orders only. We can merge equality constraints of any order into the extension we present (Equation 7.1).

The reference dynamic model for programmed motion can be viewed as a motion planner. However, motion planning in this book is defined in a different way than in most control references, for example, Li and Canny (1993) and Murray and Sastry (1993). In those works, motion planning is divided into path planning, trajectory planning, and trajectory tracking. In the approach we present, the reference dynamic model for programmed motion serves programmed motion planning with trajectory planning as a peculiar case. Programmed motion planning is defined as follows.

Definition 7.2: Programmed motion planning for a system subjected to material and programmed constraints specified by

$$B(t,q,\dot{q},...,q^{(p-1)})q^{(p)} + s(t,q,\dot{q},...,q^{(p-1)}) = 0$$

consists in finding time histories of positions and their time derivatives, that is, $q_p(t), \dot{q}_p(t),...,q_p^s(t)$, $s = 1,...,p$, along motion consistent with the constraints.

Specifically, in our formulation of programmed motion planning, trajectory planning consists of a solution $q_p(t)$ of Equation (7.1), in which a programmed constraint is the position constraint (3.76). Definition 7.2 generalizes trajectory planning to programmed motion planning.

Note that a programmed constraint has to be verified whether or not it is feasible for a given system. By feasible, we mean that the system with its actuators is capable of reaching desired positions, velocities, and accelerations needed to follow the program, and the programmed constraint does not violate any material constraint.

7.1.2 Architecture of the Model Reference Tracking Control Strategy for Programmed Motion

We employ the reference dynamic model for programmed motion (7.1) to design a control strategy for tracking programmed motions.

The control objective of programmed motion tracking is formulated as follows: Given a programmed motion specified by equations of programmed constraints and a description of a constrained system as nonlinear dynamics,

$$M(q)\ddot{q} + V(q,\dot{q}) + D(q) = Q(t,q,\dot{q}),$$

$$B(t,q,\dot{q},...,q^{(p-1)})q^{(p)} + s(t,q,\dot{q},...,q^{(p-1)}) = 0,$$

design a feedback controller that can track the desired programmed motion.

In our setting, the objective of control design is to make system positions q_i, $i = 1,...,n$, and their time derivatives follow programmed positions $q_{ip}(t)$ and their time derivatives.

To achieve our control objective, we design a tracking strategy whose architecture is presented in Figure 7.1. It is based on two dynamic models: the reference dynamic model for programmed motion (7.1) and the unified dynamic control model (5.3):

$$M(q)\ddot{q} + C(q,\dot{q})\dot{q} + D(q) = E(q)\tau,$$

$$B_1(q)\dot{q} = 0. \tag{7.2}$$

The motivation to design the tracking strategy is that a variety of programmed constraint equations disable the design of a general scheme of a tracking

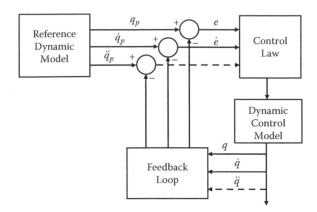

FIGURE 7.1
Architecture of the model reference tracking control for programmed motion.

controller. Instead, in this tracking strategy, we separate programmed constraints from material constraints. All constraint equations put on a system are merged into the reference dynamic model (7.1). The unified dynamic control model (7.2) is derived for the system with material constraints only. This separation of constraint types between the two dynamic models has tremendous implications. First, Equations (7.2) are equivalent to the dynamic control model that nonlinear control uses; second, the control law block may consist of both controllers that are dedicated to holonomic systems and model-based controllers for nonholonomic systems. Also, this control architecture is suitable for underactuated systems. Outputs of Equations (7.1) are inputs to a tracking controller in Equations (7.2).

From the point of view of the control theory, the model reference tracking control strategy for programmed motion may be compared with a two degrees of freedom design (van Nieuwstadt 1997). First, we generate motion that takes the system to the goal configuration and is consistent with material constraints. Then we design a controller that ensures that the system tracks this motion.

Advantages of the programmed motion tracking strategy can be summarized as follows:

- The reference dynamic model (7.1) can include arbitrary order nonholonomic constraints. In this way, any programmed motion can be planned.
- The tracking strategy separates material constraints from programmed. Then, the dynamic control model (7.2) is equivalent to model nonlinear control uses, i.e., models based on Lagrange's approach.
- The equivalence of Equations (7.2) and models based on Lagrange's approach promotes the adoption of existing control algorithms, even those dedicated to holonomic systems.

- The equivalence of Equations (7.2) and models based on Lagrange's approach enables us to use controllers that proved their good performance through laboratory tests.
- The dynamic control model (7.2) enables selecting states that are to be controlled.
- The tracking strategy takes advantage of one dynamic control model (7.2) for both holonomic and nonholonomic systems.
- The tracking strategy enables designing controllers for both motion and a force applied on the environment because the reference dynamics can be derived in the joint or task spaces.
- The tracking strategy extends "trajectory tracking" to "programmed motion tracking."
- The tracking strategy can be applied to systems whose dynamics are completely known and to systems with parametric or structural uncertainties (for details, see Section 7.3).
- The tracking strategy can be applied to underactuated systems.
- The reference dynamic model can be generated off-line and stored in a computer. A library of reference models that plan different tasks can be created. All these reference models can be applied to one dynamic control model of a specific system.

7.1.3 A Controller Design for Programmed Motion Tracking

The model reference tracking control strategy for programmed motion is developed and verified through numerical tests. In this section, we discuss practical aspects of control implementation. We are conscious of limitations of outcomes from the theoretical and simulation studies. This is why, at a controller design stage, we take into account experimental evaluations of implementation of controllers. From a variety of experiment results in tracking controllers reported in the literature (e.g., Berghuis 1993; Oriolo et al. 2002; Whitcomb et al. 1991), we wish to select controllers that are the most suitable for programmed motion tracking. Both the nature and variety of programmed constraints and requirements for accuracy of programmed motion tracking suggest the following properties of a "good" controller for programmed motion tracking:

- Reduction of on-line computation to minimum
- Elimination of the acceleration measurement
- Elimination of the need to ensure that the inverse of the mass matrix exists
- Utilization of desired programmed motion information
- Satisfaction of the criterion of the global convergence
- Application for repetitive tasks

Some works demonstrate superiority of adaptive controllers compared with conventional proportional-derivative (PD) controllers. For example, a careful evaluation and comparison of classical PD controllers with model-based non-adaptive and adaptive controllers applied to holonomic systems is reported in Berghuis (1993). Studies reported in Whitcomb et al. (1991) describe evaluation results of some adaptive passivity-based controllers. Studies in Berghuis (1993) extend these evaluations on both inverse dynamics and passivity-based controllers.

In Berghuis (1993), the simulation and experimental performance are compared based upon a performance measure, defined by position tracking errors:

$$J = \frac{1}{t} \int_0^t e^T(t)e(t)\,dt, \tag{7.3}$$

where t is the total simulation time and $e(t)$ is a vector of a position tracking error. Following this performance measure, a significant superiority of adaptive model-based controllers over classical PD controllers is demonstrated.

Other experiments show that simulation and experimentation performance measures with the same criterion J differ. Specifically, these differences are caused by unmodeled dynamic effects in control dynamics. When unmodeled dynamics becomes dominant, underrated control values are predicted by simulations. Also, experiments confirm results of many studies on friction, which conclude that unmodeled or nonadequately modeled friction may cause a significant increase in tracking errors (Armstrong–Helouvry 1991; Jarzębowska and Greenberg 2000).

In Oriolo et al. (2002), attention is paid to perturbations that act on mobile robots. They are not of equal influence on a system performance. A deviation in the direction compatible with robot mobility is not as severe as the deviation that violates nonholonomic constraints for it. In Section 5.3, for a unicycle model, we simulate small constraint violations. Our results confirm remarks reported in Oriolo et al. (2002).

Other works describe experimental designs of controllers and report limitations of simulation studies only. For instance, in Oriolo et al. (2002), it is observed that encoders become unreliable in the presence of a wheeled vehicle slippage, and the vehicle is simply lost in the environment. A design of sensory feedback for nonholonomic robots is needed; some preliminary work about a visual feedback from a fixed camera system is reported in De Luca et al. (2002). An interesting implementation comparison of kinematic-based controllers is reported in Kim and Tsiortas (2002).

Finally, one more category of comparative experiments concerns controller behavior under a wide range of operational conditions (Berghuis 1993). Different model-based adaptive and non-adaptive controllers and the PD controller were employed to track various desired trajectories, which were periodic functions with different frequencies.

The results of experimental evaluations of several model-based controllers with respect to operating conditions are recommendations and guidelines for the selection of controllers for programmed motion tracking. Specifically, we conclude the following:

- Model-based controllers outperform the classical PD controller. Adaptive model-based controllers are superior over their non-adaptive counterparts. The performance benefits are limited by the accuracy of a model that is employed to a controller design.

- The desired compensation adaptation law (DCAL) controller uniformly outperforms other model-based controllers for a wide range of frequencies and amplitudes of desired trajectories (Sadegh and Horowitz 1990, Whitcomb et al. 1991). Also, the DCAL controller appears to be the most attractive version of the passivity concept because it employs desired motion information in both the model-based compensation part and the adaptation loop.

- Predictive values of simulations decrease when unmodeled dynamics is dominant, and it affects the PD controller much more than non-adaptive or adaptive model-based controllers.

- Adaptive control algorithms are not able to learn from past experience. Tasks specified by programmed constraints may be repeatable or periodic motions. Control of such motions can be improved by application of "learnin" controllers, e.g., the repetitive control law (RCL) controller (Sadegh and Horowitz 1990).

- Robots interact with the environment when they work. Tasks like grasping, writing, scribing, painting, or object manipulating require a force regulation. To control such tasks, a hybrid force/position control is developed (Arimoto 1995; Lewis et al. 2004). For a programmed constraint that may specify a force exerted on the environment, a force controller may be desirable.

One more conclusion from this brief overview of experimental evaluation of controllers is that the performance measure (7.3) is not satisfactory in the case of programmed motion tracking. We extend it introducing the programmed motion performance measure that quantifies the tracking accuracy for the programmed motion.

Definition 7.3: The programmed motion performance measure is

$$J_P = \frac{1}{t} \int_0^t [e^T(t)e(t) + \dot{e}^T(t)\dot{e}(t) + \cdots + e^{(s)T}(t)e^{(s)}(t)] \, dt, \tag{7.4}$$

where $e(t) = q_p(t) - q(t)$ is a vector of a position tracking error, s is the constraint order, and q, q_p stand for control and programmed variables, respectively.

By the programmed variable, we mean a joint or a task space variable specified by a programmed constraint equation. For example, let e_x, e_y be position errors for the determination of Equation (7.4) for the manipulator end-effector task specified in Example 1.5. There, the end-effector coordinates x_p, y_p are programmed variables specified by the joint angles as

$$x_p = l_1 \cos \Theta_1 + l_2 \cos(\Theta_1 + \Theta_2), y_p = l_1 \sin \Theta_1 + l_2 \sin(\Theta_1 + \Theta_2), e_x = x_p - x, e_y = y_p - y. \quad (7.5)$$

To determine Equation (7.4), we need derivatives up to third-order for e_x and e_y, because third order derivatives of coordinates x, y are involved in the programmed constraint equation (1.7).

7.2 Non-Adaptive Tracking Control Algorithms for Programmed Motions

Throughout this and the next sections, two examples are detailed to illustrate the theoretical framework for the model reference tracking control strategy for programmed motion. Examples are conceptually and computationally simple, and they represent two classes of mechanical systems relevant to applications. The first is a unicycle model that is nonholonomic, and the second is a two-link planar manipulator model that is holonomic. They both are examined from the point of view of programmed motion tracking performance when different control laws are applied. Programmed motions are selected according to two criteria. First, we apply the same programs as reported in the literature and execute them using the programmed motion tracking strategy. Second, we demonstrate that our tracking strategy can be applied to programs specified by high order constraint equations, for which classical constrained dynamics and control dynamics fail.

7.2.1 Programmed Motion Tracking for a Unicycle

In Example 5.2, we demonstrated a controller design for trajectory tracking for a unicycle model. We approached the controller design using the generalized programmed motion equations (GPME). However, the controller could be designed using the classical approach, i.e., Lagrange's equations.

In this section, to demonstrate tracking motion specified by high order programmed constraints, consider the same unicycle for which the trajectory curvature profile has to get some value. The idea of imposition of this constraint is inspired by a requirement that vehicles can turn their wheels with a limited speed. This imposes a limitation on the rate of change of the trajectory curvature. Such constraint is reported in Koh and Cho (1999),

Oriolo et al. (2002), Shekl and Lumelsky (1998), and Singh and Leu (1991). However, it is not merged into a constrained dynamic model.

For the unicycle model, a constraint put on the rate of change of a trajectory curvature is formulated in Example 2.11 and is given by

$$\dot{\Phi}_1 = F_0 + \frac{\dot{x}\dddot{y} - \dddot{x}\dot{y}}{(\dot{x}^2 + \dot{y}^2)^{3/2}}. \tag{7.6}$$

Equation (7.6) is the nonholonomic constraint equation of the third order, where F_0 does not contain third order coordinate derivatives and has the form

$$F_0 = \frac{F_1 \dot{y}}{(\dot{x}^2 + \dot{y}^2)^{3/2}}.$$

First, the reference dynamic model for programmed motion specified by Equation (7.6) has to be generated for $p = 3$, according to Algorithm 2.2 from Section 2.4.2. It can be generated in the same way as it was demonstrated in Example 2.16. The dynamic control model remains unchanged.

For simulation, we select two functions for the curvature: $\Phi_1 = 5$ and $\Phi_2 = 2\sin t + 1$. They are limited and have limited derivatives. We assume that only control forces act on the unicycle, e.g., the computed torque controller with $\sigma = 3$. Figures 7.2 and 7.3 present simulation results of programmed motion tracking.

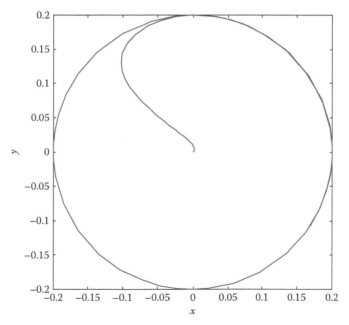

FIGURE 7.2
Programmed motion tracking for the curvature Φ_1.

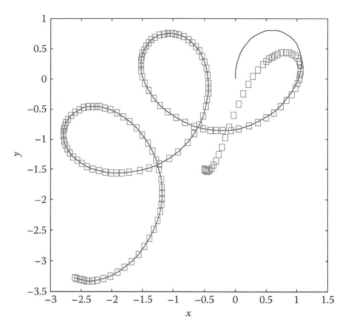

FIGURE 7.3
Programmed motion tracking for the curvature Φ_2: reference (—), controlled (□□□).

7.2.2 Programmed Motion Tracking for a Planar Manipulator

Consider a planar two-link manipulator model with two degrees of freedom described by Θ_1, Θ_2. In Example 2.16, the reference dynamics for the programmed constraint given by Equation (2.211), that is,

$$\dddot{\Theta}_2 = F_1 - F_2\dddot{\Theta}_1, \tag{7.7}$$

is derived. It is of the form

$$(b_1 - b_2F_2)\ddot{\Theta}_1 + (b_2 - \delta F_2)\ddot{\Theta}_2 + c = 0,$$
$$\dddot{\Theta}_2 = F_1 - F_2\dddot{\Theta}_1, \tag{7.8}$$

where $b_1 = \alpha + 2\beta\cos\Theta_2, b_2 = \delta + \beta\cos\Theta_2, c = -\beta\dot{\Theta}_2(\dot{\Theta}_2 + 2\dot{\Theta}_1)\sin\Theta_2 - 4/3\beta\dot{\Theta}_1^2F_2\sin\Theta_2$.

For simulation, the rate of change of the end-effector trajectory curvature is selected to be $\Phi_1 = 0.6 + 0.02t$. Motion initial conditions $y_{0r} = [1.2;0.43;-0.9;-0.5;0.35;0.35]$ satisfy Equation (7.7). Figure 7.4 shows the reference programmed motion on the (x,y) plane. We selected the programmed motion in such a way that the manipulator may follow it for some time only.

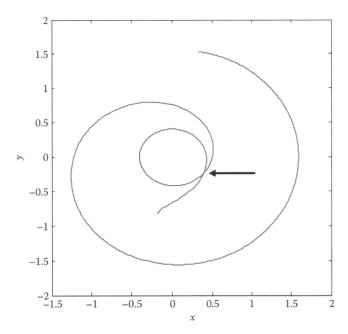

FIGURE 7.4
Reference programmed motion for the end-effector.

After reaching the position marked by the arrow, links of a given length cannot follow the program any more. This demonstrates that a programmed motion specified for any system should be analyzed by inspection of the reference motion outputs.

The dynamic control model for the manipulator is its unconstrained dynamics, that is,

$$\begin{bmatrix} \alpha + 2\beta\cos\Theta_2 & \delta + \beta\cos\Theta_2 \\ \delta + \beta\cos\Theta_2 & \delta \end{bmatrix}\begin{bmatrix} \ddot{\Theta}_1 \\ \ddot{\Theta}_2 \end{bmatrix} + \begin{bmatrix} -\dot{\Theta}_2\beta\sin\Theta_2 & -\beta\sin\Theta_2(\dot{\Theta}_1 + \dot{\Theta}_2) \\ \dot{\Theta}_1\beta\sin\Theta_2 & 0 \end{bmatrix}\begin{bmatrix} \dot{\Theta}_1 \\ \dot{\Theta}_2 \end{bmatrix} = \begin{bmatrix} \tau_1 \\ \tau_2 \end{bmatrix},$$

(7.9)

where τ_1 and τ_2 are control torques whose magnitudes are shown in Figure 7.6. The controlled motions of the end-effector, for several different initial conditions, are shown in Figure 7.5. Quite large initial torque values are due to differences in initial conditions for the controlled and programmed motions taken for illustrative purposes. As can be expected, values of the torque τ_1 that drives the lower joint are smaller than τ_2. For comparison, for the same initial conditions of the controlled motion, i.e., for $y_{0c} = [1.2;0.43;-0.9;-0.5]$, the PD controller is applied (cf., Figures 7.7 and 7.8). The PD controller gains are $k_s = 15; k_d = 25$.

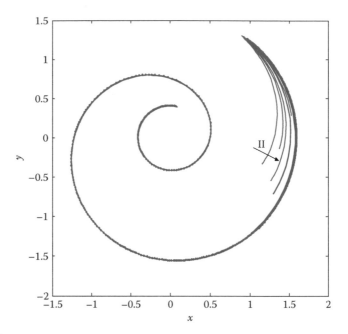

FIGURE 7.5
Programmed motion tracking for the end-effector.

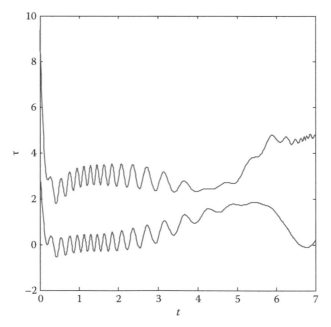

FIGURE 7.6
Computed torques in programmed motion tracking for the end-effector versus time.

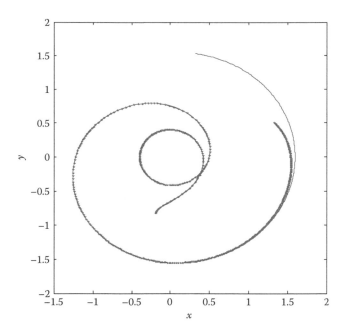

FIGURE 7.7
Programmed motion tracking by the proportional-derivative (PD) controller.

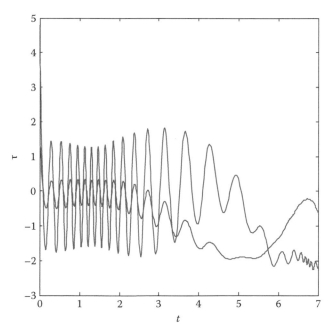

FIGURE 7.8
Proportional-derivative (PD) controller torques versus time.

7.2.3 Programmed Motion Tracking for a Two-Wheeled Mobile Robot

Consider motion of a two-wheeled mobile robot that consists of a platform and two actuated wheels as presented in Figure 1.2. The control objective for the robot is to track the reference motion:

$$B_1 = x_C^2 + y_C^2 - R^2(t) = 0, \quad R(t) = 0.2 + 0.01t. \tag{7.10}$$

It is subjected to the material constraint equations (1.1). The control dynamics for the robot is then

$$M\ddot{q}_1 + C(\dot{q}_2)\dot{q}_1 + D\dot{q}_1 = \tau,$$

$$\dot{q}_2 = B(q_2)\dot{q}_1, \tag{7.11}$$

where the following notation is introduced:

- Partition of a vector of generalized coordinates $q \in R^5$ is $q = (q_1, q_2)$, $q_1 \in R^2$, $q_2 \in R^3$, and $q_1 = (\ _r,\ _l)$ consists of driving wheel angles due to rolling, $q_2 = (x_C, y_C,\ _C)$, in which (x_C, y_C) are coordinates of the mass center of the platform, and   is the heading angle.
- The inertia matrix is

$$M = \begin{bmatrix} m_{11} & m_{12} \\ m_{21} & m_{22} \end{bmatrix} \text{ and }$$

$m_{11} = \frac{r^2}{4b^2}(mb^2 + I) + I_w$, $m_{12} = \frac{r^2}{4b^2}(mb^2 - I)$, $m_{21} = m_{12}$, $m_{22} = m_{11}$.

- The total mass consists of mass of the platform m_c and of two wheels $2m_w$, i.e., $m = m_c + 2m_w$,
- Moments of inertia are I_c, I_w, I_m – through the robot mass center, of the wheel about its axis, and about its diameter, and $I = m_c d^2 + 2m_w b^2 + I_c + 2I_m$.
- The velocity dependent matrix is

$$C = \begin{bmatrix} 0 & c_{12} \\ -c_{12} & 0 \end{bmatrix} \text{ and } \quad c_{12} = \frac{r^2 d}{2b} m_c\ .$$

- The matrix in the constraint equation is

$$B = \begin{bmatrix} 0.5r\cos & 0.5r\cos \\ 0.5r\sin & 0.5r\sin \\ 0.5r/b & -0.5r/b \end{bmatrix}.$$

- The matrix of damping coefficients is $D - diag(d_{11}, d_{22})$.
- The control torque vector is $\tau \in R^2, \tau = (\tau_r, \tau_l)$. These are real torques applied to the wheels.

Physical parameters of the robot, in SI units, are

$$m_c = 0.55, \ m_w = 0.05, \ d = 0.05, \ r = 0.026, \ b = 0.075,$$

$$I_m = 0.23 \cdot 10^{-6}, \ I_c = 0.0182, \ I_w = 0.17 \cdot 10^{-4}.$$

The reference dynamics is developed for $n = 5, k = 4, p = 1$, according to Algorithm 2.1 from Section 2.4.2. The set of equations of the reference dynamics is

$$\tilde{m}_{11}\ddot{}_r + \tilde{m}_{12}\ddot{}_l + \tilde{m}_{13}\ddot{x}_C + \tilde{m}_{14}\ddot{y}_C + \tilde{m}_{15}\ddot{} + v_1 = 0,$$

$$\dot{y}_C \cos \ - \dot{x}_C \sin \ - d\dot{} = 0,$$

$$\dot{x}_C \cos \ + \dot{y}_C \sin \ + b\dot{} = r\dot{}_r, \qquad (7.12)$$

$$\dot{x}_C \cos \ + \dot{y}_C \sin \ - b\dot{} = r\dot{}_l,$$

$$x_C^2 + y_C^2 - (0.2 + 0.01t)^2 = 0.$$

In Equations (7.12), the following notation is introduced:

$$\tilde{m}_{11} = 2I_w\left(\frac{dv}{rw} + s\right), \tilde{m}_{12} = 2I_w\left(\frac{dv}{rw} - s\right), \tilde{m}_{13} = \left[2m_w d\sin \ - (m + 2m_w)\frac{dy_C}{w}\right],$$

$$\tilde{m}_{14} = \left[(m + 2m_w)\frac{dx_C}{w} - 2m_w d\cos \ \right], \tilde{m}_{15} = I_c + 2I_m + 2m_w b^2, v_1 = \frac{4m_w d\sigma^*}{w} \cdot \ - \frac{6m_w d^2}{w} \cdot 2,$$

$$w = x_C \cos \ + y_C \sin \ , v = x_C \sin \ - y_C \cos \ , s = b/r, \sigma^* = \sigma_1 B_1 - \dot{R}(t)/2.$$

The term σ^* is introduced for simulation to stabilize the programmed constraint.

To track the programmed motion, we use the Wen–Bayard controller (Wen and Bayard 1988) whose gains are selected to be $K_d = diag(15, 15)$, $K_a = diag(15, 15)$, and $\sigma_1 = 6$. Initial conditions for the reference motion are $q_{0r} = [0;0;0.2;0;\frac{\pi}{2}]$, $\dot{q}_{0r} = [0;0;0;0.5;0]$, and for the controlled motion

$$q_{0c} = \left[0;0;0.1;0;\frac{\pi}{2}\right], \ \dot{q}_{0c} = [0;0;0;0.45;0].$$

Tracking the programmed motion (7.10) is presented in Figure 7.9. Figure 7.10 presents control torques applied to the wheels.

Based on the examples above, it can be seen that the reference model employed to control enables tracking the programmed motion using various controllers. Due to the separation of the programmed and material

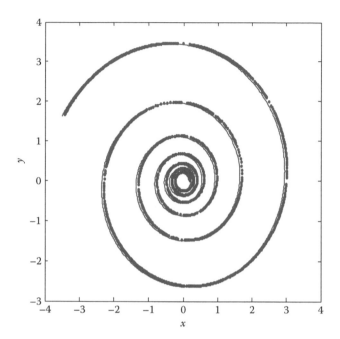

FIGURE 7.9
Reference (—) and controlled (···) robot motions within 6 minutes.

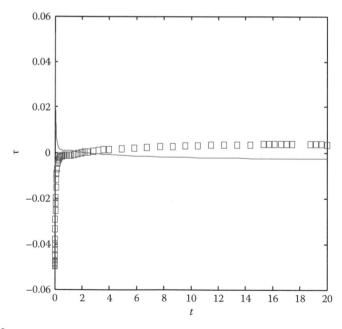

FIGURE 7.10
Torques of the Wen–Bayard controller applied to the right (□□□) and left (—) wheels.

constraints between two dynamic models, we can use controllers that are dedicated to holonomic systems.

7.3 Adaptive Tracking Control Algorithms for Programmed Motions

Dynamic models of controlled systems include many parameters that are difficult to estimate with accuracy better than 10% to 20%. Unmodeled dynamics adds uncertainty to overall system dynamics. Also, robots and manipulators are subjected to large variations in payloads. Structural and parametric uncertainties limit the correctness of tracking. In this section, we present applications of adaptive controllers to programmed motion tracking. They are selected according to guidelines presented in Section 7.1.3.

In the case when system dynamics is uncertain, approximate parameter values only can be at our disposal. These approximate values are used in the reference model to plan this motion. First, we apply the DCAL controller (Sadegh and Horowitz 1990), which we adopt to programmed motion tracking (Jarzębowska 2008b). In the DCAL, the separation of parameters from time functions is given by

$$Y_p(\cdot)\,\phi = \tilde{M}(q_p)\ddot{q}_{1p} + \tilde{C}(q_p,\dot{q}_{1p})\dot{q}_{1p} + D(q_p), \tag{7.13}$$

where $Y_p(\cdot)$ is a $(n\text{-}k) \times r$ regression matrix that depends only on known functions along the programmed motion. Utilizing Equation (7.13), the DCAL controller for programmed motion is of the form

$$\tau_p = Y_p(\cdot)\hat{\phi} + k_v r_1 + k_s e_1 + k_a \|e\|^2\, r_1. \tag{7.14}$$

In Equation (7.14), the filtered tracking error is defined as $r_1 = e_1 + \dot{e}_1$, with e_1 being the tracking error; $e_1 = q_{1p} - q_1$, k_v, k_s, k_a are scalar, constant control gains; and $\hat{\phi}$ is the $(r \times 1)$ vector of parameter estimates. The subscript p at $q_p, \dot{q}_p, \ddot{q}_p$ indicates that the controller is fed with programmed motion information. The update rule for an unknown parameter ϕ is defined as $\dot{\hat{\phi}} = \Gamma Y_p^T(\cdot) r_1$. The parameter error is $\tilde{\phi} = \phi - \hat{\phi}$, and Γ is a $(r \times r)$ positive definite diagonal constant adaptive gain matrix. For the controller (7.14), we may formulate the theorem.

Theorem 7.1

The control dynamics (5.6)

$$\tilde{M}(q)\ddot{q}_1 + \tilde{C}(q,\dot{q}_1)\dot{q}_1 + D(q) = \tau_p,$$

$$\dot{q}_2 = -B_{12}^{-1}(q)B_{11}(q)\dot{q}_1$$

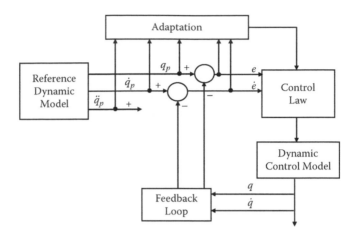

FIGURE 7.11
Architecture of the model reference tracking control for programmed motion for systems with uncertain dynamics.

with the DCAL controller (7.14) $\tau_p = Y_p(\cdot)\hat{\phi} + k_v r_1 + k_s e_1 + k_a \|e_1\|^2 r_1$ is the closed-loop dynamics that provides asymptotic stability for the position and velocity tracking errors, and the parameter error remains bounded.

The stability proof for the closed-loop control dynamics with the DCAL controller applied to the programmed motion tracking can be carried out in the same way as for the closed-loop control dynamics with the RCL, which is demonstrated in Example 3.6.

Architecture of the model reference tracking control for programmed motion with the adaptive controller is presented in Figure 7.11.

The adaptation of the DCAL for programmed motion tracking is motivated by significant reduction of the on-line computation it provides. Specifically, the regression matrix can be calculated off-line and stored in a computer. One disadvantage of the DCAL is that controller gains k_v, k_s, k_a must be rather large.

7.3.1 Adaptive Programmed Motion Tracking for a Planar Manipulator

In this simulation study, we specify the programmed motion when parameter estimates only are available. The two-link manipulator model from Section 7.2.2 is detailed, but now mass is an unknown parameter and mass estimates $m_{1ref} = m_1 = 0.2$, $m_{2ref} = m_2 = 1.9$ are used. The DCAL controller (7.14) is applied. Gains in the adaptive update rule for the unknown parameter m are selected to be $\gamma_1 = 16$, $\gamma_2 = 5$, and the controller gains are $k_v = 25$, $k_s = 15$, $k_a = 15$. Initial conditions for the reference motion are $y_{0r} = [1.2;0.43;-0.9;-0.5;0.35;0.35]$. Simulation results of programmed motion tracking by the DCAL are presented in Figure 7.12. For illustrative purposes, the manipulator controlled motion starts from rest with both angles equal to zero (cf. Figure 7.13).

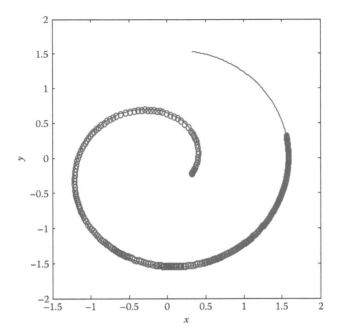

FIGURE 7.12
Reference (—) and controlled (°°°) motions for uncertain manipulator dynamics.

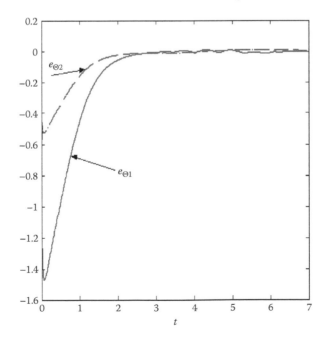

FIGURE 7.13
Angular velocity errors versus time.

This is the reason why the control torques τ_{1p} and τ_{2p} are large at the beginning of tracking (cf. Figure 7.15). These torques are real torques applied to joints 1 and 2.

In simulation studies presented so far, we have not added any friction or bounded disturbances terms into the dynamic control model (7.9). In manipulator models, friction can be characterized by the Tustin model (Armstrong–Helouvry 1991; Berghuis 1993; Jarzębowska and Greenberg 2000). It consists of a combination of Coulomb, viscous, and static friction terms. Experiments show that the static friction effects are small compared with the other two friction model components, and it is often assumed that Coulomb and viscous friction terms can approximate friction forces. We adopt the same assumption for the friction force model. Also, disturbances are added so both friction and disturbance terms applied to links 1 and 2 are

$$f_1 = 2\dot{q}_1 + 0.5\,sign(\dot{q}_1) + 0.2\sin(3t),$$
$$f_2 = 2\dot{q}_2 + 0.5\,sign(\dot{q}_2) + 0.2\sin(3t). \tag{7.15}$$

The DCAL controller performance is good in this case. Some inaccuracies are in mass correction even for higher gains taken to be $\gamma_1 = 35$, $\gamma_2 = 10$ (cf. Figures 7.14 and 7.16). Figure 7.17 presents control torques for programmed motion tracking in the presence of friction and bounded disturbances.

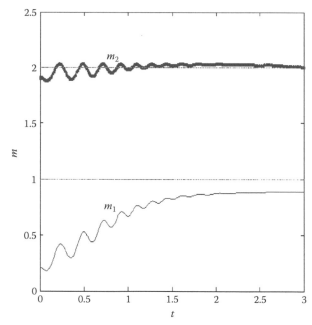

FIGURE 7.14
Mass correction for $m_1(0) = 0.2$, $m_2(0) = 1.9$, versus time.

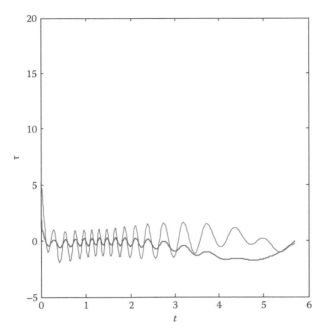

FIGURE 7.15
Control torques τ_{1p}(—) and τ_{2p}(- -) versus time.

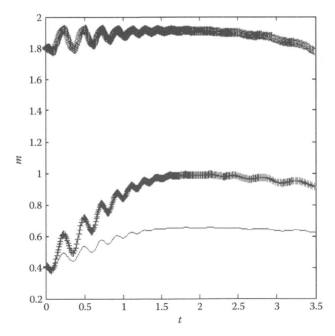

FIGURE 7.16
Mass correction when friction is added (solid line) versus time.

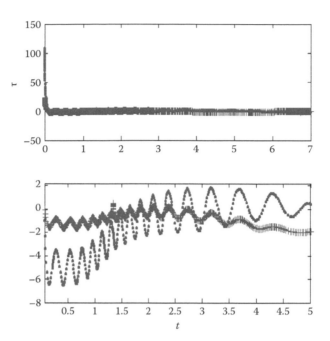

FIGURE 7.17
Control torques when friction is added to the control dynamics: τ_{1p} (ooo) and τ_{2p}(+++).

A question may be asked why any robust controller is applied in this case. The reason is that the accuracy of tracking by adaptive controllers improves with time, because the adaptation mechanism continues extracting information from the tracking error. Also, for a robust controller, even in the absence of disturbances, one cannot guarantee asymptotic stability of the tracking error.

7.3.2 Adaptive Programmed Motion Tracking for a Unicycle

In this study, we assume that the wheel mass is an uncertain parameter, and we apply the DCAL controller to track the programmed motion given by Equations (2.208) in Example 2.15:

$$\dot{x} = r\dot{\theta}\cos \quad , \dot{y} = r\dot{\theta}\sin \quad , \quad (t) = \cos(0.5\pi - t). \tag{7.16}$$

The dynamic control model for the unicycle is the same, i.e., Equations (5.25) in Example 5.2, but this time, τ_{pi}, $i = 1,\ldots,4$, are the components of the virtual torque of the DCAL controller (7.14). The controller gains are selected to be $\gamma_1 = 30$, $k_v = 15$, $k_s = 15$, $k_a = 15$. The figures below present the following simulation results: Figure 7.18—wheel mass correction, Figure 7.19—comparison of tracking performance for the RCL with $k_L = 30$ and the DCAL, and Figure 7.20—comparison of tracking performance for the DCAL and the Wen–Bayard controllers. In all figures, the reference motion is in a solid line.

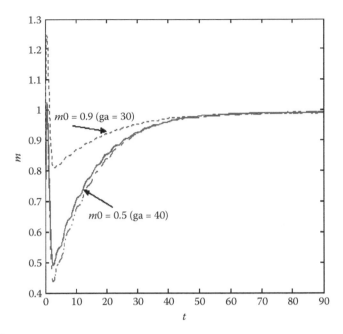

FIGURE 7.18
Wheel mass correction versus time.

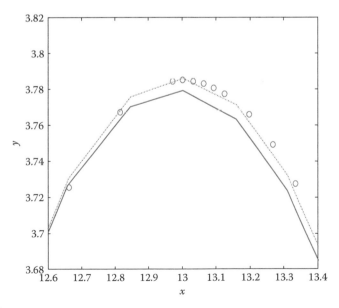

FIGURE 7.19
Tracking by the repetitive control law (RCL) (ooo) and desired compensation adaptation law (DCAL) (---) controllers.

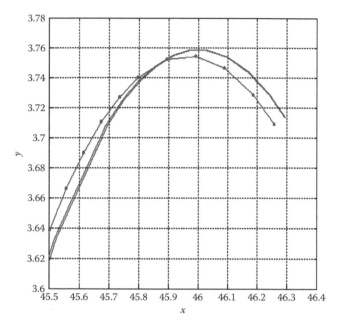

FIGURE 7.20
Tracking by the Wen–Bayard (o-o-o) and desired compensation adaptation law (DCAL) (---)
controllers.

Magnitudes of virtual control torques for the DCAL and Wen–Bayard controllers are compared in Figure 7.21. It can be seen that for the same sets of initial
conditions close to the reference, magnitudes of torques generated by the two
controllers are similar. The tracking performance by the Wen–Bayard controller
is a little worse. The Wen–Bayard controller can be recognized as "better" when
we can assume that parameters of a system are measured with a satisfactory
accuracy, because it requires less computation than does the DCAL. The focus
of this book is not the examination of properties of various controllers, and we
do not provide more comparisons of controllers with respect to other criteria.

In conclusion, for each controller we apply the programmed motion performance measure (7.4) and determine tracking errors for programmed variables.
Tables 7.1 and 7.2 summarize these results. The first three columns in both
tables contain the performance measures (7.3) determined for position errors.
The last columns contain the programmed motion performance measure (7.4).

Initial conditions for the manipulator motion are $y_{0r} = [1.2;0.43;-0.9;0.35;$
$-0.8;0.35]$ and $y_{0c} = [1.2;0.2;-0.9;0]$. Initial conditions for the unicycle motion
are $y_{0r} = [1;2;1;0;0;0;0;0]$ and $y_{0c} = [1;2;1;0;0.8;0;1;0]$.

The controller gains were the same as indicated in simulation studies.

Based on data from Tables 7.1 and 7.2, conclusions are as follows:

    1. Initial motion conditions for the controlled motion of the manipu-
       lator are selected to be close to reference ones. In this case, the

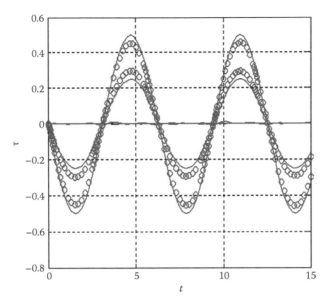

FIGURE 7.21
Magnitudes of control torques for desired compensation adaptation law (DCAL) (---) and Wen–Bayard (o-o-o) controllers versus time.

TABLE 7.1

Performance Measures for the Manipulator Motion Tracking

Manipulator	$J_{\Theta 1}$	$J_{\Theta 2}$	$J_{\Theta 1+\Theta 2}$	J_p
PD	0	0.014	0.014	0.0039
DCAL	0	0.023	0.023	0.0022
Computed torque ($\sigma = 10$)	0	0.021	0.021	0.0017

TABLE 7.2

Performance Measures for the Unicycle Motion Tracking

Unicycle	J_{x+y}	J_φ	J_Θ	J_p
PD	0	0	0.098	0.098
DCAL	0	0	0.058	0.062
RCL	0	0	0.086	0.092
W-B	0	0	0.115	0.139

Note: proportional derivative (PD); desired compensation adaptation law (DCAL); repetitive control law (RCL); Wen–Bayard (W-B);.

programmed motion performance measures obtained for the three tested controllers are similar. However, the PD controller performs worse than the two others.

2. Initial motion conditions for the controlled motion of the unicycle are selected to be the same or close to the reference ones. The DCAL controller outperforms the three other controllers we test.

3. It can be verified based on the manipulator example that even though the performance measure for joint variables was quite poor, the programmed motion performance measure was satisfactory. Thus, we can see the difference between the performance measures as in Equations (7.3) and (7.4).

7.4 Learning Tracking Control Algorithms for Programmed Motions

In many applications, mechanical systems perform the same tasks repeatedly, i.e., programs may be repetitive by their nature. The unicycle program given by Equation (7.16) is repetitive, and its programmed trajectory on the (x,y) plane is a periodic function. The repetitive program may be viewed as a sequence of trials, and it can be desirable to improve tracking performance from trial to trial. Let us employ a repetitive controller (Sadegh et al. 1990). The underlying philosophy for derivation of the repetitive control law (RCL) adopted to encompass programmed motions tracking is as follows (Jarzębowska 2008b). If a programmed motion is periodic, the dynamics

$$u_p = M(q_p)\ddot{q}_p + C(q_p,\dot{q}_p)\dot{q}_p + D(q_p) + p(\dot{q}_p) \tag{7.17}$$

are repeatable. Even if an unknown constant parametric quantity $p(\dot{q}_p)$ is present in Equation (7.17), the signal of the control vector u_p is periodic or at least repeatable. We may assume that the program is periodic with period T, and it yields that $u_p = u_p(t-T)$. For the repeatable dynamics (7.17), the RCL is formulated as

$$\tau_p = \hat{u}_p(t) + k_v r + k_s e + k_a \|e\|^2 r, \tag{7.18}$$

where $\hat{u}_p(t)$ is a $(n \times 1)$ vector of a learning term that compensates for the repeatable dynamics $u_p(t)$, and it is updated from trial to trial by the learning update rule:

$$\hat{u}_p(t) = \hat{u}_p(t-T) + k_L r, \tag{7.19}$$

with k_L being a positive scalar control gain. Other quantities in Equation (7.19) are as follows: r is a filtered tracking error defined as $r = e + \dot{e}$, with e being a tracking error and $e = q_p - q$; k_v, k_s, k_a are scalar, constant, control gains. The learning update rule (7.19) can be written in terms of the learning error, which is defined as

$$\tilde{u}_p(t) = u_p(t) - \hat{u}_p(t). \tag{7.20}$$

Relation (7.19) can be presented in the form

$$u_p(t) - \hat{u}_p(t) = u_p(t) - \hat{u}_p(t - T) - k_L r,$$

and utilizing the assumption that the programmed motion is repeatable, the learning error update rule is

$$\tilde{u}_p(t) = u_p(t - T) - \hat{u}_p(t - T) - k_L r = \tilde{u}_p(t - T) - k_L r. \tag{7.21}$$

For the RCL controller (7.18) adopted for programmed motion tracking, we formulate Theorem 7.2.

Theorem 7.2

The control dynamics (5.6)

$$\tilde{M}(q)\ddot{q}_1 + \tilde{C}(q, \dot{q}_1)\dot{q}_1 + D(q) = \tau_p,$$

$$\dot{q}_2 = -B_{12}^{-1}(q)B_{11}(q)\dot{q}_1$$

with the RCL controller (7.18) $\tau_p = \hat{u}_p(t) + k_v r_1 + k_s e_1 + k_a \|e_1\|^2 r_1$ is the closed-loop dynamics that makes position and velocity tracking errors asymptotically stable.

The proof of the theorem is demonstrated in Example 3.6.

The RCL controller applied to the strategy for programmed motion tracking is presented in Figure 7.22.

The RCL controller seems to be attractive because it improves its performance from one trial to the next, and it requires very little information about a system we control, and very little on-line computation. It satisfies then requirements we formulated for programmed motion tracking controllers. Figures 7.23 and 7.24 present tracking the programmed motion (7.16) by the RCL and DCAL controllers.

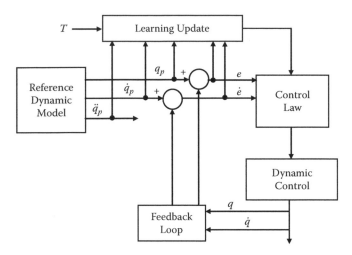

FIGURE 7.22
Architecture of the model reference tracking control for a repetitive programmed motion.

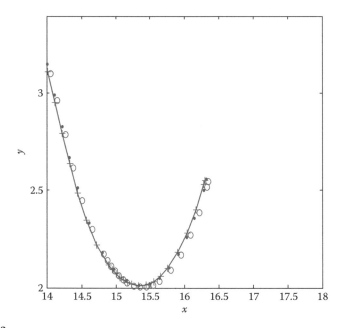

FIGURE 7.23
Tracking performance comparisons: computed torque (···); repetitive control law (RCL) (ooo); Wen–Bayard (+++) controllers.

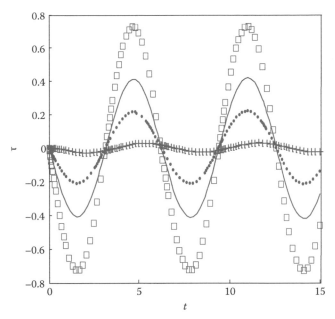

FIGURE 7.24
Repetitive control law (RCL) torques: τ_3 (□□□), τ_4 (+++), desired compensation adaptation law (DCAL) torques: τ_3 (—), τ_4 (···) versus time.

7.5 Tracking Control Algorithms for Programmed Motions Specified in Quasi-Coordinates

In Section 2.4.3 we derived the GPME specified in quasi-coordinates, i.e., Equations (2.231), which are the reference dynamics for control purposes. The constraint equations specified in quasi-coordinates have the form as in Equation (2.228), that is, $\omega^p = \Omega_\beta^p\left(t, q_\sigma, \omega_\sigma, \omega_\sigma, ..., \omega_\mu^{(p-1)}, \omega_\mu^p\right)$.

The control goal is as follows: Given a programmed motion specified by the constraints (2.228) and the system reference dynamics (2.231), design a feedback controller to track the desired programmed motion.

The unified dynamic control model has the form of Equations (2.151):

$$M(q)\dot{\omega} + C(q, \omega) + D(q) = \tilde{\tau},$$

$$B(q, \omega) = 0.$$

$$(7.22)$$

where control torques are added in the right-hand side.

Equations (7.22) are the GPME in quasi-coordinates for $p = 1$. They consist of $(n\text{-}b)$ equations of motion and b equations of material constraints or conservation laws.

7.5.1 Tracking Control of the Unicycle Model Specified in Quasi-Coordinates

In Example 2.17 we illustrated the application of the constrained systems modeling framework in quasi-coordinates based on the unicycle model. The control dynamics model according to Equations (7.22) is

$$(mr^2 + I_\theta)\dot{\omega}_3 = \tau_3,$$

$$I\,\dot{\omega}_4 = \tau_4,$$

(7.23a)

with the kinematic relations

$$\dot{x} = r\omega_3 \cos \quad , \quad \dot{y} = r\omega_3 \sin \quad , \quad \dot{\theta} = \omega_3, \quad \dot{} = \omega_4. \tag{7.23b}$$

To track the program specified by Equations (2.232), we use the computed torque controller whose components are

$$\tilde{\tau}_3 = (mr^2 + I_\theta)\tilde{u}_3 \quad \text{and} \quad \tilde{\tau}_4 = I\,\tilde{u}_4 \tag{7.24}$$

with $\tilde{u}_3 = \dot{\omega}_{3p} + 2\sigma_1(\omega_{3p} - \omega_3) + \sigma_1^2(\theta_p - \theta)$, $\tilde{u}_4 = \dot{\omega}_{4p} + 2\sigma_2(\omega_{4p} - \omega_4) + \sigma_2^2(\ _p -\ )$.

The subscript p stands for the programmed variable in the reference equation, σ_1 and σ_2 are the convergence rates selected for the specific program. For our simulation study, they are both set equal to 20. Simulation results are presented in Figures 7.25 and 7.26.

It can be seen that the equations of the Boltzmann–Hamel type are first order differential equations. They are in the reduced-state form and may offer a fast way to obtain equations of motion either for the reference or control blocks. Also, it is convenient to select the quasi-velocities, since they satisfy both the material and programmed constraints.

7.5.2 Tracking Control of the Planar Manipulator Model Specified in Quasi-Coordinates

A two-link planar manipulator is a holonomic system, which we make nonholonomic by an imposition of the nonholonomic constraint on it. Formulating the same program for the manipulator end-effector as in Example 2.18, that is, it has to move along a trajectory for which its curvature changes according to some function $\Phi^* = \frac{d\Phi(t)}{dt}$, we obtain its specification in quasi-velocities as

$$\ddot{\omega}_2 - (1 - F_2)\frac{l_2}{l_1}\ddot{\omega}_1 - F_1 l_2 = 0. \tag{7.25}$$

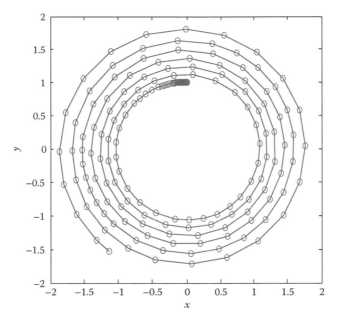

FIGURE 7.25
Programmed motion tracking: reference (~), controlled motion (ooo).

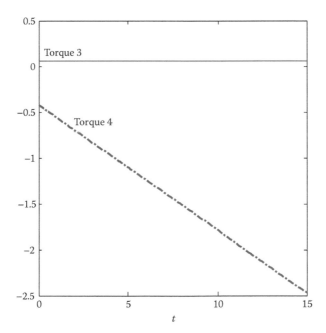

FIGURE 7.26
Control torques versus time: Torque3—heading angle, Torque4—rolling angle.

The reference dynamics of the manipulator is given by Equations (2.237):

$$\frac{b_1 - b_2 - F_2(b_2 - \delta)}{l_1}\dot{\omega}_1 + \frac{b_2 - F_2\delta}{l_2}\dot{\omega}_2 + c = 0,$$

$$\ddot{\omega}_2 - \frac{(1 - F_2)l_2}{l_1}\ddot{\omega}_1 - F_1 l_2 = 0, \qquad (7.26)$$

$$\omega_1 = \dot{\Theta}_1 l_1,$$

$$\omega_2 = (\dot{\Theta}_1 + \dot{\Theta}_2)l_2.$$

The control dynamics (7.22) become

$$\dot{\omega}_1 = u_1,$$

$$\dot{\omega}_2 = \left[1 - \frac{1}{\delta}(\delta + \beta\cos\Theta_2)\right]\frac{\dot{\omega}_1 l_2}{l_1} - \frac{\beta}{\delta}\sin\Theta_2\frac{\dot{\omega}_1^2 l_2}{l_1^2}, \qquad (7.27)$$

$$\omega_1 = \dot{\Theta}_1 l_1,$$

$$\omega_2 = (\dot{\Theta}_1 + \dot{\Theta}_2)l_2.$$

Tracking the programmed motion specified by Equation (7.25) is achieved in the same way as in Example 5.3.

The control-oriented modeling and the controller design for the manipulator model developed in quasi-coordinates result in the compact reference and control dynamics. The simulation is faster, and no numerical stabilization of the constraint equations is needed.

7.6 Tracking Control Algorithms for Programmed Motions with the Velocity Observer

Mobile robots may be equipped with position and velocity measurement devices; however, signals obtained in this way can be noise sensitive. Velocity signals can be obtained by differentiation of position signals, but when calculated for low or high velocities, they can be inadequate. To reduce weight and costs, robots are often not equipped with velocity sensors. These limitations can be circumvented by a velocity observer design. The main challenge in designing an observer-controller feedback for nonholonomic systems is the presence of the Coriolis term in their dynamic control model, which results in quadratic cross-terms of unmeasured velocities. For this reason, observer designs proposed for manipulators cannot be directly applied to nonholonomic systems (Fossen and Nijmeijer 1999). In Do et al. (2003), a velocity observer is designed for output feedback tracking for a unicycle-type robot based on two-level tracking control architecture, that is, on kinematic and dynamic models derived in generalized coordinates.

In this section, we present a model-based tracking control strategy extended by an incorporation of a nonlinear observer. The observer is designed to enable only position and orientation measurement for feedback tracking. The strategy, as presented in Section 7.1, is originally developed for programmed motion tracking when the full state of a system is available for measurement.

The velocity observer is designed for car-like vehicles whose kinematics is as that of a unicycle. For this reason, our approach cannot be directly applied to a general class of nonholonomic systems. However, kinematics of many mobile robots employed in practice are equivalent to that of the unicycle.

Consider a mobile robot model that consists of a platform and two actuated wheels as presented in Figure 1.2. A dynamic control model for the robot developed in generalized coordinates has the form of Equations (7.11).

Notice that the linear and angular velocities $v_r = \frac{r}{2}(\,\cdot\,_r + \,\cdot\,_l)$ and $\omega_r = \frac{r}{2b}(\,\cdot\,_r - \,\cdot\,_l)$ of the robot are not used in Equations (7.11) as control variables. Instead, the angular velocities of wheels are the control variables in (7.11), so its form is convenient for designing a velocity observer.

For control applications, it is more advantageous to develop the control dynamics (7.11) in quasi-coordinates using the Boltzmann–Hamel equations. To this end, we introduce a quasi-velocity vector $\Omega = (\omega, \omega^*)$, $\omega \in R^m$, $\omega^* \in R^k$, where $m = n - k$ is the number of degrees of freedom of a system, and k is the number of non-material, i.e., task-based constraints and material nonholonomic constraint equations. Assume that the material constraints on the robot are of an order not higher than one and have the form

$$\omega_\beta^* = \omega_\beta^*(t, q_\sigma, \dot{q}_\sigma) = 0. \quad \beta = 1, ..., k \tag{7.28}$$

The generalized Boltzmann–Hamel equations for a system with the constraints (7.28) have the form (2.151) and together with n kinematic relations (2.150) are a set of $(n\text{-}k)$ equations of motion.

Due to the introduction of the quasi-velocities, k of them are identically satisfied, and the rest $(n\text{-}k)$ may be selected arbitrarily by a designer. These $(n\text{-}k)$ quasi-velocities may reflect control variables in a system. What follows is a form of the final equations of motion, which depend upon ω, and ω^* may be canceled after the equations are derived, as it was presented in Example 2.10

where the roller-racer motion dynamics is governed by the equations of the Boltzmann–Hamel type.

The material nonholonomic constraints on the robot motion due to rolling its wheels without slipping are

$$\dot{x} = \frac{r}{2}(\dot{}_r + \dot{}_l)\cos \quad , \quad \dot{y} = \frac{r}{2}(\dot{}_r + \dot{}_l)\sin \quad . \tag{7.29}$$

Due to the robot design, quasi-velocities may be selected as

$$\omega_1 = \dot{}_r, \quad \omega_2 = \dot{}_l,$$

$$\omega_3^* = \dot{x} - \frac{r}{2}(\dot{}_r + \dot{}_l)\cos \quad , \quad \omega_4^* = \dot{y} - \frac{r}{2}(\dot{}_r + \dot{}_l)\sin \quad . \tag{7.30}$$

The control dynamics (7.11) derived in quasi-coordinates is of the form

$$M\dot{\omega} + C(q_2, \omega)\omega + D\omega = \tau,$$

$$\dot{q}_2 = B(q_2)\omega. \tag{7.31}$$

The matrices M, C, D, and B remain unchanged comparing to Equations (7.11).

The control objective is to design a model-based controller that tracks a reference motion provided that only position and orientation, i.e., $q_2 = (x, y, \ )$ are available for measurement. The reference motion for the robot is specified by a programmed constraint equation.

To design a velocity observer, we aim at elimination of the term with velocity products in the first equation in (7.31). To this end, we transform coordinates as

$$z = A(q_2)\omega, \tag{7.32}$$

and we assume that $A(q_2)$ is a globally invertible (2×2) matrix with bounded elements, which has to be determined. We apply Equation (7.32) to the first of equations (7.31) to obtain

$$\dot{z} = [\dot{A}(q_2)\omega - A(q_2)M^{-1}C(\dot{q}_2)\omega] + A(q_2)M^{-1}(\tau - D\omega). \tag{7.33}$$

We would like to select $A(q_2)$ in such a way that the square bracket in the right-hand side of Equation (7.33) vanishes. To this end, we use the second of equations (7.31) and solve partial differential equations with respect to elements of $A(q_2)$. They result in

$$a_{11} = s_{11} \sin w + s_{12} \cos w, a_{21} = s_{21} \sin w + s_{22} \cos w,$$

$$a_{12} = \frac{s_{12}\sqrt{m_{11}^2 - m_{12}^2} + s_{11}m_{12}}{m_{11}} \sin w - \frac{s_{11}\sqrt{m_{11}^2 - m_{12}^2} - s_{12}m_{12}}{m_{11}} \cos w,$$

$$a_{22} = \frac{s_{22}\sqrt{m_{11}^2 - m_{12}^2} + s_{21}m_{12}}{m_{11}} \sin w - \frac{s_{12}\sqrt{m_{11}^2 - m_{12}^2} - s_{22}m_{12}}{m_{11}} \cos w,$$

and arbitrary constants are selected to be

$$s_{11} = s_{22} = 0, s_{12} = s_{21} = \frac{m_{11}}{m_{11}^2 - m_{12}^2},$$

and $w = d \ / \sqrt{m_{11}^2 - m_{12}^2}$.

Equations (7.31) in the new coordinates (z, q_2) are

$$\dot{z} = A(q_2)M^{-1}\tau - E(q_2)z,$$
$$\dot{q}_2 = B(q_2)A^{-1}(q_2)z \tag{7.34}$$

where $E(q_2) = A(q_2)M^{-1}DA^{-1}(q_2)$. They are linear in the unmeasured state z. Based on the structure of Equations (7.34), we apply a passive observer that needs q_2 only, i.e., $(x, y, \ )$ for feedback:

$$\dot{\hat{z}} = A(q_2)M^{-1}\tau - E(q_2)\hat{z} + K_1(q_2 - \hat{q}_2),$$
$$\dot{\hat{q}}_2 = B(q_2)A^{-1}(q_2)\hat{z} + K_2(q_2 - \hat{q}_2), \tag{7.35}$$

where $\hat{z}$, $\hat{q}_2$ and $\dot{\hat{z}}$, $\dot{\hat{q}}_2$ are the estimates of z, q_2, and velocities, respectively.

Gain matrices K_1, K_2 in Equations (7.35) are selected as

$$E^T(q_2)P_1 + P_1E(q_2) = Q_1, K_2^T P_2 + P_2K_2 = Q_2, (B(q_2)A^{-1}(q_2))^T P_2 - P_1K_1 = 0, \tag{7.36}$$

where Q_1, Q_2 and P_1, P_2 are positive definite matrices. To detail stability of the observer (7.55), we formulate the following theorem.

Theorem 7.3 (Jarzębowska 2011a)

The control dynamics (7.31) with the velocity observer (7.35) is globally exponentially stable at the origin $(\tilde{z}, \tilde{q}_2) = 0$.

Proof: We write the observer estimation error dynamics, based on Equations (7.31) and (7.35), to obtain

$$\dot{\tilde{z}} = -E(q_2)\tilde{z} - K_1\tilde{q}_2,$$
$$\dot{\tilde{q}}_2 = B(q_2)A^{-1}(q_2)\tilde{z} - K_2\tilde{q}_2, \tag{7.37}$$

where $\tilde{z} = z - \hat{z}$, $\tilde{q}_2 = q_2 - \hat{q}_2$ are estimation errors. Let us take the Lyapunov function candidate as

$$V(\tilde{z}, \tilde{q}_2) = \tilde{q}_2^T P_2 \tilde{q}_2 + \tilde{z}^T P_1 \tilde{z}. \tag{7.38}$$

Matrices P_1 and P_2 are positive definite, so $V(\tilde{z}, \tilde{q}_2) \geq 0$. The Lyapunov function derivative along the solution of Equations (7.37) yields

$$\dot{V} = -\tilde{q}_2^T Q_2 \tilde{q}_2 - \tilde{z}^T Q_1 \tilde{z}. \tag{7.39}$$

Based on properties of quadratic forms, we can write

$$\alpha_{ql} \|\tilde{q}_2\|^2 \leq \tilde{q}_2^T Q_2 \tilde{q}_2 \leq \alpha_{qu} \|\tilde{q}_2\|^2, \quad \alpha_{zl} \|\tilde{z}\|^2 \leq \tilde{z}^T Q_1 \tilde{z} \leq \alpha_{zu} \|\tilde{z}\|^2,$$

and hence, $\dot{V}$ can get an upper bound as

$$\dot{V} \leq -\alpha \left(\|\tilde{q}_2\|^2 + \|\tilde{z}\|^2 \right) \tag{7.40}$$

where $\alpha = \min(\alpha_{ql}, \alpha_{zl})$. In the same way, we show that V is upper bounded, that is,

$$V \leq \beta \left(\|\tilde{q}_2\|^2 + \|\tilde{z}\|^2 \right) \tag{7.41}$$

where $\beta = \max(\beta_{qu}, \beta_{zu})$. Both (7.40) and (7.41) yield

$$V(t) \leq V(0) \exp(-\alpha t / \beta), \tag{7.42}$$

and we conclude that the Lyapunov function V exponentially converges to zero. Based on properties of quadratic forms and on (7.42), we find that

$$\beta_{zl} \|\tilde{z}\|^2 \leq \tilde{q}_2^T P_2 \tilde{q}_2 + \tilde{z}^T P_1 \tilde{z} = V(t) \leq V(0) \exp(-\alpha t / \beta)$$

and hence,

$$\|\tilde{z}\| \leq \sqrt{\frac{V(0)}{\beta_{zl}}} \exp(-\alpha t / 2\beta).$$

Since $z = A(q_2)\omega$, the estimated velocity exponentially converges to $\dot{q}_1$, i.e., to ω. Similarly, it can be verified that

$$\|\tilde{q}_2\| \leq \sqrt{\frac{V(0)}{\beta_{ql}}} \exp(-\alpha t / 2\beta).$$

q.e.d.

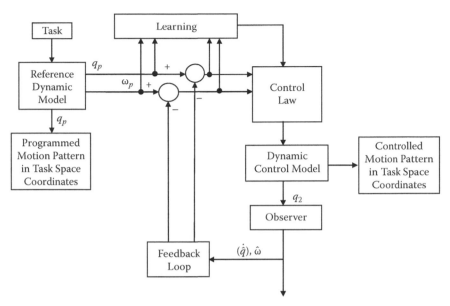

FIGURE 7.27
Controller-observer system within the model reference tracking control strategy for programmed motion.

If a damping term is absent in the control dynamics (7.31), we can prove that applying the observer (7.35) $\tilde{z}$ and $\tilde{q}_2$ are bounded.

Architecture of the model reference tracking strategy with the velocity observer is presented in Figure 7.27.

Assume that the robot is to perform a repetitive motion. Then, the block of specialized terms to the control law consists of a "learning" term. For the repetitive motion, we apply a repetitive control low (RCL) (7.18).

Assume that the robot is subjected to position and first order constraints only. Then, its reference dynamics can be derived using the Boltzmann–Hamel equations (2.151), and it is of the form

$$M(q)\dot{\omega} + V(q,\omega) + G(q) = 0,$$
$$B_1(t,q)\omega + b_1(q) = 0, \tag{7.43}$$

where $V(q,\omega)$ and $G(q)$ are $(n\text{-}k)$-dimensional vectors. The matrix $M(q)$ dimension is $(n\text{-}k) \times n$, and $B_1(t,q) - (k \times n)$. Relations between vectors of generalized coordinates q and quasi-velocities ω have to be added to Equations (7.43). In (7.43), quasi-velocities may be selected suitably to the programmed constraint equations and may differ from their selection in (7.31).

To illustrate the theory, let us design a controller-observer system for the mobile robot from Section 7.2.3. Control design parameters are $P_1 = P_2 = K_2 = diag\,(1,1)$, $K_1 = (B(q_2)A^{-1}(q_2))^T$. A programmed constraint is specified as

$$(t) = \cos(0.5\pi - t). \tag{7.44}$$

The robot motion is also subjected to the material constraints (7.29). The quasi-velocities are selected as

$$\omega_1^* = \dot{x} - \frac{r}{2}(\dot{\ }_r + \dot{\ }_l)\cos\ , \quad \omega_2^* = \dot{y} - \frac{r}{2}(\dot{\ }_r + \dot{\ }_l)\sin\ ,$$

$$\omega_3^* = \dot{\ }(t) - \sin(0.5\pi - t), \quad \omega_4 = \frac{r}{2b}(\dot{\ }_r - \dot{\ }_l) = \omega_r. \tag{7.45}$$

Equations of the reference dynamics (7.43) have the form

$$I\,\dot{\omega}_4 = \frac{-\dot{\ }^2(t)(mr^2 + I_\theta)(x\sin\ + y\cos\ )}{4r^2(x\cos\ + y\sin\ )^3},$$

$$\dot{x} = \omega_4\cos\ , \tag{7.46}$$

$$\dot{y} = \omega_4\sin\ ,$$

$$\dot{\ } = \sin(0.5\pi - t).$$

Tracking of the programmed motion (7.44) is presented in Figure 7.28. Figure 7.29 shows control torques as time functions that are applied to the right and left wheel to track the desired motion.

7.7 Other Applications of the Model Reference Tracking Control Strategy for Programmed Motion

7.7.1 Hybrid Programmed Motion-Force Tracking

Industrial robots perform a variety of tasks that are realized by their end-effectors. Many of these tasks cannot be defined in terms of the end-effector motion only. These are, among others, painting, writing, scribing, or grinding. They all demand some specified contact force between the end-effector and the surface on which it works. The surface is viewed as a constraint

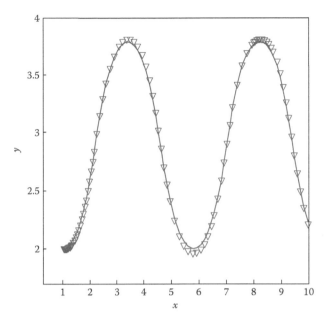

FIGURE 7.28
Reference (——) and controlled (∇) robot motions in (x,y) plane.

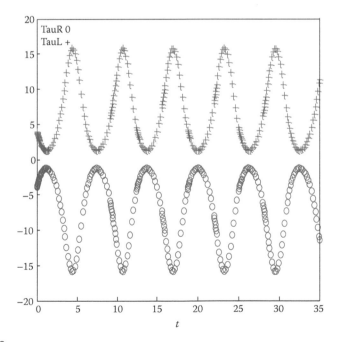

FIGURE 7.29
Torques (in *Nm*) produced by the repetitive control law (RCL) controller: TauR (o)—right wheel,
TauL (+)—left wheel.

for the robot motion, and contact forces we wish to control are referred to as constraint forces. A dynamic control model to control contact forces is a model with a holonomic constraint equation in the direction normal to the surface the force is exerted on. As we demonstrated in Section 7.1, using the model reference tracking control strategy for programmed motion, we can track any programmed motion. Specifically, we can write or scribe anything on a surface with the end-effector, but we cannot control the constraint force directly using this strategy. This is due to the absence of constraint reaction forces in the unified dynamic control model (5.2). In this section, we demonstrate that architecture of the tracking strategy is flexible, and motion tracking can be combined with force tracking. From now on, we use the term *programmed motion/force controller* or just *motion/force controller*, since we incorporate a force controller into the model reference tracking control strategy. The force part may be any force controller adopted from the literature.

Many control design approaches to force feedback control are proposed (see, e.g., Arimoto 1995; Lewis et al. 2004; McClamroch and Wang 1988). If a system dynamics is assumed to be completely known, a feedback-linearizing control scheme can be applied. It enables developing a linear position-force controller. First, the joint space control formulation (5.2) has to be replaced by the task space formulation for the purpose of hybrid programmed motion/force control. It means that for a task space vector x, $x \in R^n$ we have to select a relation $x = h(q)$, where $h(q)$ is found from kinematics and relationships between joint and task spaces. Differentiating this relation twice with respect to time, we obtain $\ddot{x} = J(q)\ddot{q} + \dot{J}(q)\dot{q}$, where J is the nonsingular $(n \times n)$ task space Jacobian matrix. Solving it for $\ddot{q}$ results in

$$\ddot{q} = J^{-1}(q)[\ddot{x} - \dot{J}(q)\dot{q}]. \tag{7.47}$$

Substituting (7.47) into (5.2) yields

$$M^*(q)J^{-1}(q)\left[\ddot{x} - \dot{J}(q)\dot{q}\right] + C^*(q,\dot{q}) + D^*(q) + J^T(q)f = \tau, \tag{7.48}$$

where f is a $(n \times 1)$-vector of contact forces and torques in task space. The corresponding feedback-linearizing control for the dynamics (7.48) is

$$\tau = M^*(q)J^{-1}(q)[v - \dot{J}(q)\dot{q}] + C^*(q,\dot{q}) + D^*(q) + J^T(q)f, \tag{7.49}$$

where v is a new $(n \times 1)$ vector of control inputs such that $v = \left[(v_T)^T, (v_N)^T\right]$, and v_T is the tangent component of v, which is the programmed motion control input, and v_N is the normal component of v, which is the force control input. It can be seen from Equation (7.49) that $\ddot{x} = v$, so task space motion is globally linearized and decoupled. We can design programmed motion and force controllers separately. The v_T component can be designed as $v_T = \ddot{q}_p + 2\sigma\dot{e} + \sigma^2 e$, with $e = q_p - q$ as the position tracking error, and σ as a positive convergence rate. The tracking error satisfies the equation $\ddot{e} + 2\sigma\dot{e} + \sigma^2 e = 0$ and converges to zero exponentially. The v_N component is designed assuming that the surface an end-effector moves on is a spring, the force exerted upon it is $f_o = k_e(z\text{-}z_o)$, k_e is the surface stiffness, and z denotes the normal direction. The force that the end-effector is to exert on the surface is f_p. Then

$$v_{Ni} = \frac{1}{k_{ei}}[\ddot{f}_{pi} + k_{di}(\dot{f}_p - \dot{f}_o) + k_{si}(f_p - f_o)], \qquad (7.50)$$

where k_{ei} is the i-th component of the surface stiffness, and k_{di}, k_{si} are the i-th components of control gains. Task space forces are needed for implementation of this control law. Measured forces are usually not task space forces needed for Equation (7.50). To obtain task space forces, a transformation can be made to relate them to forces measured by sensors mounted on the end-effector wrist. Also, this force control law requires a task space force derivative, which is usually unavailable. The stiffness equation $f_o = k_{ei}(z\text{-}z_o)$ can be used to get the i-th task space force derivative. Other force controllers based on different models of the environment and dynamic interaction between the end-effector and the environment can be employed for our tracking strategy.

For constrained systems, we are interested in model-based controllers that incorporate material holonomic or nonholonomic constraint effects. To the best of the author's knowledge, a theoretical framework that incorporates effects of constraint forces into a robot controller by using classical results in dynamics is proposed in McClamroch and Wang (1988). This framework is known as a reduced-state position-force control and is based upon the observation that when the robot end-effector contacts environmental constraints, which are mostly surfaces, reaction forces between the end-effector and the environment develop. Hence, the end-effector positional freedom is reduced, which means that the degrees of freedom are reduced.

We present a modification and certain redefinition of the reduced-state position-force controller for some classes of robot tasks. Specifically, we extend it to a programmed motion-force controller (Jarzębowska 2007).

Let us first recall that in order to design a reduced-state position/force controller, a reduced-state dynamics in terms of constraint space coordinates have to be obtained. In McClamroch and Wang (1988) and You and Chen (1993), where other force controllers are designed in the same spirit, Equations (5.2) are a starting point. One way to obtain equations of motion in a reduced state is to find a transformation $q = h(w)$ that relates the $(n\text{-}a) \times 1$ dimensional constraint space vector w to the $(n \times 1)$ dimensional joint space vector q. For specific problems, to find $h(w)$, we can use relations given by constraint equations and a robot kinematics. This $(n \times 1)$ vector function must be selected such that

1.

$$\left[\frac{\partial h(w)}{\partial w}\right]^T A^T(q)\big|_{q=h(w)} = 0, \quad A(q) = \frac{\partial \phi(q)}{\partial q}, \quad \phi(q) = 0$$

is a holonomic constraint equation, and the matrix $A^T(q)$ contains a linearly independent columns over the joint space.

2. The $n \times (n-a)$ Jacobian

$$J(w) = \frac{\partial h(w)}{\partial w}$$

contains $(n-a)$ independent rows along the constraint space motion given by w.

The choice of $h(w)$ is not unique, and the conditions above are related to the reduced-state model. Condition (1) is used to ensure that constraint forces can be decoupled from the constraint space motion represented by w. Condition (2) ensures that the decoupled model consists of $(n-a)$ independent equations.

To develop the reduced-state position/force controller, we differentiate $q = h(w)$ once and twice with respect to time and substitute q, $\dot{q}$, and $\ddot{q}$ to Lagrange's equations with multipliers (2.90). We obtain the reduced-state dynamics described by $(n-a)$ equations, which are decoupled from contact forces. Based on this dynamics, the reduced-state position/force feedback-linearizing controller is designed as

$$\tau = M(w)J(w)(\ddot{w}_p + k_v\dot{e} + k_s e) + C(w, \dot{w}) + A^T(w)(\lambda_p + k_f e_f), \tag{7.51}$$

where k_v, k_s are diagonal, positive-definite $(n-a) \times (n-a)$ control gain matrices, and k_f is a diagonal positive-definite $(a \times a)$ control gain matrix. The constraint position tracking error is $e = w_p - w$, and we assume that the desired force λ_p is a known bounded function, from which the force tracking error is defined as $e_f = \lambda_p - \lambda$.

In our approach to force tracking control, we do not have to generate the reduced-state dynamic model, because we have it. Our concern is what control forces we would like to retrieve. Second, the reduced-state equations (5.6) can be derived in any set of coordinates. In order to retrieve a contact force in a specific direction, we transform the constraint equation accordingly. Assume that a constraint function $\phi(q) = 0$ satisfies the following: There is an open set $\Omega \quad R^{n-a}$ and a function $f : \Omega \quad R^a$ such that $\phi(q_1, f(q_1)) = 0$ for all $q_1 \in \Omega$. This assumption holds according to the implicit function theorem for some constant vector $\tilde{q} \in R^n$ that satisfies $\phi(\tilde{q}) = 0$, if rank $J(\tilde{q}) = a$ in some neighborhood of $\tilde{q}$. Let this assumption hold with $\Omega = R^{n-a}$. We make a transformation so that the constraint equation can be presented in a simpler form. To this end, partition q into $q = (q_1, q_2)$, where $q_1 \in R^{n-a}$ and $q_2 \in R^a$. Define the nonlinear transformation $g(q): R^n \to R^n$ given by

$$g(q) = x = \begin{bmatrix} q_1 \\ q_2 - f(q_1) \end{bmatrix} \tag{7.52}$$

which is differentiable and has a differentiable inverse transformation $h(x): R^n \to R^n$ such that

$$h(x) = q = \begin{bmatrix} x_1 \\ x_2 + f(x_1) \end{bmatrix} \tag{7.53}$$

and the vector x partition is $x = (x_1, x_2)$, and $x_1 \in R^{n-a}, x_2 \in R^a$. The Jacobian matrix of the inverse transformation is given by

$$J(x) = \begin{bmatrix} I_{n-a} & 0 \\ \partial f(x_1)/\partial x_1 & I_a \end{bmatrix}. \tag{7.54}$$

The equations of motion (5.6) with the holonomic constraint equation $\phi(q) = 0$ expressed in terms of the variable x have the form

$$M(x_1)\ddot{x}_1 + C(x_1, \dot{x}_1)\dot{x}_1 + D(x_1) = \tau_1, \tag{7.55}$$

$$x_2 = 0.$$

Now, select a computed torque controller in the form

$$\begin{bmatrix} \tau_1 \\ \tau_2 \end{bmatrix} = \begin{bmatrix} M(x_1)[\ddot{x}_{1p} + k_v \dot{e}_1 + k_s e_1] + C(x_1, \dot{x}_1)\dot{x}_1 + D(x_1) \\ A^T(x_2)(\lambda_p + k_f e_f) \end{bmatrix}, \tag{7.56}$$

where e_1, e_f are the position and force tracking errors, and the desired force λ_p is a known bounded function. Because the Equations (5.6) are equivalent to Lagrange's equations in the reduced-state form, the type of stability for the position and force tracking errors can be determined following Lewis et al. (2004) and McClamroch and Wang (1988). Both the position error and the force tracking error admit an asymptotic stability result. The variable $x_2 \in R^a$ in Equations (7.55) can be viewed as a new hypersurface on which motion described by the variable vector x takes place. In other words, x can be viewed as a vector of some virtual task coordinates. In many robot applications, x becomes a vector of real task space coordinates.

Our development of the motion-force controller can be summarized as follows:

1. The reduced-state position/force controller comes in a natural way and a coordinate transformation is needed to transform the constraint equation only. Instead of position tracking control, we can develop programmed motion tracking.

2. For some tasks and control forces, we can decompose motion and force controllers. It is possible due to the selection of the constraint space coordinates. The reduced-state dynamics (5.6) enables this operation.

3. Based on (2), if motion control and force control can be decomposed, there is no need to derive the control dynamics (5.6) in task space coordinates, viewed either as virtual or real.

4. Based on (2) and (3), one can directly apply a force controller to regulate the contact force.

5. Based on (2), (3), and (4), one can apply different force controllers than the one designed by Equation (7.51).

The form of the motion/force controller (7.56) is convenient when the transformations $g(q)$ and $h(x)$ can be presented in the forms $g(q) = (x_1, x_2) = (q_1, 0)$, $q_1 \in R^{n-a}$, and $h(x) = (q_1, 0) = (x_1, x_2)$, where $x_2 = 0$ is a constraint equation. For systems constrained by nonlinear constraint equations $\phi(q) = 0$, motion and force control decomposition may not be possible. Also, the implicit function theorem ensures the existence of a local function $f(\cdot)$ such that $\phi(q_1, f(q_1)) = 0$ for all $q_1 \in \Omega$. As mentioned earlier, it may not be possible to obtain such a function. On the other hand, many dynamic control models (5.6) for real systems are either unconstrained or holonomically linearly constrained models. This is why there may be many applications of the controller (7.56).

To illustrate how we can retrieve constraint forces and regulate them, take the two-link planar manipulator dynamics (7.9) and incorporate constraint effects. For the manipulator model, select the following vectors: $q = (q_1, q_2)$, with $q_1 = (\Theta_1, \Theta_2)$, $q_2 = 0$, and $x = (x_1, x_2)$, with $x_1 = (x, y)$, $x_2 = z$. Assume that we want to regulate a contact force that develops between the surface and the end-effector that operates on it. The transformation $g(q)$ from q to x is

$$x = l_1 \cos \Theta_1 + l_2 \cos(\Theta_1 + \Theta_2), \ y = l_1 \sin \Theta_1 + l_2 \sin(\Theta_1 + \Theta_2), \ z = 0.$$

Then, the constraint equation is $x_2 = z$. For the planar manipulator remarks (1) through (5) hold. It means that its dynamic control model can be developed in joint space coordinates and has the form of Equations (7.9).

Select the programmed constraint detailed in Section 7.2.2, for which the rate of change of the end-effector trajectory curvature is specified to be $\Phi_1 = 0.6 + 0.02t$. Using the development presented above, the force controller can be added and decomposed from the motion controller. As a result, we can track both the programmed motion and the force. Let us apply the force part of the controller (7.56) with the assumption that surface stiffness is $k_e = 1000$, and z denotes the normal direction. The force the end-effector is to exert upon the surface is f_p, and $f_p = 1 - exp(-t)$. Then

$$\tau_2 = \frac{1}{k_e}[\ddot{f}_p + k_{d1}(\dot{f}_p - \dot{f}_o) + k_{s1}(f_p - f_o)],\qquad(7.57)$$

where $k_{d1} = 10$, $k_{s1} = 10$. The reference force f_p and the control force exerted by the end-effector are shown in Figure 7.30. Figure 7.31 presents the end-effector vertical motion and velocity.

We demonstrated in this section that for certain tasks, it was possible to separate motion tracking from force tracking and to control the desired force. Other force control laws may replace the one we used. All of them, however, may be employed within one motion/force tracking strategy in the same way.

7.7.2 Application of a Kinematic Model as a Reference Model for Programmed Motions

The reference model for programmed motion introduced in Section 7.1.1 is a dynamic model, in which the number of equations of constraints is less than the number of degrees of freedom of a system. It may happen that

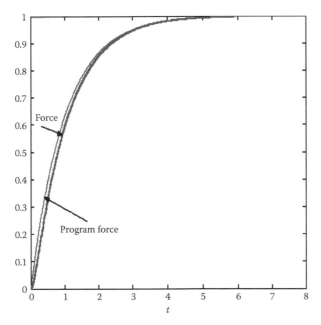

FIGURE 7.30
Reference and control forces exerted on the surface by the end-effector versus time.

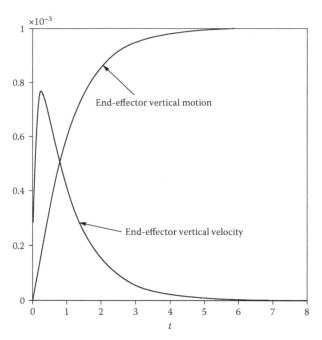

FIGURE 7.31
End-effector vertical position and velocity versus time.

the number of equations of constraints is equal to the number of degrees of freedom. Then the system motion is fully specified provided that the constraint equations are not mutually exclusive. In this case, the reference model consists of equations of material and programmed constraints. This is a kinematic model that specifies motion, and it becomes a reference model. Such a situation may occur when several motion requirements are specified. For example, for a manipulator end-effector, one may want to specify both trajectory and velocity changes. The possibility of velocity control can be significant, since one may want to terminate motion when the program is accomplished.

The reference model, which is kinematic, is in the form of Equation (3.78), but now B is a $(n \times n)$ matrix, with n being the number of coordinates describing motion, i.e.,

$$B(t, q, \dot{q}, \ldots, q^{(p-1)})q^{(p)} + s(t, q, \dot{q}, \ldots, q^{(p-1)}) = 0. \tag{7.58}$$

The kinematic reference model is a motion planner now, and its outputs are inputs to the unified dynamic control model (5.3). Implementation of the kinematic reference model that replaces the dynamic reference model does not change anything in the tracking strategy architecture developed in Section 7.1, and it can extend the scope of its applications.

Example 7.1

Let us consider the two-link planar manipulator model from Section 7.2.2 and formulate a kinematic reference model for it. We desire the manipulator end-effector to move in such a way that its trajectory curvature changes according to Equation (7.7), and it has a specified velocity—that is, $\sqrt{\dot{x}^2 + \dot{y}^2} - v_p = 0 = B_1$. For these two constraint equations, the kinematic reference model is

$$
\begin{bmatrix} F_2 & 1 \\ (a_9 - a_{10})F_2 & 0 \end{bmatrix} \begin{bmatrix} \dddot{\Theta}_1 \\ \dddot{\Theta}_2 \end{bmatrix} = \begin{bmatrix} F_1 \\ \dot{v}_p^2 + v_p \ddot{v}_p - a_{10}F_1 - A_3 - S \end{bmatrix},
\tag{7.59}
$$

where F_1, F_2, are defined in Example 1.5, the programmed velocity is $v_p = 1$, and the trajectory curvature $\Phi(t)$ is assumed to be constant. Other quantities in Equation (7.59) are

$$
A_3 = a_7^2 + a_8^2 + a_5 A_1 + a_6 A_2, \quad a_9 = a_5(a_1 + a_2) - a_6(a_3 + a_4), \quad a_{10} = a_5 a_2 - a_6 a_4,
$$

and a_i, $i = 0,1,\ldots,8$, are defined in Example 1.5.

The function S consists of stabilizing terms, i.e., $S = \alpha \dot{B}_1 + \beta B_1$, where α, β are rates of convergence of the differentiated constraint to the original and are selected to be $a = 20$, $b = 3$. In our example, S has the form

$$
S = \alpha(a_5 a_7 + a_6 a_8 - v_p \dot{v}_p) + \beta\left(\sqrt{a_5^2 + a_6^2} - v_p\right).
\tag{7.60}
$$

The reference motion according to Equations (7.59) and the programmed motion tracking by the PD controller, for which gains are selected to be $k_s = 15$, $k_d = 25$, are presented in Figures 7.32 and 7.35. Initial motion conditions are as follows:

reference motion—$y_{0r} = [-0.2;0.43;-0.5;0;-0.4;0]$;

controlled motion—$y_{0c} = [0.5;0.2;-0.5;-0.6]$.

Figure 7.33 presents position tracking errors, and Figure 7.34 presents accelerations of both joints in time.

Example 7.2

Consider a task of steering a unicycle along a desired trajectory to a rest position achieved by a programmed motion tracking. To this end, supplement the equation of the constraints (7.16) by one equation that specifies termination of motion after a predefined time. Select an initial velocity, say $v = 10$ m/s, and terminate motion after 20 seconds, i.e., $v = f(t)$. The kinematic reference model (7.58) is

$$
\begin{bmatrix} 1 & 0 & 0 & 0 \\ 0 & 1 & 0 & 0 \\ 0 & 0 & 1 & 0 \\ 0 & 0 & 0 & 1 \end{bmatrix} \begin{bmatrix} \dot{x} \\ \dot{y} \\ \dot{\theta} \\ \cdot \end{bmatrix} - \begin{bmatrix} f(t)\cos \\ f(t)\sin \\ f(t)/r \\ \sin(0.5\pi - t) \end{bmatrix} = 0
\tag{7.61}
$$

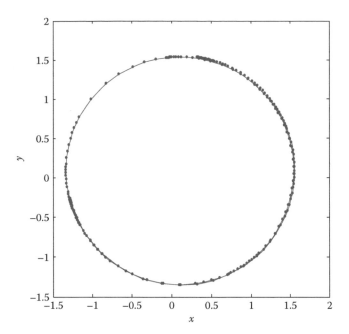

FIGURE 7.32
Reference (—) and controlled (ooo) motions of the end-effector.

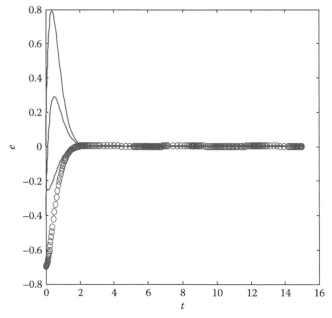

FIGURE 7.33
Tracking errors: e_{θ_1} (ooo), e_{θ_2} (--), $e_{\dot{\theta}_1}$ (...), $e_{\dot{\theta}_2}$ (---) versus time.

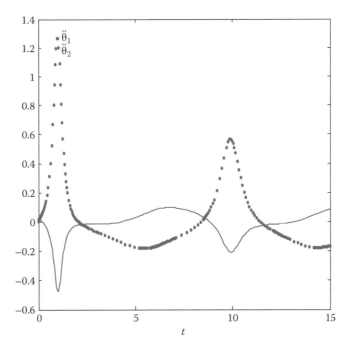

FIGURE 7.34
Joint accelerations $\ddot{\theta}_1$ (ooo), $\ddot{\theta}_2$ (—) versus time.

Figures 7.36 and 7.37 present tracking of the programmed motion specified by Equation (7.61). We apply the Wen–Bayard controller with control gains $k_d = 20$, $k_s = 15$. Initial conditions for the reference and controlled motions are $y_{0r} = [1;1;0;0]$ and $y_{0c} = [1;1;0.9;0]$.

7.7.3 Robot Formation Control

The idea of an application of the model reference tracking control strategy for programmed motion to the robot formation control is based upon the strategy properties. One of its unique properties is that different controllers may be plugged into it based on a task to perform, a system design, or to preserve a desired tracking precision. The other is its modular architecture that enables adding new blocks like other robot control blocks and changing them from mission to mission. In order to extend the strategy to a leader–follower system, some concepts have to be redefined and new assumptions and concepts adopted. This section presents a preliminary study of an extension of the model reference tracking control strategy for programmed motion to a leader–follower system (Jarzębowska 2011b). A control theoretic leader-to-follower framework that enables a group of robots performing task-based missions is developed. It enables tracking not only trajectories but also other task-based motions. Its modular architecture enables switching between controllers for a leader as well as changing followers based on a specified mission.

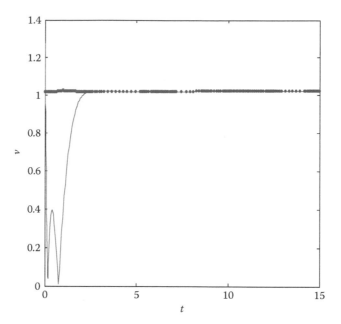

FIGURE 7.35
Tracking of the end-effector velocity versus time.

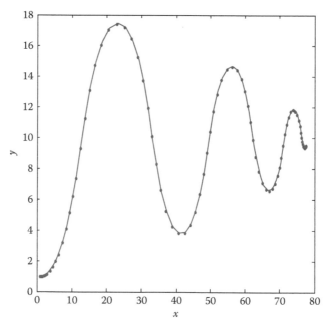

FIGURE 7.36
Programmed motion tracking by the Wen–Bayard controller (···).

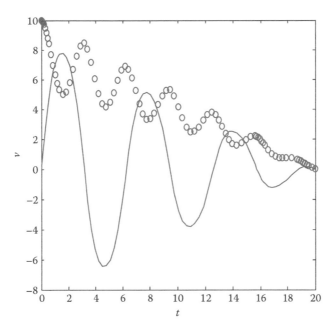

FIGURE 7.37
Unicycle velocity components v_x(—), v_y(ooo) versus time.

A leader–follower system may consist of two or several mobile vehicles that are coordinated in order to achieve complex tasks such as search or rescue in, for example, hazardous environment, distributed manipulation, exploration, and transportation of large objects (Jennings et al. 1997; Nguyen et al. 2003). The leader–follower system or a robot formation may work more efficiently in a shorter time than a single robot. The formation members may be low-complexity robots, because each one can be specialized in one specific task, e.g., navigation or manipulation.

There is a variety of control problems that may be formulated for robot formations. They determine and motivate ways in which these problems are investigated. Usually, by formation control, the problem of controlling the relative positions and orientations of robots in the group is meant. At the same time, the group may move as a whole. Such problems in formation control include moving into formation, maintaining the formation shape (Schneider et al. 2000), or switching between formations (Fierro et al. 2002). In Das et al. (2002), a framework for multirobot coordination is developed to maintain or change formation while following a specified trajectory and to perform cooperative manipulation tasks like grasping objects.

Different control strategies that may be applied to formation control offer a diversity of control goals that can be achieved as well as different degrees of the formation reliability and task precision (Renaud et al. 2004). There are several methodology approaches to robotic formation control, such as behavior-based control (Balch and Arkin 1998), generalized coordinates (Spry and Hedrick 2004), virtual structures, or leader–follower using local sensory information and vision-based feedback (Cruz et al. 2007; Das et al. 2002; Dierks and Jagannathan 2009; Mariottini et al. 2005; Michaud et al. 2002). The most popular approach is the leader–follower method where followers stay at a specified distance and bearing from a leader (Cruz et al. 2007; Das et al. 2002; Dierks and Jagannathan 2009). Also, the most popular in the leader–follower approach is the design of a kinematic controller that assumes a perfect velocity tracking and neglects follower dynamics (Balch and Arkin 1998; Cruz et al. 2007; Das et al. 2002; Li and Chen 2005; Mariottini et al. 2005). The control laws are motivated by ideas from the nonlinear control theory of nonholonomic systems, e.g., from the input–output linearization as in Das et al. 2002. A few works develop a framework that may incorporate the formation dynamics into the control strategy (Dierks and Jagannathan 2009). The control laws there are based on the backstepping in order to include the formation dynamics. Their model-based controller is developed based on the traditional two-level tracking control architecture for nonholonomic systems, and a multilayer neural networks is used to learn the dynamics of a leader and followers. In Dierks and Jagannathan (2009), the formation leader follows the virtual leader, and the virtual leader's velocity is defined by a time function that is differentiable.

Herein, we are interested in a robot formation tracking predefined motions, not necessarily trajectories. They may be combined with force control as well. These motions, which are task-based missions, may serve some search, transportation, or cleaning tasks. By formation control, tracking a specified motion by a formation of robots is understood.

The extension of the tracking strategy for programmed motion requires an introduction of a tracking protocol that consists of new concepts and assumptions. The basic ones are as follows:

C1. Formation control is understood as tracking a predefined motion by all of the formation members.

C2. Predefined motion is attributed to a leader, and we refer to it as a reference motion. The reference motion specifies a task-based mission.

C3. The task-based mission is a high-level command that is generated at the leader level.

C4. Any task-based mission may be plugged to the leader controller that is based on the model reference tracking control strategy for programmed motion.

C5. Control of a formation shape may be a task-based mission.

A1. A leader is a mobile robot that may measure its position and orientation as well as velocities when needed. It is a reference robot for followers.

A2. A follower is a mobile robot, and it may measure its position and orientation with respect to a leader and to the environment, and its velocity components.

A3. The formation members follow the leader.

A4. The shape of the formation may also be predefined.

A5. The follower may work within the centralized or decentralized control coordination. We assume the decentralized control mode.

A6. Each follower computes and executes its own controller based on the mission at hand.

A7. The followers do not measure their relative positions—they may keep the initial distance specified for them.

A8. No obstacle avoidance is incorporated into the follower tracking algorithm. It is possible, however, to include it into the follower control module.

The coordination of the formation tracking is presented in Figure 7.38 for the case of a leader and two followers. The leader is provided with a task-based mission. Its motion is controlled within the model reference tracking control strategy for programmed motion. The outputs of the reference motion are positions and velocities needed to conduct the mission. They are inputs to the follower controller. Each follower may compute and execute its own controller to track the mission-based inputs.

For the purpose of a leader–follower tracking control design for task-based missions, we adapt the strategy presented in Section 7.1. This strategy is for a leader control, and the outputs are passed on to the followers. For control purposes, we introduce definitions.

Definition 7.4: The dynamic model of a constrained leader (7.1) is referred to as a reference dynamic model for a mission—the reference mission dynamics.

The reference mission dynamics (7.1) serves mission planning. It is defined as follows.

Definition 7.5: Mission planning for a leader subjected to the task-based constraints (3.78) consists in finding time histories of positions $q_m(t)$ and their time derivatives in motion consistent with the constraints.

Specifically, in this formulation, planning a mission, which is a trajectory to follow, consists in obtaining a solution $q_m(t)$ of Equations (7.1), in which a task-based constraint equation is algebraic. Solutions of Equations (7.1) also serve as verification of whether a task-based constraint is reasonable for a leader.

The reference mission dynamics (7.1) is employed to design the model reference tracking control strategy for task-based missions, shortly referred to as the strategy for mission tracking.

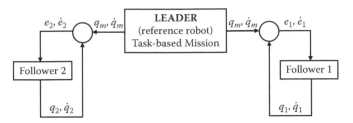

FIGURE 7.38
Coordination of the leader–follower tracking control mission.

A control goal for a leader is as follows: Given a task-based mission specified by the constraints (3.78) and the reference mission dynamics (7.1), design a feedback controller to track motion specified by the mission.

A follower dynamic control model is the same as in Equations (5.3). The follower dynamics may be substituted by its kinematic control model when needed.

A control goal for a follower is as follows: Given a control dynamics (5.3) and the reference mission inputs as positions $q_m(t)$ and velocities $\dot{q}_m(t)$, design a feedback controller to track the reference mission inputs.

Following the leader, which is in fact tracking a desired mission, is carried out according to the scheme presented in Figure 7.38.

The control theoretic framework presented in Section 7.1 and the leader-to-follower tracking control protocol enable performing a mission. The leader may track a variety of task-based missions, and followers may be changed depending upon the mission. Also, tracking controllers may be selected for each mission for both leader and follower control blocks. A mission may be changed and not affect the architecture of the strategy for mission tracking. Only the reference mission dynamics changes. The leader dynamics may be uncertain and adaptive controllers may be applied (Jarzębowska 2008b). A distance between a leader and a follower may be selected a priori, and there is no typical bearing parameter for the follower. Instead, the reference orientation is passed on to the follower from the reference mission dynamics.

Example 7.3

Take a leader, which is the same mobile robot as in Section 7.2.1. Its mass is $m = 1$, a wheel radius is $r = 1$, and it rolls without slipping on a plane surface. Nonholonomic material constraints for the robot are

$$\dot{x} - r\dot{\theta}\cos = 0, \dot{y} - r\dot{\theta}\sin = 0. \tag{7.62}$$

We would like the leader to move a zigzag reconnaissance trajectory specified by the constraint

$$(t) = \cos(0.5\pi - t). \tag{7.63}$$

We assume that only control forces act on the leader. The reference mission dynamics for the robot is developed in Example 2.15 by Equations (2.210). The dynamic control model of the leader is given by Equations (5.25) from Example 5.2.

The mission motion specified by Equation (7.63) is repetitive with period $t = 2\pi$. This suggests applying a learning controller.

Assume that a follower is a robot of the same kinematics as the leader. Its wheel radius is $R = 1$ and mass $M = 1$, and it rolls without slipping on a plane surface. Its motion describes φ—heading angle measured from the axis x; and θ—rotation angle due to rolling, measured from a fixed reference. Coordinates of the wheel contact point with the ground are (x,y). The control dynamics for a follower is the same as for the leader, but this time different control algorithms for torques τ_i may be selected, e.g., the PD.

In Figure 7.39, tracking the task-based mission specified by Equation (7.63) by the leader and the followers is presented. The tracking controller for the leader is the RCL controller, and the PD controllers are applied to the followers. The followers move at specified distances behind the leader, and a distance between them is also specified. MATLAB is applied to all simulation studies. Dimensions in figures are in SI units, and the reference task-based motion is in a solid line.

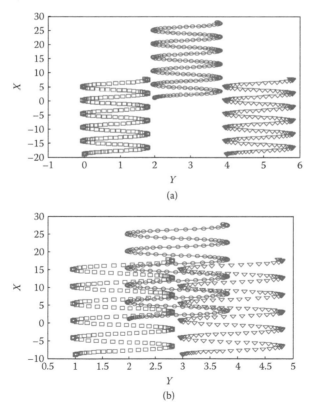

(a)

(b)

FIGURE 7.39
Reconnaissance mission tracking by the leader and two followers that move at some specified distances as in (a) and (b). O – leader motion, □,Δ - followers motion.

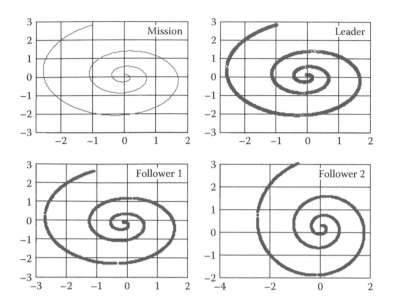

FIGURE 7.40
Search mission tracking by the leader and two followers.

We may change a mission to a kind of search of the environment and request followers to follow a leader along a trajectory

$$x^2 + y^2 = R^2, \qquad R = 0.08t + 1. \qquad (7.64)$$

Figure 7.40 presents tracking motion according to the task-based mission specified by Equation (7.64). Some specified distance is kept between the followers, and they do not collide (see Figure 7.41).

To demonstrate possibilities of specifications of task-based missions, consider a task that is motion not just along a specified trajectory but along a trajectory of a specified change of its curvature profile. For illustrative purposes, take two curvature profiles: one constant $\kappa = 5$ and the second $\kappa = 2\sin t + 1$. In the first case, the task is motion along a circle, and in the second the robot motion has to satisfy the constraint of Equation (2.162b) written in the form

$$\dot{\kappa} = F_0 + \frac{\dot{x}\dddot{y} - \dddot{x}\dot{y}}{(\dot{x}^2 + \dot{y}^2)^{3/2}}, \qquad (7.65)$$

where F_0 consists of terms without third-order time derivatives of the variables.

Select the computed torque controller to track motion according to Equation (7.65):

$$\tau_p = \ddot{q}_{1p} + 2\sigma(\dot{q}_{1p} - q_1) + \sigma^2(q_{1p} - q_1) \qquad (7.66)$$

with $\sigma = 3$. Figures 7.42 and 7.43 present navigation toward a mission specified by the two curvature profiles. In Figure 7.43, robots do not collide. Their trajectories intersect, but the same positions are reached at different time instants.

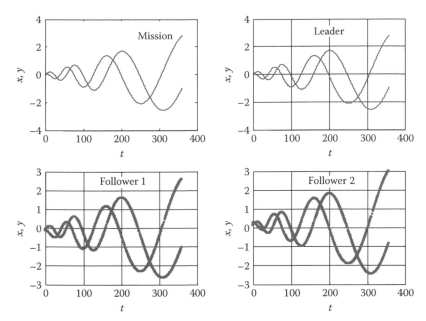

FIGURE 7.41
Time histories of positions for the leader and the followers.

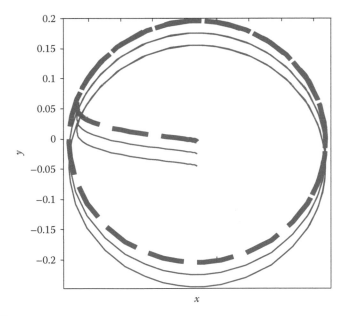

FIGURE 7.42
Navigation of a robot formation toward the task specified by a change of a trajectory curvature $\kappa = 5$. Dashed line-leader motion; solid line-motion of the followers.

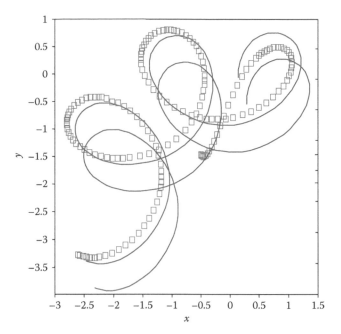

FIGURE 7.43
Navigation of a robot formation toward the task specified by a change of a trajectory curvature
$\kappa = 2\sin t + 1$. Squared line—leader motion; solid line—motion of the followers.

Other task-based missions specified by the constraint equations can be handled in the same way. Other control algorithms can be implemented to navigate the formation toward the task-based mission, and no changes are required in the strategy architecture.

Problems

1. Derive the reference leader dynamics from Example (7.3) when it is subjected to the constraint of Equation (7.64).
2. Derive the reference leader dynamics from Example 7.3 when it is subjected to the constraint of Equation (7.65).
3. Prove Theorem 7.1, that is, the stability for the closed-loop dynamics (5.6) with the DCAL controller (7.14).

References

Arimoto, S. 1995. Fundamental problems of robot control: Part II: A nonlinear circuit theory towards an understanding of dexterous motions. *Robotica* 13:111–122.

Armstrong-Helouvry, B. 1991. *Control of machines with friction*. New York: Kluwer Academic.

Balch, T., and P. Arkin. 1998. Behavior-based formation control for multirobot teams. *IEEE Trans. Automat. Contr.* 15:926–939.

Berghuis, H. 1993. Model-based robot control: From theory to practice, PhD diss., University of Twente, Netherlands.

Cruz, D., J. McClintock, B. Perteet, O. Orqueda, Y. Cao, and Fierro, R. 2007. Decentralized cooperative control. *IEEE Contr. Syst Mag.* Vol 58–78.

Das, A.K., R. Fierro, V. Kumar, J.P. Ostrowski, J. Spletzer, and C.J. Taylor. 2002. A vision-based formation control framework. *IEEE Trans. Robot. Automat.*, 18(8):813–825.

De Luca, A., G. Oriolo, L. Paone, and P.R. Giordano. 2002. Experiments in visual feedback control of a wheeled mobile robot. In *Proc. IEEE Conf. Robot. Automat.* 2073–2078. Washington, DC.

Dierks, T., and S. Jagannathan. 2009. Asymptotic adaptive neural network tracking of nonholonomic mobile robot formations. J. Intell. Robot Syst. 56: 153–176.

Do, K.D., Z.P. Jiang, and J. Pan. 2003. A global output-feedback controller for simultaneous tracking and stabilization of unicycle-type mobile robots. In *Proc. 42nd IEEE Conf. Decision Contr.* 3852–3857. Maui, HI.

Fierro, R., P. Song, A.K. Das, and V. Kumar. 2002. Cooperative control of robot formations. In *Cooperative control and optimization*, ed. R. Murphey and P. Pardalos, 73–93. Dordrecht, Netherlands: Kluwer.

Fossen, T.I., and H. Nijmeijer (Eds). 1999. *New directions in nonlinear observer design. Lecture notes in control and information science 244.* London: Springer-Verlag.

Jarzębowska, E., and J. Greenberg. 2000. On an improved vehicle steering system friction force model. Techn. Report No. AJ 441, *Ford Technical Journal*.

Jarzębowska, E. 2007. *Model-based tracking control strategies for constrained mechanical systems.* Palermo: International Society for Advanced Research.

Jarzębowska, E. 2008a. Quasi-coordinates based dynamics modeling and control design for nonholonomic systems. *Nonlinear Anal.* 16(16):1741–1754.

Jarzębowska, E. 2008b. Advanced programmed motion tracking control of nonholonomic mechanical systems. *IEEE Trans. Robot.* 24(6):1315–1328.

Jarzębowska, E. 2011a. A velocity observer design for tracking task-based motions of unicycle type mobile robots. *Commun. Nonlinear Sci. Numer. Simulat.* 16:2301–2307.

Jarzębowska, E. 2011b. Leader-follower tracking control design for task-based missions. *Int. J. Aut. Veh. Syst.* 9(3/4): 203–218.

Jennings, J., G. Whelan, and W. Evans. 1997. Cooperative search and rescue with a team of mobile robots. In *Proc. IEEE Int. Conf. Advanced Robot* 193–200.

Kim, B.M., and P. Tsiortas. 2002. Controllers for unicycle-type wheeled robots: Theoretical results and experimental validation. *IEEE Trans. Robot. Automat.* 18(3):294–307.

Koh, K.C., and H.S. Cho. 1999. A smooth path tracking algorithm for wheeled mobile robots with dynamic constraints. *J. Intell. Robot. Syst.* 24:367–385.

Lewis, F.L., C.T. Abdallah, and D.M. Dawson. 2004. *Control of robot manipulators.* New York: Macmillan.

Li, Y., and X. Chen. 2005. Dynamic control of multi-robot formation. In *Proc. IEEE Int. Conf. Mechatronics.* 352-357, Taiwan.

Li, Z., and J. Canny. 1993. *Nonholonomic motion planning.* Norwell, MA: Kluwer.

Mariottini, G.L., G. Papas, D. Prattichizzo, and K. Danilidis. 2005. Vision-based localization of leader-follower formations. In *Proc. IEEE Conf. Decision Contr.* 635–640, Seville, Spain.

McClamroch, H.N., and D. Wang. 1988. Feedback stabilization and tracking of constrained robots. *IEEE Trans. Automat. Contr.* 33(5):419–426.

Michaud, F., D. Letourneau, M. Guilbert, and J.-M. Valin. 2002. Dynamic robot formations using directional visual perception. In *Proc. Int. Conf. Robot. Syst.* 2740–2745. Switzerland.

Murray, R.M., and S.S. Sastry. 1993. Nonholonomic motion planning: Steering using sinusoids. *IEEE Trans. Automat. Contr.* 38(8):700–716.

Nguyen, H., N. Pezehkhian, M. Raymond, A. Gupta, and J.M. Spector. 2003. Autonomous communication relays for tactical robots. In *Proc. IEEE Int. Conf. Adv.Robot.* 35–40. Portugal.

Oriolo, G., A. De Luca, and M. Vendittelli. 2002. WMR control via dynamic feedback linearization: Design, implementation, and experimental validation. *IEEE Trans. Contr. Syst. Techn.* 10(6):835–852.

Renaud, P., E. Cervera, and P. Martinet. 2004. Towards a reliable vision-based mobile robot formation control. In *Proc. IEEE/RSJ Int. Conf. Intell. Rob. Syst.* 3174–3181. Japan.

Sadegh, N., and R. Horowitz. 1990. Stability and robustness analysis of a class of adaptive controllers for robotic manipulators. *Int. J. Robot. Res.* 9(3):74–92.

Sadegh, N., R. Horowitz, W.-W. Kao, and M. Tomizukn. 1990. A unified approach to the design of adaptive and repetitive controllers for robotic manipulators. *Trans. ASME* 112:618–629.

Schneider, F.E., D. Wildermuth, and H.L. Wolf. 2000. Motion coordination in formations of multiple mobile robots using a potential field approach. In *Distributed autonomous robotic systems*, ed. L.R. Parker, G. Bekey, and J. Barhen, 305–314. Tokyo: Springer-Verlag.

Shekl, A.M., and V.L. Lumelsky. 1998. Motion planning for nonholonomic robots in a limited workspace. In *Proc. IEEE/RSJ Intl. Conf. Intell. Robot. Syst.* 1473–1478.

Singh, S.K., and M.C. Leu. 1991. Manipulator motion planning in the presence of obstacles and dynamic constraints. *Int. J. Robot. Res.* 10(2):171–187.

Spry, S., and J.K. Hedrick. 2004. Formation control using generalized coordinates. In *Proc. IEEE Int. Conf. Dec. Control.* 2441–2446. Bahamas.

van Nieuwstadt, M.J. 1997. Trajectory generation for nonlinear control systems. *Techn. Report CDS-96-011.* Department of Mechanical Engineering, California Institute of Technology, Pasadena.

Wen, J., and D.S. Bayard. 1988. A new class of control laws for robotic manipulators. Part 1: Non-adaptive case. *Int. J. Contr.* 47(8):1361–1385.

Whitcomb, L.L., A.A. Rizzi, and D.E. Koditchek. 1991. Comparative experiments with a new adaptive controller for robot arms. In *Proc. IEEE Conf. Robot. Automat.* 2–7. Sacramento, CA.

You, L.-S., and B.-S. Chen. 1993. Tracking control designs for both holonomic and nonholonomic constrained mechanical systems: A unified viewpoint. *Int. J. Contr.* 58(3):587–612.

Yun, X., and N. Sarkar. 1998. Unified formulation of robotic systems with holonomic and nonholonomic constraints. *IEEE Trans. Robot. Automat.* 14(4):640–650.

8

Concluding Remarks

This book offers control engineers and researchers a unified framework for dynamic modeling and tracking control design of constrained mechanical systems. The final application of the framework is the development of a new tracking strategy, which is referred to as the model reference tracking control for programmed motion. It is dedicated to tracking tasks specified by equations of programmed constraints. This tracking strategy provides a unified approach to motion tracking for both holonomic and nonholonomic mechanical systems. It is based on two dynamic models. One is a reference model, which is a constrained dynamic model for a system subjected to arbitrary order nonholonomic constraints, and the second is a dynamic control model that includes material constraints only.

The constrained dynamics model is the first contribution of the book content to the area of modeling mechanical systems. It originates from an observation made in the context of applications for control that mechanical systems such as robots are designed to perform a variety of tasks and are subjected to motion limitations at the same time. Usually, for each task, a tracking control strategy is designed separately. An exception is trajectory tracking for which unified control approaches are worked out. Many of the constraints put on robot motions as well as design or control requirements for them can be specified by algebraic or differential equations. All constraints that can be specified in the form of equations can be put together into a unified constraint formulation, which we presented in Chapter 2. For systems subjected to such constraints, the method of the generation of equations of motion, referred to as the generalized programmed motion equations (GPME), was developed in Chapter 2. The second dynamic model is a unified dynamic control model, and it is used to develop a model-based tracking control strategy in Chapter 7. This control model is a unified one for both holonomic and nonholonomic systems.

In most control schemes currently used, the first step before a control strategy is designed is a reduction of the state of equations of control dynamics, which is derived using the Lagrange approach. Effort is made to eliminate constraint reaction forces. A dynamic control model already in the reduced-state form is provided in this book. The contribution in modeling constrained mechanical systems can be summarized as follows. The GPME provide a unified modeling capability, i.e., dynamics of a system subjected to constraint equations of arbitrary order can be generated. Lagrange's equations with multipliers are a peculiar case of the GPME.

The second contribution of this book relates to dynamic control models for constrained systems that control theory uses. There are no other models than the ones based on Lagrange's formulation and its modifications. The book introduces theoretical results in modeling of constrained systems to nonlinear control theory. To the best of the author's knowledge, for the first time, a possibility of modeling systems with nonholonomic constraint equations of arbitrary order and subsequent tracking motions specified by these constraints is demonstrated. The new methodology is based on matching results in analytical dynamics of constrained systems with nonlinear control theory tools. This is the model reference tracking control strategy for programmed motion. It conforms to current trends of an interdisciplinary approach to control and contributes to a stream of active research in the area of control of nonholonomic systems. It develops a new approach to tracking and extends the scope of possible applications of nonlinear control theory.

The new tracking strategy, i.e., the model reference tracking control strategy for programmed motion offers a couple of advantages. First, it extends trajectory tracking to programmed motion tracking. It enables applications of control algorithms dedicated to holonomic systems. A library of reference models can be created for one dynamic control model of a system in question. Such a tracking control design enables a unified approach to a variety of constraints control engineers have to deal with. All constraints, motion, and design requirements that can be specified by equations can be put together, and a constrained dynamic model for a system can be generated.

The unified dynamic control model is free of constraint reaction forces; however, we can retrieve them, and using the model reference tracking control strategy, we can develop a hybrid programmed motion/force tracking controller. Force controllers developed by other researchers can be applied within this tracking strategy architecture.

Flexibility of the tracking strategy architecture enables an incorporation of velocity observers to accommodate practical requirements of reducing costs and weight of sensor devices.

The tracking strategy can be applied to designing tracking controllers for underactuated systems. They are usually treated as a separate and special class of nonholonomic systems with second order nonholonomic constraints. We show that using our tracking strategy, no special treatment is necessary as far as tracking control is concerned. Also, we obtain one more new result—we additionally subject underactuated systems to the programmed constraints.

Modeling parametric and structural uncertainties of systems can be taken into account within the tracking strategy presented in this book.

Finally, a kinematic model of a constrained system that completely specifies the programmed motion may serve as a reference model, and no modification of architecture of the tracking strategy is necessary.

The tracking strategy we develop and simulations we present are not free of disadvantages. One group of difficulties may consist in the derivation and simulation of the reference dynamics, which may be time consuming and may require numerical stabilization of the solutions of constrained dynamics. Another group of problems may concern robustness of the tracking controllers we design to large disturbances. We demonstrate in the examples that they can cope well with small disturbances, but large disturbances may cause problems in getting back to track motions consistent with nonholonomic material constraints.

Based on the way the model reference tracking control strategy for programmed motion is designed, several additional options that are not presented in the book can be taken into account. They are modeling options that do not change or enrich the tracking methodology we present. The first is actuator dynamics that can be included in the reference and control models. In this way, a generalization of a control problem to electromechanical systems can be obtained. Second, other friction and disturbance models, adequate for the systems in question, can be included in both reference and control dynamics. Finally, compliance and flexibility in modeling mechanical systems might be included in modeling and control design. This latter option, however, is usually challenging as far as control design is concerned.

Index

T - #0383 - 071024 - C3 - 234/156/14 - PB - 9780367381103 - Gloss Lamination